Adaptation under Fire

BRIDGING THE GAP

Series Editors
James Goldgeier
Bruce Jentleson
Steven Weber

Adaptation under Fire

How Militaries Change in Wartime

DAVID BARNO & NORA BENSAHEL

OXFORD
UNIVERSITY PRESS

OXFORD
UNIVERSITY PRESS

Oxford University Press is a department of the University of Oxford. It furthers
the University's objective of excellence in research, scholarship, and education
by publishing worldwide. Oxford is a registered trade mark of Oxford University
Press in the UK and certain other countries.

Published in the United States of America by Oxford University Press
198 Madison Avenue, New York, NY 10016, United States of America.

Library of Congress Cataloging-in-Publication Data
Names: Barno, David W., author. | Bensahel, Nora, 1971– author.
Title: Adaptation under fire : how militaries change in wartime /
David Barno & Nora Bensahel.
Description: New York : Oxford University Press, [2020] |
Series: Bridging the gap | Includes index.
Identifiers: LCCN 2019054868 (print) | LCCN 2019054869 (ebook) |
ISBN 9780190672058 (hardback) | ISBN 9780190672072 (epub) |
ISBN 9780190937348 (online)
Subjects: LCSH: Military art and science. | Military doctrine—United States. |
Operational art (Military science)—Case studies. | Tactics—Case studies. |
Adaptability (Psychology)
Classification: LCC U104 .B365 2020 (print) | LCC U104 (ebook) |
DDC 355/.033273—dc23
LC record available at https://lccn.loc.gov/2019054868
LC ebook record available at https://lccn.loc.gov/2019054869

1 3 5 7 9 8 6 4 2

Printed by Sheridan Books, Inc., United States of America

To the men and women of the US military, who will face the challenges of fighting the nation's future wars

Contents

Introduction: Why Military Adaptation?

THE US MILITARY plans and thinks incessantly about wars and conflict—yet it inevitably fails to foresee what comes next. Former secretary of defense Robert Gates once quipped that, since Vietnam, the United States has a perfect record in predicting the next war: "We have never once gotten it right."[1] In late 2000, for example, virtually no US military planners would have envisioned that 12 months later the United States would be involved in a large-scale unconventional campaign in the mountains of remote, landlocked Afghanistan—no contingency plans existed for such an eventuality. The 2014 Russian invasion and annexation of Crimea, the rise of the Islamic State, and the global dispersion and rebirth of al-Qaeda since 2001 have each taken military planners by surprise. The future will assuredly hold more of the same. Resurgent great power competition, nuclear-armed rogue nations, and evolving unconventional threats all are now combining in unpredictable ways, exacerbated by unprecedented rates of global change. These factors massively compound the complexities that US military planners face.

Why does the US military have such a hard time envisioning the types of conflict it will confront in the future? The answer, of course, is that the problem does not lie solely with the US military—the problem is with prediction itself. As the famous saying goes, "It is difficult to make predictions—especially about the future."[2] Phil Tetlock, a political scientist who has long studied forecasting, made headlines in 2005 with a study concluding that the average expert making a prediction was "roughly as accurate as a dart-throwing chimpanzee."[3] Though his subsequent work identifies ways to make better predictions,[4] the human ability to predict the future will always remain limited. The future is simply too complicated, too dependent on unidentifiable causes, too subject to human frailties and emotions. The history of war is littered with examples, as Gates suggests, of nations and their militaries laboring incessantly in peacetime to imagine how the opening days of the next conflict will unfold—only to be taken aback by real events. These failures to accurately predict are sometimes labeled after the fact as failures of imagination. Despite the available evidence, the French could not imagine a German armored blitzkrieg in 1940, nor could the Americans credibly imagine a crushing air attack on Pearl Harbor in 1941. And, in their aftermath, the attacks of September 2001 were described in just this same way.[5] Yet even if militaries do imagine the next war accurately, the opening battles often unfold in spectacularly unexpected ways—with even well-trained armies often taken by surprise.[6]

Adaptability, therefore, is a crucial task for any organization—military and civilian, private sector and public sector alike. The environment in which every human enterprise operates is always changing, often in completely unpredictable and unexpected ways. Organizations will always try to make their best guesses about what the future will hold, but they will inevitably be wrong most if not all of the time. What matters most, then, is the *ability to successfully adapt to unforeseen circumstances as they arise*. This is even more true for militaries than other types of organizations, since the consequences can be existential. As the eminent military historian Sir Michael Howard once argued about doctrine:

> I am tempted to declare dogmatically that whatever doctrine the Armed Forces are working on now, they have got it wrong. I am also tempted to declare that it does not matter that they have got it wrong. What matters is their capacity to get it right quickly when the moment arrives.[7]

Given the impossibility of accurate prediction, how can the US military best prepare for its future conflicts?

Without question, the US military must continue to plan. Militaries have historically made detailed plans for possible future wars, for two reasons. First, they need to make decisions about how to train, organize, and equip their forces today. Those decisions simply cannot be deferred until the next war unfolds. Second, the planning process itself has value. As Dwight Eisenhower once remarked, "Plans are worthless, but planning is everything."[8] Through planning, militaries gain important insights about their adversaries, the terrain upon which they might have to fight, and the uses of the weaponry they may need to employ. Even if imperfect, prewar planning usually provides a better starting point for entering the next conflict than no plans at all.

Still, the clash of arms always reveals the unexpected. First, battles are often punctuated by one and sometimes both antagonists reeling from the new and unforeseen. Surprising uses of weaponry, outmoded tactics, wrongheaded strategy, inflexible leadership, or an unanticipated farrago of these factors frequently produce disruptive shocks on the battlefield. In concert, these opening blows shape the character of the war in ways that untested prewar assumptions often fail to anticipate. Militaries risk defeat if they cannot successfully adapt to the battlefield changes unfolding around them—and defeat carries a heavy, immediate, and at times existential price for armies and nations alike.

Adaptability is, and has always been, an essential attribute of successful military forces. But there are good reasons to expect that adaptability will become even more important during future conflicts. First and foremost, we are living in a time of great strategic uncertainty. For the first time in decades, the United States is facing the rise of major power competitors, some of whom are armed with vast economic resources, advanced military and civilian technologies, and the willingness to invest in first-rate military power. At the same time, both states and nonstate actors are increasingly operating in the space between peace and war,[9] in what is often called the gray zone. Gray zone conflicts involve some aggression or use of force, but their defining characteristic is ambiguity—about the ultimate objectives, the participants, possible violation of international treaties and norms, and the role that military forces should play in response.[10] Gray zone adversaries can employ subversion, malicious social media, and sophisticated indirect attacks through cyber disruptions and radicalized individuals to undermine adversaries and achieve political objectives.[11]

War may be leaping out of its traditional boundaries in the gray zone, but the rise of revanchist great powers and a cluster of rogue states such as Iran and North Korea also presents complex conventional threats unlike anything the United States has faced since the Cold War. Any future US war that responds to Russian or Chinese armed aggression, or defends against Iranian or North Korean provocations, will involve more than just the traditional clash of conventional arms. The long-standing delineations that bounded conventional wars, like geography, distance, legal strictures, and international norms, are now eroding, if not disappearing altogether. Any major conventional war in the future will undoubtedly involve widespread attacks rippling across space and cyberspace, which could unhinge both civil and military capabilities in the United States.

Furthermore, we are living in a time of exponential change, unfolding at unprecedented speeds. Different authors offer different explanations for why this is happening, but they almost all agree that the combination of rapid technological advances, globalization, increased connectivity, and urbanization is reshaping our world in profound ways.[12] One study estimated that the impact of these combined forces on the global economy is 3,000 times higher than the impact of the Industrial Revolution.[13] To understand why, consider how quickly new technologies are being adopted. It took 38 years for radio to reach an audience of 50 million users. Television reached an audience of that size in 13 years. In the information age, however, even those timelines seem absurdly slow. Apple's first iPod reached the same 50-million mark as radio in just four years; the internet, in three years; Facebook, in one year; Twitter, in nine months; and Pokémon Go, in only 19 days.[14] Or consider the explosive growth of mobile phones. By 2020, more people in the world will have a mobile phone than will have access to either electricity or running water.[15] Today's smartphones already possess vastly far more computing power than the Apollo 11 mission that put the first astronaut on the moon.[16] And the computers aboard *Voyager I* and *II*—the only man-made spacecraft to enter deep space—provide approximately 235,000 times less memory and 175,000 times less speed than a smartphone with 16 gigabytes of storage, far less than most of us carry in our pockets today.[17]

These are just a few examples of the many rapidly changing technologies that are revolutionizing and disrupting all aspects of government, business, and society—and inevitably militaries around the world as well. The technologies that are emerging from what has been called the Fourth Industrial Revolution are also transforming the ways in which wars are fought.[18] New standoff munitions, unmanned systems, cyberweapons, hypersonic aircraft, and directed energy weapons all present strikingly new and complex challenges—and we cannot even begin to foresee the new weapons technologies that will rapidly evolve to replace them. The rise of artificial intelligence, machine learning, gene editing, robotics, and autonomous systems complicates that picture even further. As new and unimaginable technologies continue to shape the world we live in, the character of conflict will inevitably shift as well.

In such a complex world where changes can occur at light speed, the ability of the US military to predict the size, scope, and character of the next war will remain extraordinarily difficult. The battlefields of tomorrow will be far more expansive than those of the past, stretching the traditional boundaries of war. The distant high ground of space, the once unreachable depths of the sea floor, and the unbounded cyber domain that reaches into every corner of Americans' homes and lives all may become fighting fronts in the next conflict.

This brings us to the central question of the book: *Is the US military adaptable enough to prevail in the wars of the 21st century?* To begin to answer that question, Part I of the book explores what adaptability involves. Chapter 1 starts by defining what we mean by adaptation, and identifies the three critical components of wartime adaptability—doctrine, technology, and leadership—which frame the rest of our analysis and recommendations. Chapters 2, 3, and 4 examine each of these components in more detail and provide some historical examples to illustrate the dynamics of both successful and unsuccessful adaptation. Although important examples exist throughout military history, our illustrations in these chapters come from the 20th century, which ushered in the advent of modern warfare. We also include at least one instance in each chapter of adaptation in foreign militaries as well. Although most of this book focuses on the US experience, these dynamics have broader applicability and will significantly affect other nations' armed forces as well.

Part II uses the analytic framework of doctrine, technology, and leadership to understand US military adaptation (or the lack thereof) during the recent wars in Iraq and Afghanistan. The sections on Iraq in Chapters 5 through 8 are generally longer than those on Afghanistan, because much more analysis has been done, and far more published about that war. But the experiences in Afghanistan are equally important for examining the future, since both of these wars have cast a deep shadow over America's armed forces. The US military's track record in adapting to these prolonged, unconventional conflicts is decidedly mixed. It includes some stunning successes, but also some painful failures that unnecessarily cost the lives of many US soldiers in both wars. In the case of Iraq, it brought the United States to the brink of outright defeat.

Part III then returns to our central question, examining whether the US military is sufficiently adaptable for the wars of the future. Chapter 9 examines the future threat environment that the US military is likely to confront. We argue that the exponential rates of change will make it even more difficult to prepare effectively for future conflicts, which means that the gap between the war that was predicted and the real war that erupts—what we call the *adaptation gap*—will only continue to grow. Chapter 10 assesses the adaptability of the US military today in each of the three key areas of doctrine, technology, and leadership, and delivers some sobering findings. Chapter 11 concludes with some recommendations that we offer to help improve US military adaptability.

Although effective adaptation poses a profound challenge for all of the military services, this book focuses exclusively on land forces. We made this decision for two reasons, one practical and one theoretical. First, the outcomes of war are almost always decided on land. Wars are always fought for political ends, which are nearly always wrapped up in control of boundaries, territory, and populations. Populations also entirely reside on land, requiring militaries to control and dominate the land domain in order to win wars. The air and sea domains have only rarely proven decisive in past conflicts, and we believe that will likely remain the case into the future as well. Even though fighting in future wars will take place at sea and in the air, outer space, and cyberspace, most will still be won or lost in battles to control the land and its people. Thus, this book focuses solely on the adaptability of US ground forces—the army, the marines, and special operations forces—because their operations will likely be the most decisive, as well as the most bloody.

Our second reason to focus on land forces is theoretical. Military adaptation on land is by far the hardest form of adaptation. Land warfare can involve literally millions of individuals, decisions, and communications nodes that all must be coordinated and synchronized for the military to be able to fight successfully and win. Adaptation in this domain requires identifying, understanding, and disseminating needed changes effectively throughout this vastly complex system—a daunting challenge at any time, let alone amid the incredible pressures and stresses of combat. Naval and air operations, by contrast, usually involve coordinating the interaction of *platforms*—scores, or, in the largest battles, perhaps hundreds or a few thousand airplanes and ships. Moreover, the world's burgeoning populations and the countless man-made structures of civilization do not usually directly affect war at sea or war in the air, with each domain remaining relatively open and unhindered by man or his edifices. But the architectures of civilizations and the millions who inhabit them do exert a tremendous influence on land campaigns, making them even more complex. Adaptation in the air and sea domains is by no means easy, but it nevertheless remains substantially less complicated than adaptation on land. That said, we believe the framework and insights in this book will also apply to adaptation at sea and in the air. We leave it to future scholars to explore whether that is correct.

Most importantly, however, this book is designed to provide a deeper understanding of why adaptability matters. We aim to help policymakers improve adaptability so that the US armed forces, starting with those who fight on land, are as prepared as possible for the challenges they will face in their future wars. Our analysis and suggestions should be relevant for civilian decision-makers in the White House, on Capitol Hill, and in the Pentagon, as well as senior US military leaders. We hope that our analytic framework, analysis, and findings will also prompt a robust discussion on needed changes among those who serve throughout the military today. They will be both the ultimate change agents and beneficiaries of this work.

PART I

A Framework for Analysis

I

Understanding Adaptation

Defining Adaptation

What is adaptability? Clearly it involves more than just change, since organizations change all the time.[1] They change when they hire new employees, purchase new software, or redesign their letterhead—but none of these changes necessarily help an organization adapt to unforeseen circumstances. According to Amy Zegart, adaptability involves two other elements as well. First, the changes must be of sufficient *magnitude* to transform what an organization does or how it does it. Second, and perhaps more importantly, the changes must lead to an *improved fit* between the organization and the environment in which it operates: the organization must "change in ways that keep pace with environmental demands, whatever those might be and however fast they might develop."[2]

Keeping pace, however, means that adaptability is not a one-time thing. An organization cannot simply adapt to one changed circumstance and declare victory, because by the time it has done so, the environment will likely have changed again and more adaptation will be required. In business, companies continue to change and improve to gain market share on their competitors. In warfare, opponents continue to change and improve in order to fix problems identified in battle and to gain advantages over their adversaries.

But as two organization theorists argue, "the worst of all possible worlds" is for organizations to change themselves frequently, with all the upheaval that entails, and yet still fail to keep up with their environment.[3] Adaptability is therefore a dynamic process that must keep up with the *rate of change* in its environment. That is far more easily said than done. As one former US intelligence official lamented, "There's no point in saying we're going at half the speed of Moore's Law when the world is going at Moore's Law . . . It's not how fast we've changed. It's how fast we've changed compared to the world."[4]

What happens to organizations that cannot adapt effectively? Private sector companies that fail to adapt go out of business, which happens remarkably often.

More than 500,000 firms fail each year in the United States—which means that an American company goes out of business approximately once every minute.[5] Most of these are small businesses, but more than 194 large public companies filed for bankruptcy between 2010 and 2016, including former Fortune 500 powerhouses Eastman Kodak, Blockbuster, and Borders.[6] These companies failed to adapt to competition from digital cameras, Netflix, and Amazon, respectively. Government agencies that do not adapt rarely face that same fate because they are extremely hard to eliminate,[7] but politicians can punish them in other ways, such as slashing their budgets and personnel billets. National security agencies that fail to adapt put their citizens at risk, as Zegart argues about the US intelligence community before the 9/11 attacks.[8] And militaries that cannot adapt to unexpected battlefield conditions risk losing their wars, or winning at devastating costs. Williamson Murray, who has spent much of his career studying military effectiveness, argues that because militaries cannot accurately predict the next war, "one of the foremost attributes of military effectiveness must lie in the ability of armies, navies, or air forces to recognize and adapt to the actual conditions of combat, as well as to the new tactical, operational, and strategic, not to mention political, challenges that war inevitably throws up."[9] In fact, adaptation is so important that scholars Eliot Cohen and John Gooch identify the failure to adapt as one of the three distinct types of military failure.[10] And it is firmly embedded in US military doctrine, which repeatedly stresses the need for adaptive leaders, forces, and training.[11] US Army doctrine, for example, states that the army "stresses the importance of adaptability due to the rapid pace of world events and the dynamic change that occurs across related military operations."[12] It also "acknowledges that societal change, evolving security threats, and technological advances require adaptability."[13]

Why Military Adaptation Is So Difficult

Military adaptation, however, is much easier to talk about in principle than to achieve in practice. Why is this the case? A large part of the reason is that militaries are, by necessity, large bureaucratic organizations, and bureaucracies are notoriously resistant to change. But there are several additional factors that result from the unique mission of fighting and winning wars that make military adaptation an especially challenging problem.

Organizational Resistance to Change

As Stephen Rosen has noted, "Almost everything we know in theory about large bureaucracies suggests not only that they are hard to change, but that they are *designed not to change*."[14] At the turn of the last century, the great sociologist

Max Weber argued bureaucracies became the dominant form of administration because of their "technical superiority over any other form of organization."[15] They offer predictability, continuity, and clarity, among other advantages, and make decisions more quickly than less formal forms of administration or those where decisions require reaching consensus.[16]

Taken together, these characteristics enable bureaucracies to simplify an extraordinarily complex environment by *reducing uncertainty*. The world is a complicated, unpredictable place. For private sector firms, markets change rapidly, innovation makes products obsolete, and staff members are constantly joining and leaving the company. Organizations in the public sector are not subject to the relentless pressure of market forces, of course, but face other challenging uncertainties like which political party will win the next election and how that will affect their future budgets and priorities. Bureaucracies of all types therefore try to minimize uncertainty to the greatest extent possible, so they protect their ability to act rationally and achieve their stated objectives.[17]

Bureaucracies reduce uncertainty in many ways, such as focusing on short-term problems rather than long-term strategies that require making predictions about the future.[18] But their primary mechanism for doing so involves establishing standard operating procedures (SOPs). These rules and routines enable even very large organizations to coordinate the collective work of their individual members and to make sure that they respond to similar inputs, challenges, and problems with reliably consistent actions. Standardization has its benefits. To be effective, however, SOPs must be simple, clear, and easy to both teach and learn. In many ways, they are "the memory of an organization,"[19] because they collectively represent all of the decisions the organization has made in the past. Organizations do not have to waste time and resources trying to solve a problem they have had in the past; SOPs enable them to simply do what they did last time. They also ensure consistency despite constant employee turnover, enabling newly hired employees to perform their jobs in the same ways as their predecessors.[20] As a result, bureaucracies excel at routine, at repetition, and at uniformity across people, places, and time.

The trade-off is that the characteristics that reduce uncertainty also limit a bureaucracy's ability to adapt to new and changing circumstances. As Zegart writes, "The very routines, policies, and norms that improve an organization's reliability and stability reduce prospects for adaptation."[21] Bureaucracies resist change because change increases the uncertainty that they are *deliberately designed to avoid*. Thus, they usually prefer to stick to the way that they have always done things, even when change could help them function more effectively.[22] This does not mean that adaptation and innovation never occur. An organization that remained exactly the same, no matter what, would not be able to keep up with an inevitably changing environment. A private firm would go out of business; a public agency would face budget cuts and curtailed missions from its political overseers; militaries would lose wars. But it does mean that change is slow, difficult, usually incremental, and frequently resisted.[23] As Graham Allison argues, "The best explanation of an organization's behavior at t is $t - 1$; the best prediction of what will happen at $t + 1$ is t."[24]

This natural organizational tendency is also compounded by individual psychology and social dynamics. Over time, employees develop vested interests in keeping things the way that they are—not just to maintain their own personal power, though that does often occur, but also because they internalize the bureaucracy's norms of behavior. Again, there are positive benefits to this dynamic, because it can create a powerful esprit de corps.[25] But every advantage has a corresponding disadvantage, and the trade-off here is that change can become a moral and political problem rather than simply a technical adjustment.[26] As Zegart compellingly notes, "With employees viewing proposed changes in moral terms ('this is wrong!') rather than technical terms ('this is not efficient!'), organizational changes stand little chance."[27]

The Additional Challenges Facing Militaries

Military organizations suffer from these challenges just like other types of organizations. Since the armed forces that fight most wars are large organizations, militaries necessarily face the problems of bureaucratic adaptation.[28] Yet they also face some additional challenges that go beyond the general problems of bureaucracies. Militaries have a unique purpose that profoundly differentiates them from private companies and even other organizations in the public sector. They are designed to fight and win wars, which involve chance and friction; their wars involve strategic interaction with an adversary; their wartime environments are often radically different from their training environments; and they face the potentially existential costs of failure. Taken together, these characteristics make military adaptation a particularly vexing challenge.

Chance and Friction

As the great military theorist Carl von Clausewitz wrote in his masterpiece, *On War*, wars defy logic and rational prediction because of two related factors: chance and friction.[29]

Chance, according to Clausewitz, is an inescapable element of war. "No other human activity," he writes, "is so continuously or universally bound up with chance,"[30] or "has such incessant and varied dealings with this intruder."[31] Chance involves factors and dynamics that cannot be controlled, and in many cases, are simply unknowable. Weather, for example, is fundamentally a matter of chance. An invasion planned for temperate weather can turn into a disaster with unexpected torrential rains and mud that impede otherwise easy military movements. As Hitler found out during his invasion of the Soviet Union in 1941, the early arrival of the severe Russian winter halted his previously unchecked momentum. His heretofore victorious troops found themselves unprepared for wintry conditions on the outskirts of Moscow, and many froze to death in their summer uniforms.[32] Chance also played a critical role in the Pacific theater of World War II. If the US aircraft carriers of the Pacific Fleet had not happened to be away from their home port at Pearl Harbor on December 7, 1941, they might have been sent to the bottom alongside the fleet's battleships during the surprise Japanese attack. Such an outcome would have prolonged the war and exposed Australia and even Hawaii to grave risk from further Japanese advances.[33] Chance means that military leaders cannot control many aspects of their wartime environment, no matter how hard they try; all of their efforts to fight the enemy are "aimed at probable rather than certain success."[34]

On the battlefield, military leaders also experience uncertainty through what Clausewitz calls friction. In an 1812 essay, he wrote, "The conduct of war resembles the workings of an intricate machine with tremendous friction, so that combinations which are easily planned on paper can be executed only with great effort."[35] When he wrote *On War* years later, he expanded on this idea: "Everything in war is very simple, but the simplest thing is difficult. The difficulties accumulate and end by producing a kind of friction that is inconceivable unless one has experienced war."[36] Murray describes friction as "those almost infinite number of seemingly insignificant incidents and actions that can go wrong, the impact of chance, and the horrific impression of combat on human perceptions."[37] And as the army's chief military historian during the Korean War so eloquently observed, "One gun properly manned, one mine properly placed, one range properly measured, one registration properly executed, one bit of light, one cloud, one piece of ice, one puddle, the dust of a road—all can go to make or break the success of a small engagement and may mean life or death."[38]

Many factors can cause friction, including poor or nonexistent information, rumors, fear, and the presence of danger.[39] Friction cannot be avoided. It affects both adversaries, but in unpredictable and often unequal ways. It is not only an inherent part of the battlefield, but the armies that Clausewitz describes as an "intricate machine"[40] are made up of millions of people, connections, and communication points—and the potential for friction exists within each one. That is why Clausewitz argues that friction distinguishes war in theory from war in practice.[41] It is fairly easy, for example, to plan how a platoon will take a hill. But it can be enormously challenging for the platoon to actually take that hill amid deadly incoming fire, smoke and deafening noise, equipment that breaks down, and mounting casualties. If soldiers are sufficiently adaptable, however, they can overcome some of the inevitable problems that friction causes. During the D-Day invasion of June 1944, for example, US paratroopers found themselves dropped widely across Normandy, with units intermingled and often miles from their intended drop zones. Enemy fire and confusion at night caused the inexperienced pilots of the transport aircraft to wildly scatter their troops without regard to long-planned targets. The carefully developed plan for units to seize specific objectives had to be radically modified by paratroopers from different divisions banding together and improvising once they landed.[42]

Taken together, chance and friction lead Clausewitz to estimate that "three quarters of the factors on which action in war is based are wrapped in a fog of greater or lesser uncertainty."[43] One recent interpretation of Clausewitz uses insights from chaos theory to expand on this point. Alan Beyerchen argues that Clausewitz fundamentally understood the concept of nonlinearity, because he maintained that "the core cause of analytic unpredictability in war is the *very nature of interaction itself.*"[44] Wars are enormously complicated endeavors that involve dynamic interactions not only between adversaries, but at every level down to that of the individual soldier. Clausewitz stresses "the disproportionately large role of the *least* important of individuals and of minor, unforeseeable incidents. 'Friction' conveys Clausewitz' sense of how unnoticeably small causes can become amplified in war until they produce macroeffects, and that one can never anticipate those effects."[45]

The need to function effectively in an environment characterized by chance and friction leads militaries to resist change even more than other types of bureaucracies. All organizations create standard operating procedures to manage uncertainty, but militaries must function in one of the most uncertain areas of human endeavor. They therefore seek to impose as much order on the relentless chaos of war as possible. They rely heavily on standard operating procedures—the higher forms of which are formalized in doctrine—so that soldiers and leaders reflexively know how to respond to problems even amid the death, destruction, and confusion of the battlefield. Militaries create strong hierarchies, codified in a formal rank structure, so that everyone in uniform knows who is in charge and whose orders to follow. They require strict discipline, even in ways that can seem trivial in peacetime, to ensure that their troops follow orders effectively even when their lives are in grave danger. As Murray writes:

> The demand of discipline and rigid respect for one's superiors—on which cohesion in battle depends—are antithetical to the processes of

adaptation, which require a willingness on the parts of subordinates to question the revealed wisdom of their superiors. *It is this inherent tension between the creation of disciplined, obedient military organizations, responsive to direction from above, and the creation of organizations adaptive to a world of constant change that makes military innovation in peacetime and adaptation in war so difficult.*[46]

Strategic Interaction with the Adversary

As noted already, wars present an environment for adaptation that is very different from any other form of human endeavor. Businesses may face fierce competitors, but not adversaries aiming to kill their employees and destroy their homes and cities. The stakes at war are thus immeasurably higher. At the same time that one military is struggling to understand, change, and adapt, its adversary is doing exactly the same. War is not a one-way arcade shooting gallery, but, as Clausewitz describes, a "duel on a larger scale,"[47] where the opponent dodges, shoots back, and continually looks for other advantages. Adversaries in war can be expected to use every bit of their wiles and energy to outwit and defeat their opponent, again and again.

Dealing with a capable, thinking adversary makes wartime adaptation more important and infinitely more difficult than any sort of peacetime competition. Following the 2003 US invasion of Iraq, for example, Iraqi insurgents sought out effective ways to attack the well-armed and equipped US forces occupying the country. Leftover munitions from the demobilized Iraqi army became a ready source of explosives that could be converted into roadside bombs, called improvised explosive devices (IEDs), to attack American troops and their vehicles. As the United States worked to overcome this new threat—eventually spending billions of dollars on largely ineffective countermeasures—the insurgents continued to adapt the size, placement, and means of detonating their IEDs. These relatively cheap weapons inflicted thousands of American casualties and stymied a vast array of US counteradaptations for years.[48] IEDs became the signature weapons of the Iraq War, and yet they were entirely unanticipated before the conflict. As the United States rediscovered in Iraq, war never stands still, and both combatants continually adapt until the last shots are fired.

The Radical Difference between Peacetime and Wartime

Most organizations are designed to do what they actually do every day. Private companies seek to make profits by selling goods or services, making deals with other companies, and deciding when and how to borrow money or invest their profits. Similarly, public agencies seek to make policy by interacting with stakeholders, lobbyists, and the elected officials who oversee them. Their environments continually provide feedback, telling them what they are doing well (or badly) through indicators like rising (or falling) profits.

Militaries, by contrast, are designed to do something that they do only rarely. Their raison d'être is to fight and win wars,[49] but fortunately wars do not occur very often. That means that unlike most organizations, militaries spend the vast majority of their time preparing for their ultimate purpose but not actually performing it. Even though "the battle is the payoff" for militaries, as the title of one American soldier's account of World War II accurately notes,[50] many militaries go years and sometimes even decades without fighting a battle. As Michael Howard so eloquently writes, the profession of a soldier "is almost unique in that he may have to exercise it only once in a lifetime, if indeed that often. It is as if a surgeon had to practice throughout his life on dummies for one real operation; or a barrister appeared only once or twice in court towards the close of his career; or a professional swimmer had to spend his life practicing on dry land for an Olympic championship on which the fortunes of his entire nation depended."[51] These challenges become even more difficult at the organizational level. Militaries do not, and cannot, know when or sometimes even whom they will fight, what battlefield conditions will look like, how new technologies will be used, or the political and sociological conditions that will form their strategic context.[52] They do their best to prepare, innovate, and adapt through constant exercises, experiments, and simulations—but they cannot truly know how ready they are until they face the crucible of actual combat. There is no way for militaries to credibly evaluate their preparations, according to Murray, "until the audit of war itself, in which fear, chaos, ambiguity, and uncertainty dominate."[53]

Existential Costs of Failure

No organization wants to fail, but the risks of failure are far more profound for militaries than for any other type of organization. As we have noted, organizations resist change for a rational reason: change increases the uncertainty that they are designed to avoid, which may put them at a serious competitive disadvantage. Organization theorist James March suggests that "an institution eager to adopt innovative proposals will survive less luxuriantly and for shorter periods than others . . . being the first to experiment with a new idea is not likely to be worth the risk."[54] For a private company, the worst consequence of failing is bankruptcy, which may cause investors to lose a great deal of money. For a public agency, the worst consequence of failing is being eliminated, or more commonly, facing severe budget cuts and being rendered irrelevant.

For militaries and for their nations, however, the stakes can be existential. Losing a war can cause a state to lose some or all of its territory as well as its capacity for self-governance. Militaries are extremely conservative about change precisely because the stakes are so high—both for the state and for the men and women in uniform. Tradition weighs heavily on militaries because, as Murray observes, "The approaches that succeeded on earlier battlefields were often worked out at a considerable cost in blood."[55] As a result, those who advocate for dramatic change face an uphill battle. As one military officer has written, "Evidence for change must be sufficiently convincing to overcome the understandable caution inherent in the military. The burden of proof in an argument for change clearly rests with the advocate of reform."[56]

Military commanders also tend to resist changes that they believe will put the lives of their soldiers at undue risk. Unlike the leaders in nearly any other field, commanders have the authority to send their people into battle knowing that they may be wounded or killed. That profound responsibility shapes everything that they do, starting with peacetime training and extending all the way to the battlefield. Commanders are trained to put achieving the mission above the welfare of their soldiers, while always trying to balance the needs of both. They typically have a sound sense of the risks involved when dealing with things they have done before, but change introduces uncertainty and unknown risks. As Suzanne Nielsen argues, "Military leaders may hesitate to abandon 'tried-and-true' weapons systems, organizations, or tactics in favor of new approaches that may—in their view—unnecessarily put lives at risk."[57]

Despite these very rational reasons for resisting change, militaries must nevertheless find ways to remain adaptable. The impossibility of predicting future battlefield conditions and the inescapable roles of chance and friction mean that a military that does *not* adapt to battlefield conditions risks losing a war just as much as a military that makes too many bold and imprudent changes. Military adaptation is thus a constant balancing act, an ongoing search for the equilibrium of being able to change just enough, but not too much. And unlike Goldilocks, who only had to find the right bowl of porridge once, the best equilibrium today may not be the best equilibrium tomorrow.

Existing Approaches to Military Adaptation

Many scholars have explored the causes of military change—and how militaries manage to overcome all of these diverse challenges. For most of the 20th century, these scholars tended to focus on historical narratives, operational histories, and individual case studies of bureaucracy and politics.[58] In the 1980s and early 1990s, however, two factors led to a resurgence of interest in this particular subject. First, the end of the Cold War meant that the United States and its European allies had suddenly and unexpectedly defeated the Soviet Union without firing a shot, and entered into an era of relative peace. Historians saw some interesting emerging parallels with the interwar period—the years after World War I when the victorious militaries had to prepare for a possible future war with limited resources and dwindling public interest. Second, a wave of new positivist research on military change emerged, which inspired many social scientists to delve more deeply into this subject. Unlike the historians, who sought to explain single cases, the social scientists sought to identify the broader causes of military change that could apply more generally across space and time.

These two distinct approaches continue to uneasily coexist in more recent work on military change. But before exploring them in more detail, we first need to define *innovation* and *adaptation*, since both historians and social scientists unfortunately use these terms in inconsistent and sometimes contradictory ways.[59] Some scholars use the term "innovation" to describe all forms of military change;[60] others define innovation as applying only to peacetime change.[61] Some argue that innovation requires fundamentally new approaches, while adaptation adjusts existing approaches—regardless of whether the military is at peace or at war.[62] Some argue that there is a sliding scale between innovation and adaptation rather than a bright dividing line between them;[63] others argue that adaptation is one of two types of innovation.[64] Such conceptual confusion makes it challenging to grasp the essential characteristics of these processes and to compare the works of different authors.

We believe that Williamson Murray offers the clearest and most compelling definitions of these terms, and have chosen to use them throughout this book. Murray argues that innovation occurs during peacetime, while adaptation occurs during wartime. He readily acknowledges the similarities between them: both require a willingness to change, imagination about the possibilities of change, and an effective flow of ideas from lower to higher ranks.[65] Yet despite these similarities, Murray emphasizes that the environments in which innovation and adaptation occur are profoundly different. His eloquent discussion of these differences is worth quoting at length:

> In peacetime, time poses few significant challenges to the innovator; he may lack significant resources, but he has time to form, test, and evaluate his ideas and perceptions. The opposite is true in war. There, those involved in combat usually possess a plethora of resources, but time is not one of them; those pursuing serious changes in doctrine, technology, or tactics in the midst of a conflict have only a brief opportunity to adapt. Adding to their difficulties is the fact that as their organization adapts, so,

too, will the enemy. Moreover, in peacetime those who will innovate in the emerging strategic environment will have to consider indeterminate opponents who, more often than not, represent an idealized, rarely changing depiction of potential enemies. In war, however, the enemy is real, and as a human entity, he, too, adapts to the conditions he confronts and, more often than not, in a fashion that may be largely unexpected by his opponents. Thus, *adaptation demands constant, unceasing change because war itself never remains static* but involves the complexities thrown up by humans involved in their attempt to survive.[66]

The Historical Approach to Adaptation

The historical approach has dominated scholarly work on military change until relatively recently. As Adam Grissom notes, the study of military innovation extends at least as far back as Thucydides.[67] The vast majority of the scholars who followed in his footsteps have utilized the same type of historical approach: identifying an interesting example of military change (or lack thereof) and then exploring all of the factors that combined to produce (or inhibit) that change. Such scholars do not seek prescriptive lessons for the future, because history is too inexact, too messy, and too contingent to offer such clarity. Instead, they search for insights and patterns that shed light on current problems.[68]

In recent years, Williamson Murray and Allan Millett have been the most prominent scholars continuing this tradition. After the end of the Cold War, they identified a number of notable similarities between the emerging strategic environment and the interwar period—a time when "military institutions had to come to grips with enormous technological and tactical innovation during a period of minimal funding and low resource support."[69] In order to explore these parallels, they followed the same method that produced their extraordinary three-volume study of military effectiveness several years earlier.[70] They brought together a number of fellow historians to examine these similarities, which led to their 1996 edited volume called *Military Innovation in the Interwar Period*.[71]

The innovation project directly influenced their later work and led to a breakthrough on military adaptation. In 2010, they published a new edition of *Military Effectiveness*, which included a brief essay reflecting on the original work. In it, they write that their work on innovation "helped clarify some of the unfinished business" in the original three volumes.[72] One of the things that they had missed was that "part of the effectiveness of military forces must be their ability to adapt to the actual conditions of combat."[73] They did not discuss adaptation any further in that essay, but they did include a tantalizing footnote stating that Murray was in the process of writing a book focusing solely on that subject.

Murray published the promised book in 2011, which remains the most detailed study of military adaptation to date.[74] After explaining why military adaptation is so important for success on the battlefield, he examines why it has nevertheless proven so challenging from ancient times through today. Most of the book consists of five superb case studies of military adaptation from the 20th century, some of which we draw upon in subsequent chapters. He identifies some recurring patterns, especially that "most military organizations and their leaders attempt to impose prewar conceptions on the war that they are fighting, rather than adapt their assumptions to reality."[75] (This unfortunate pattern reflects how often militaries struggle with one of Clausewitz's most important dictums: "The first, the supreme, the most far-reaching act of judgment that the statesman and commander have to make is to establish . . . the kind of war on which they are embarking; neither mistaking it for, nor trying to turn it into, something that is alien to its nature.")[76]

Murray makes a tremendous contribution to the literature by focusing exclusively on military adaptation during wartime. Yet, as good as it is, it does not go far enough in helping answer the question at the heart of this book: how can the US military become more adaptable so it can respond effectively to future challenges? The fundamental limitation of the historical approach is that it deliberately avoids asking, much less answering, such prescriptive questions. It provides rich, detailed descriptions of cases and frequent patterns and offers invaluable context for those interested in current policy problems. But it does not provide an analytic framework that identifies which factors are most important, which can then provide a guide for action.

Social Science and the Positivist Approach

Unlike historians, social scientists explicitly try to identify causal mechanisms that apply to a wide range of cases and that can lead to clear policy prescriptions. Their approaches fall squarely within the positivist tradition—an approach that uses scientific forms of inquiry in the social sciences (and beyond).[77] Positivists seek to provide generalized explanations of complex social phenomena by coming as close to the scientific method as possible. In contemporary political science and strategic studies, this usually involves linking variables and hypotheses into theories that can be tested, using either qualitative or quantitative data.[78]

In 1984, Barry Posen published the first major social science approach to understanding military innovation. In *The Sources of Military Doctrine*, he argues that militaries do not innovate on their own; instead, innovation occurs when political leaders intervene to make it happen.[79] Posen's work inspired a new wave of social science research on military change. A few years later, Stephen P. Rosen countered that innovation and adaptation can result from competition within the military services rather than from civilian intervention.[80] During the 1990s, many scholars engaged in this debate, adding their own refinements or counters to these two seminal works, but largely remaining within the parameters of their basic models.[81] In the late 1990s and early 2000s, Theo Farrell and his collaborators developed a new approach, focusing on the role of institutional culture in promoting (or hindering) innovation.[82]

Despite their notable differences, these models of innovation share one common limitation: they all conceptualize military change as led from the top.[83] Farrell himself has noted that the main debate among social scientists is not about whether change occurs from the top down; it instead focuses on whether that top-down change is led by military or civilian leaders.[84] The models assume, implicitly or explicitly, that militaries require senior leaders to overcome the problems discussed earlier. If organizations resist change and military organizations particularly resist change, then strong leadership must be required to break through the inertia.

Starting in the mid-2000s, however, some scholars started to examine how military change could occur from the bottom up. In part, their interest was spurred by Grissom's 2006 article reviewing the innovation literature, which explicitly noted this gap. After identifying numerous historical examples of innovation and adaptation that top-down approaches do not explain, he concludes that the "door is open for an individual or a group of scholars to make a major contribution to the field by developing the empirical and conceptual basis for explaining cases of bottom-up innovation."[85]

Even more importantly, however, the ongoing wars in Iraq and Afghanistan provided numerous real-world examples of adaptation that occurred without the involvement of senior leaders. Several scholars have studied small-unit adaptation in these wars, often at the battalion level and occasionally delving down to the company level as well.[86] They provide powerful examples of what historians have long understood: troops on the front lines of combat often adapt effectively to the challenges they face. Yet they also identify some of the limitations with bottom-up adaptation. Nina Kollars, for example, argues that even though soldier-led adaptations can be helpful in specific situations, they can also be "idiosyncratic, unrefined, and narrow in application" and that they are "prone to being forgotten or lost within the organization."[87] Furthermore, efforts to disseminate and institutionalize these adaptations often face strong resistance from other organizations within the units' service, especially those focused on long-term issues such as force development and acquisition. As a result, these wartime adaptations tend to involve bureaucratic bypasses and workarounds that cannot be easily replicated elsewhere on the battlefield.[88] But as we have noted, adaptation requires more than just change; change must also be of sufficient magnitude to transform how an organization operates. Yet few, if any, of the adaptations in these case studies fundamentally changed how the higher headquarters or the service operated. These studies also do not address broader questions of how larger formations adapt to unanticipated battlefield challenges, or how these bottom-up adaptations may interact with top-down change to produce (or inhibit) more systemic adaptation throughout a combat force.

To use an imperfect analogy, then, the current state of the adaptation literature is like looking at the tactical and strategic levels of war without examining the operational level that connects the two.[89] Like bottom-up approaches, the tactical level of war focuses on specific military techniques and procedures, and how fielded forces operate together in specific battles with the enemy. Like top-down approaches, the strategic level of war focuses on broad questions about how wars are conducted and how the military is used to achieve national political objectives. But there is no parallel to the operational level of war, that critical junction where commanders link the two together and utilize large fielded forces to conduct campaigns over a period of time.

Our approach seeks to fill this important gap. In the rest of this chapter, we provide an analytic framework for understanding military adaptation that focuses on three key factors: the degree to which a military's doctrine, technology, and leadership promotes (or hinders) its ability to adapt to battlefield challenges and realities.

Analyzing Adaptation: The Roles of Doctrine, Technology, and Leadership

Doctrine, technology, and leadership are the three most critical components of military adaptability. They are important individually because they shape how military forces conduct their operations, the weapons and systems with which they fight, and the critical decisions leaders make in combat. Yet they are even more important collectively because they interact in ways that often decide the fates of armies and the outcomes of wars. Rigid doctrine, inflexible technology, and dogmatic leaders are a recipe for disaster, given the uncertainty, chaos, and surprises that characterize every war. Militaries with adaptable doctrine, technology, and leadership have a critical wartime advantage that may spell the difference between victory and defeat.

Doctrine

Doctrine, according to military historian J. F. C. Fuller, is "the central idea of an Army."[90] It is "the cement which holds an army together,"[91] by providing a common fabric of understanding and language for all of its members. It permits a military organization to impose order on the chaos of war through what Barry Posen calls "institutional principles of how to fight."[92] Despite generations of junior officers' cynicism, doctrine is not simply "the opinion of the senior officer in the room." Instead, it helps reduce the uncertainty of war by providing soldiers with a shared understanding of how they should accomplish their missions and a common framework of how to approach unforeseen problems.

As we have noted, Posen's *The Sources of Military Doctrine* was the first major social science approach to understanding military change, and it remains one of the most influential works on the subject today. It focuses on doctrine as an element of grand strategy, specifically examining whether doctrine is offensive, defensive, or deterrent in nature; the degree to which doctrine is coordinated with foreign policy objectives; and the degree to which it is innovative. He concludes that balance-of-power theory explains these aspects of doctrine better than organization theory does.[93] The book prompted a wave of new scholarship that pushed back against Posen's realist conclusion, arguing that institutional[94] and cultural[95] theories provided more explanatory power and that militaries are capable of far more independent innovation than Posen posits.[96]

Because all of these works directly respond to *The Sources of Military Doctrine*, however, they share its focus on doctrine as an element of grand strategy.[97] While important, that remains a very limited perspective on what doctrine entails, as Posen himself has readily acknowledged.[98] Scholars have paid far less attention to the other elements of military doctrine—especially those at the operational and tactical levels, which are the ones that consume the attention of military professionals on a daily basis. Indeed, the grand strategic issues that Posen and his critics address are often barely recognizable as doctrinal issues to those in the military. Instead, they view doctrine as a guide to *action on the battlefield*.[99] While he was serving as the army chief of staff, General Gordon Sullivan wrote, "Doctrine is the *how* in the way the Army expects to conduct its operations; it is the accepted way we conduct our missions."[100] Michael Evans, a highly respected military scholar, further described doctrine as "the foundation of military professional knowledge. Doctrine is to soldiers what blueprints are to architects or briefs to lawyers."[101]

From this perspective, doctrine is critically important because it reduces the pervasive uncertainty of the battlefield. As noted earlier, Clausewitzian fog and friction are an inevitable and inescapable part of war.[102] Military doctrine helps address this uncertainty by providing soldiers with a shared understanding of how they should accomplish their missions. According to *The Oxford Companion to Military History*, doctrine "can simplify the decisions that have to be made under the duress of combat, by limiting the range of choices that are deemed relevant to given circumstances."[103] It provides well-understood means of approaching typical battlefield problems, such as launching an air attack on an adversary's airbase or executing the ground defense of a river line. Doctrine is the uniquely military manifestation of the general bureaucratic tendency to manage uncertainty through standard operating procedures—except it must endure in life-and-death situations amid the chaos of battle.

Doctrine also helps reduce the friction and fog of warfare by facilitating rapid comprehension and understanding on the battlefield. As one US Army major recently noted, "doctrine provides a common language for military professionals to communicate with each other. This is particularly important under fire when information must be quickly and accurately transferred—and universally understood."[104] Perhaps most critically, it provides a common framework of how to approach unforeseen problems, reducing the need for detailed orders when time is of the essence and communications are frequently limited. Harald Høiback writes that "doctrine can facilitate tacit conversations in the heat of battle by pre-arranging salience and distinguishing relevant precedents. If everybody in a unit comprehends the situation in a similar way and trusts everyone else to do it, countermeasures can be taken more swiftly without cumbersome and time-consuming coordination."[105]

There is an inevitable tension inherent within military doctrine, however, that is particularly relevant for this book. J. F. C. Fuller described this tension almost a century ago as the "dual conception of doctrine."[106] First, doctrine must be, in his words, "a fixed method of procedure" that promotes the understanding and unity of effort described previously.[107] As a result, doctrine is often written down—to standardize common practices and to easily disseminate them throughout the force.[108] The US Army, for example, does this through an extensive collection of field manuals and other doctrinal publications. In June 2019, for example, the "Doctrine and Training" section of the Army Publishing Directorate's website listed 1,340 different documents divided into 11 different categories—totaling tens of thousands if not hundreds of thousands of pages.[109] Although most soldiers will never read most of these publications, they will often know the ones related to their specific military occupations fairly well. Taken together, they represent the army's collective approach to conducting military operations.

Fuller's second conception of doctrine, however, seems directly at odds with the first. He writes that it must be flexible: it "must be looked upon as power to formulate a correct judgment of circumstances and to devise a course of procedure which will fit conditions."[110] It must be able to change according to circumstances in order to be effective. Otherwise, as he writes in the florid language of his times, "the danger of a doctrine is that it is apt to ossify into a dogma, and to be seized on by mental emasculates who lack virility of judgment, and who are only too grateful to rest assured that their actions, however inept, find justification in a book."[111] More recently, General James Mattis (who commanded US Marines in Iraq in 2004 and later served as secretary of defense) often expressed a similar sentiment by paraphrasing Oscar Wilde: "Doctrine is the last refuge of the unimaginative."[112]

Doctrine must therefore be fixed but flexible. It must enable all members of a military organization to approach problems in a similar way, but without stifling imagination. Doctrine must simultaneously promote unity of effort and individual judgment. In short, it must be adaptable.

Technology

Technology, even in the most primitive forms, has always played a significant role in warfare. Once human beings evolved from battling one another with their bare hands to pummeling each other with rocks or sticks, the race for employing the most advanced and deadly technologies in war began. Ever since, belligerents have wrestled with ways to employ weapons and equipment, the raw materiel of war, ever more effectively. Rocks and sticks evolved to bows and arrows, which begat firearms and artillery, leading to machine guns and long-range rockets, and now today's precision munitions and remotely piloted armed drones. The destructive power of technology at war has continued to increase, which makes it even more important for adversaries to adapt effectively to its ever-changing employment.

For the purposes of our study, we define technology as the weapons, equipment, and accoutrements of battle employed in armed conflict. These may include civil capabilities pressed into military use such as wireless radios or trucks, or may be unique military items such as main battle tanks or artillery pieces. All form a sweeping menu of diverse technology from which military leaders can pick and choose before and during wars, assembling new means of improving the battlefield effectiveness of their forces. Moreover, the demands of war itself can catalyze the demand to develop new, yet unimagined or undiscovered, technologies that provide a wartime advantage. Nearly every war produces such new developments, whether it be the tank, the proximity fuse for artillery shells, or the atom bomb. Finding and fielding the most advanced technologies for a military in peacetime offers the potential for war-winning advantage on a future battlefield. Every army is seized with the importance of keeping up with if not surpassing its potential enemies in this never-ending contest. Arms races between nations are very much the product of this urge, and reflect how important it is to have a better, faster, and more lethal set of military capabilities than a prospective future adversary.

As a result, military history often focuses heavily on the role of technology in shaping the conduct of war. This was particularly true in the 20th century, as noted scholar Stephen Biddle has observed: "The World War I trench stalemate was seen as the product of the industrial revolution in the machine gun, new artillery, and mass munition production. World War II was seen as a war of movement brought on by the tank, the airplane, and the radio. Postwar conflict was seen as overshadowed by the atom bomb."[113] This view is consistent with a broader school of thought called *technological determinism*. It views technological change as something that naturally happens through scientific research, which then affects society in some way.[114] The causal arrow, in this view, points in a straight line. New weapons systems are developed by researchers that function in specific ways, and those ways shape how wars are fought. This view is commonly held among many military scholars, particularly those in the neorealist tradition.[115]

More recent approaches, however, suggest that this linear approach is both incomplete and misleading. Starting in the 1980s, scholars developed an alternative approach called the *social shaping of technology*. They critiqued many aspects of technological determinism, but for our purposes, the most important critique argues that human agency shapes how technologies are used just as much as, if not more than, the other way around.[116] In other words, users can figure out new and creative uses for existing technologies that give them different purposes and have notable consequences.[117] Take, for example, the 9/11 attacks. Before that date, very few if any people had thought to use an airplane as a missile in a terror attack. That was a diabolically clever way to use an existing technology in an entirely new way, the results of which have profoundly shaped US foreign policy for almost two decades.[118]

How does this help us understand military adaptation? When a war begins, militaries are stuck with the weapons and systems they already have. During the opening battles, they do not have time to develop new ones to address the unanticipated battlefield challenges they encounter. Instead, front-line troops must figure out how to modify or adjust their existing technologies to address their tactical challenges, or find completely new ways to utilize them. If the war extends past days and weeks into months and years, then the military may have enough time to design, procure, and field new weapons and other types of equipment. Yet, as we discuss in Chapter 4, this requires adaptability at the institutional level—the ability to think creatively about possible technological solutions, and to overcome the inevitable resistance within the military bureaucracy to rapidly produce new systems and get them to the troops in combat as quickly as possible.

Leadership

Leadership is a central element of any collective human endeavor. At its core, leadership is the art of persuading others to take on and accomplish tasks that they would otherwise not do. Leaders also inspire and motivate their followers to achieve things together that none could do alone. Military leaders execute their authority through a chain of command that links the most senior generals to the privates and sergeants on the battlefield. In times of war, military leaders are responsible for sending men and women into harm's way, often on missions that could lead to grievous injury or death. There is no greater human responsibility.

Effective military leaders bring their willpower, intellect, and physical vigor to the battlefield. They set the example for others to follow, and infuse energy and determination into those around them. They can rally dispirited troops facing defeat and motivate them to fight on. They can reinvigorate exhausted units, faltering on the cusp of victory, to prevail through sheer force of will. At higher levels of command, the power of their confidence and determination can not only change the course of battles, but reverse the course of entire wars. Exceptional senior leaders can even inspire entire armies, as great commanders including Hannibal, Napoleon, and Eisenhower have demonstrated.

Ineffective leaders can be devastating in any environment, but their effect upon military operations is disproportionately large. In war, bad decisions by military leaders can have devastating consequences for their followers and, at times, their nations. Leaders who fail to grasp the true character of the battlefield, who are wedded dogmatically to failing tactics or strategies, and who are unable or unwilling to seek out good ideas from their subordinates can wreak grievous damage. The failings only multiply with increasingly catastrophic effects at higher levels of command. Weak leaders can undermine their troops' strengths, and even precipitate disaster or outright defeat.

Beyond the battlefield, senior military leaders also have immense responsibilities that are mirrored by their outsized organizational impact. Before the war, they make many of the key decisions on how an army is to be trained, organized, and outfitted. They make far-reaching judgments and approve key assumptions about the character of the next war, and are the principal architects of the force that will fight that war, whenever it comes. When wars break out, these leaders often make key decisions that directly affect the fighting forces—judgments to develop new military capabilities or modify old ones, keep or abandon established doctrine and procedures, and potentially shape the strategic direction of the war. These types of decisions have far-reaching impacts before wars are fought, and even greater effects once wars erupt.

Leadership is the most important of the three factors in this framework. Strong and adaptive leaders can overcome problems with doctrine and technology, but weak and inflexible leaders can undermine even the best examples of each. For this reason, developing adaptive leaders in peacetime, who are ready to adjust to the always unpredictable demands of war, may be one of the most important objectives of any military. A military whose leadership meets that high standard will be far better prepared for its future wars than one whose leaders fall short.

A Note on Culture

One possible criticism of our analytic framework is that we do not include organizational culture. Scholars have been long been interested in understanding the extent to which different countries have different strategic cultures,[119] and how the cultures among and within the military services shape their behavior.[120] More recently, cultural approaches have been used to understand broad strategic phenomena, from the sources of national power and military effectiveness, to more specific questions about the behavior of individual military organizations.[121] And, as we noted earlier, Theo Farrell and his colleagues have used cultural approaches to help explain some of the processes of military change that are so central to this book.

Why, then, do we not include organizational culture as a fourth critical component of adaptability alongside doctrine, technology, and leadership? We believe that culture is an integral element of each of these three components, rather than a separate component with independent explanatory power. In other words, we focus on how organizational culture *expresses itself* through a military's doctrine, its utilization of technology, and its leadership. The approaches mentioned earlier all focus on how cultures and norms shape organizational choices and decisions—but we believe that understanding adaptability requires understanding the *effects* of those choices and decisions rather than simply their original sources.

Applying the Framework

The next three chapters examine each element of our framework in more detail. Chapters 2, 3, and 4 identify specific characteristics of adaptability in doctrine, technology, and leadership, respectively, and provide examples from 20th-century conflicts to illustrate how those characteristics either helped or hindered wartime adaptability. Chapters 5 through 8 then use these characteristics to assess how adaptable US ground forces were during the wars in Iraq and Afghanistan. (We divide the discussion of leadership adaptability into two chapters to make it easier to follow.) Chapters 9 and 10 use these characteristics to help answer the key question of this book: whether the US military is adaptable enough for the types of conflicts it will face in the future. Chapter 11 concludes with a series of recommendations on how to make today's US military more adaptable.

2

The Role of Doctrine

ALL MILITARIES ENTER wars with some degree of relatively established doctrine, which is the conceptual and practical basis for how they will operate in battle. As discussed in Chapter 1, doctrine provides one of the most important ways that militaries try to impose a degree of order upon the utter chaos that is war. It provides commanders and soldiers with a common road map to help navigate anticipated combat missions, and helps them think about how to address a wide range of contingencies. Good doctrine also helps prepare soldiers to respond effectively to unanticipated developments on the battlefield—Clausewitz's fog and friction—long before the first shots are fired.

There are many bases upon which to judge a military's doctrine. Doctrine developed and refined between wars is in many ways a bet at the roulette table, because no military, no matter how prescient, can fully anticipate the character of its next war. Strategists and planners relentlessly study the campaigns of distant wars, dissect the recent wartime experiences of other armies, and carefully reflect upon their own battlefield victories and defeats as they rethink and refine their doctrine in preparation for the next war, whenever it may come.

Some militaries can predict their likely future adversaries, such as a hostile neighboring state, and thus have a good understanding of the opposing military capabilities they may face. They may also know the geography and terrain upon which they are likely to fight, such as the deserts or mountains, forests or cities of their home country or immediate region. Other militaries may face far greater uncertainty as they prepare for future conflicts. They may have to consider a wide range of potential adversaries, and thus face far more unknowns when anticipating opposing military capabilities, climates, and geography. In either case, however, Michael Howard's enduring observation about doctrine remains sound. It does not matter, he argues, whether armies develop the wrong doctrine: "What matters is their capacity to get it right quickly when the moment arrives."[1]

No matter how confident they are in their prewar doctrine, militaries need to ensure that their doctrine can adapt to the unexpected realities of the battlefield. In order to adapt their doctrine effectively, militaries must

- **Ensure that doctrine is flexible and not dogmatic.** Doctrine should be more descriptive than prescriptive. Militaries that view doctrine as a rigid formula are inevitably brittle in battle and can be quickly defeated by more agile foes. Effective doctrine requires decentralized application by tactical leaders who work from broad principles rather than detailed rules in the fog and friction of war.
- **Rapidly assess battlefield performance and identify when doctrine needs to change.** Continuous, objective self-criticism of doctrinal performance during combat is essential to identify shortcomings and correct them. The entire chain of command should be engaged in this process, with candor and ruthless critiques expected from all levels. Senior headquarters cannot be exempt from criticism and have a special responsibility to lead and empower self-criticism at all levels.
- **Search out and incorporate effective new doctrinal suggestions from the entire chain of command.** Changing doctrine must occur both from the bottom up and the top down. Junior leaders, midgrade officers, and senior commanders must all share a commitment to improving doctrine based upon battlefield results. Organizations must have both a culture and mechanisms that encourage the best ideas to flow up the chain of command.
- **Rapidly disseminate and implement doctrinal changes throughout the force.** Doctrinal adaptation does not result from good ideas alone. Newly devised changes must be effectively disseminated throughout the force and accompanied by any necessary training. Processes that facilitate these important follow-through steps must be regularly practiced before the war begins. After a war starts, doctrinal adaptation often involves rapid retraining during lulls in combat or between major campaigns.

Chapters 3 and 4 distinguish adaptability at the tactical level from adaptability at the theater or institutional level, and analyze each one separately. But adaptation in doctrine only occurs at the institutional level, almost by definition. Squads and platoons can innovate and experiment with doctrine, but their new approaches do not *become* doctrine unless and until they are approved by senior military leaders, who often serve in military headquarters far from the battlefield. These leaders are ultimately responsible for the doctrine that their forces use in combat.

	Adaptable Doctrine	Rigid Doctrine
Accurate Prewar Doctrine	German Army, 1939–1940	Egyptian Army, 1973
Mistaken Prewar Doctrine	Israeli Army, 1973	French Army, 1939–1940

FIGURE 2.1 Examples of Doctrinal Adaptation during Wartime

As Figure 2.1 shows, this chapter examines four examples of wartime doctrinal adaptation from two different conflicts: Germany and France in 1939–1940, and Israel and Egypt in 1973. The German and Egyptian armies both developed prewar doctrine that proved largely accurate in their planning, which gave them considerable advantages over their opponents. France developed prewar doctrine that was both mistaken in its expectations and rigid, a catastrophic combination that significantly contributed to France's defeat by the Germans in 1940. In 1973, Israel entered the war with mistaken doctrine, but its doctrine was inherently adaptable enough to overcome that disadvantage. Its rapid battlefield adaptation allowed it to defeat Egypt, which began the war with accurate doctrine, but whose doctrine was ultimately too rigid to adapt on a fast-changing battlefield.

Germany and France, 1939–1940: Accurate and Adaptable Doctrine Defeats Mistaken and Rigid Doctrine

Few battles demonstrate the importance of adaptable doctrine better than the opening battles of World War II in Europe, which culminated in the fall of France in June 1940. The war began on September 1, 1939, when Germany launched a surprise attack against Poland. German prewar doctrine was largely accurate for the types of battles it faced at the start of the war, a powerful advantage not shared among its early adversaries. The German invasion of Poland, joined by the Soviets just 17 days later, quickly overwhelmed the capable but outmaneuvered Polish military. After the rapid Polish defeat, German military leaders could reasonably conclude that their army had performed exceptionally well, and that its now combat-proven doctrine would prove effective in a looming clash on its western front with France. However, the German High Command did exactly the opposite. It spent the intervening months learning from its experiences in Poland and aggressively adapting, testing, and refining its offensive doctrine. France, by contrast, remained steadfastly committed to the defensive doctrine it had developed after World War I. The French High Command refused to reexamine its doctrine even though the fall of Poland should have taught the French that Germany was fighting much differently than it had three decades earlier. When Germany invaded France on May 10, 1940, it immediately became clear that the French prewar doctrine was not only mistaken, but also far too rigid to adapt to the German onslaught. France was completely defeated in six short weeks. The cost of French doctrinal failure was high: the German occupation that followed lasted for nearly four more years.

German Army Doctrine: Accurate and Adaptable

German army doctrine at the outset of World War II is a striking example of the best-case scenario for any military preparing to go to war. Its doctrine proved largely accurate in assessing the shape of the next war, and at the same time, remained very adaptable. Between World Wars I and II, the German army transformed itself from a defeated force to an entirely new organization with a powerful and original new doctrine. Moreover, its ability to rapidly adapt and improve this new doctrine was demonstrated after the 1939 Polish campaign. This combination made it doubly dangerous and set the stage for the stunning victories it would inflict on the Allies in the following years.

The German military drew substantially different battlefield lessons from World War I than the Allied powers did. In the last two years of the war, Germany had pioneered a new doctrine called "infiltration tactics" that upended years of static trench warfare thinking. It relied upon small squads of specially trained "storm troops" who were equipped with light machine guns, flamethrowers, and hand grenades that allowed them to move rapidly across the battlefield. Most importantly, their tactics emphasized bypassing enemy strongpoints to plunge ever deeper into the enemy's rear to sow confusion and chaos.[2] The German army developed this new doctrine and retrained its combat forces to employ it in a matter of months in the midst of ongoing combat operations, a truly remarkable feat.[3] This new doctrine, and the way in which it was developed and infused into the force in the midst of combat operations, made a deep impression on German army leaders and remained a key element of German thinking after the war.

The 1919 Treaty of Versailles forced the defeated German army to absorb strict limitations on its size, its officer corps, and its equipment. Ironically, these limitations may have catalyzed innovative thought and creative doctrinal development between the wars. Defeat animates militaries in ways that victory often does not, and the German army was able to leverage its defeat in 1918 into a series of aggressively pursued reforms and innovations that led it to become arguably the most capable military in the world by 1939. Its bold new offensive doctrine, derived from its intense study of World War I, was perhaps the foremost of those.

As the German army endured the massive cutbacks stipulated by Versailles, its leaders saw the opportunity presented by austerity. The size of the German army was slashed from a wartime high of 13.5 million to only 100,000 troops, and the German officer corps was limited to just 10,000. But instead of cutting the force haphazardly, German military leaders selected those who would be allowed to remain through a competitive process. Many combat-experienced officers became noncommissioned officers and stayed in uniform. The very best officers were retained, oftentimes at ranks far junior to their previous wartime positions. Colonels became captains or even lieutenants, and even some generals became colonels once more. This competitive selection created a small army of unparalleled quality, and one that retained some of the best minds of the wartime force for the years of peacetime thinking and practice to come.

Versailles also tightly limited German military equipment. No tanks or armored vehicles were allowed, and the air force was abolished entirely. Limits were also placed on the numbers and types of artillery, machine guns, and even light weapons. But these restrictions in weaponry—like the personnel limits—had perversely positive effects. Unlike their French counterparts, German officers did not have to develop effective ways to employ large stocks of aging equipment left over from the last war. There were no stockpiles of obsolete artillery or tanks to maintain, no recurrent costs of training large forces in tactics dictated by aging weapons that were still on hand. German military leaders were free to think conceptually, to imagine what might be possible and to envision combinations of doctrine and emerging capabilities to meet challenges that a new war might bring.

By the early 1930s, German military thinkers recognized that their innovative World War I infiltration tactics could be effectively combined with evolving new technologies in communications, aviation, and the internal combustion engine to create entirely new battlefield capabilities. Instead of using well-armed, nimble infantry detachments on foot to strike deep into the enemy's rear, they saw an opportunity to update and accelerate the effects of their infantry assault doctrine. German officers modernized these trench warfare tactics by adding fast-moving armored vehicles and replacing concentrated artillery with far-reaching air power. The new doctrine also encouraged commanders to move well forward on the battlefield, operating from newly designed, mobile, wireless command vehicles, so they could see the battlefield and make critical on-the-spot decisions.

These intellectual breakthroughs enabled a clean break with the legacy weaponry and doctrine that encumbered the postwar thinking of the victorious Allied powers. It also reflected the agility of thought of the much smaller and intensely capable German officer corps, which was catalyzed by defeat in ways that the victorious Allies were not. The Germans thought, discussed, debated, and wrote about this emerging doctrine actively during the interwar years.[4]

The German army also strongly emphasized the role of junior officers and NCOs in making independent decisions on the battlefield, unencumbered by rigid doctrine. The principle of decentralized operations was deeply ingrained in the German military, with junior leaders expected to use initiative and seize opportunities on the battlefield without waiting for orders. The German army traditionally emphasized the notion of *Führungsprinzipien*, leadership principles, and its companion concept of *Auftragstaktik*, or mission orders.[5] These concepts involved empowering leaders at all levels to act with bold initiative within their higher commanders' intent. In fact, junior officers could disobey orders from above if they felt that alternate approaches would better accomplish their assigned missions.

This emphasis on decentralized, independent battlefield actions, long a part of German military thinking, once again became a central tenet of German army doctrine in the modest force of the post-Versailles period. Mission orders were regularly emphasized and practiced during peacetime training exercises. Moreover, the German army relentlessly critiqued the performance of its leaders and units in exercises and war games. Commanders and staff officers at all levels were expected to do so candidly and objectively, without regard to personal embarrassment or potential career damage. This candor extended to critiquing the performance of senior officers and higher headquarters as well. These principles made German doctrine inherently adaptable in the face of battle.

By 1939, the newly expanded German military had a new and novel offensive doctrine enabled by sizable numbers of aircraft and armored vehicles with modern wireless communications. Although other Western armies dabbled in mechanized operations, the Germans made it the centerpiece of their operations. Unlike French and British tanks, all German tanks could receive new orders by radio while on the march. This created substantial agility by sharply reducing the time it took German commanders to make battlefield decisions and issue orders to forces on the move—down to minutes or even seconds. German aircraft also played a central role in ground operations. They had air-to-air and limited air-to-ground communications, which enabled them to coordinate with advancing tanks and provide immediate close air support to break down enemy centers of resistance.

The German army that invaded Poland in September 1939 was entirely transformed from the army that had been defeated in November 1918. Moreover,
its doctrine, training, and leadership principles had been entirely modernized,
whereas its potential adversaries continued to operate much as they had in the
past. The Polish campaign provided the first test of its new doctrine, and enabled
it to further refine its tactics and new weaponry. Using its new form of mechanized, combined-arms warfare, the German army smashed the large and capable
Polish military in just five weeks. Its fast-moving armored formations quickly
overwhelmed enemy resistance. These mechanized spearheads sliced through the
Polish defenses and struck deep into their rear areas, surrounding and rapidly
defeating large Polish formations. This new doctrine quickly became known in
the Western press as *Blitzkrieg*, or "lightning war."

Not only had the Germans largely succeeded in developing accurate doctrine before the war, but that doctrine remained inherently adaptable. Within
weeks of the Polish surrender, the German army immediately started studying the lessons of the Polish campaign. The German High Command ordered
detailed after-action reports from all units that had been involved, and over the
following months, it produced sharply self-critical accounts of its battlefield performance. These noted a myriad of deficiencies, including problems coordinating air support, poor small-unit camouflage, insufficient junior leader initiative,
and poor physical fitness among reservists.[6] The critiques became harsher as the
self-assessment moved up the chain of command. Senior-level headquarters were
even more self-critical of their own combat performance than were lower-level
units, a highly unusual phenomenon in military hierarchies.[7]

The lessons learned through this process led the German army to revise its
doctrine once again. These adaptations were tested, refined, and polished in a
series of tactical exercises and war games of increasing scale and complexity during the winter of 1939–1940, which were designed to prepare the force for the
invasion of France and western Europe the following spring.[8] Moreover, the exercise program contained its own process of critique and evaluation to ensure that
the doctrinal changes were being effectively implemented and adapted where
needed.[9] By the spring, the Germans were well prepared to fight the Allied powers in the west.

French Army Doctrine: Mistaken and Rigid

The French army, victorious at the end of World War I, drew substantially different lessons from that long and bloody conflict than the defeated Germans. By
1917, relying upon detailed orders to orchestrate massive and complex offensive
operations had become deeply embedded in French military practice. Thus, the

French army struggled to adapt during the last year of the war, when mobile operations became more commonplace. As a senior French commander observed at the time, "The command no longer knows how to shake off the formula of position warfare, to break away from the detailed prescriptions and multiple calculations which it has occasioned. When confronted by the unexpected, it remains confused; its action is marked by slowness and hesitation; thus, in most cases, it allows the most favorable opportunities for action to escape. It no longer knows how to act with speed."[10] The French army's experience in World War I, which was dominated by the intricate planning required for "methodical battle" and the central role of firepower, continued to have an outsized impact on its interwar thinking and planning.[11]

France ended World War I supremely confident in its army and its generals. French generals believed that their wartime military doctrine had been validated by the Allied victory in 1918 and should therefore be retained. The French army remained a large and potent force after the war, in contrast to the militaries of nearly all of the other belligerents.[12] Five million men remained on the active and reserves rolls, and the army retained vast stockpiles of wartime weapons and equipment. In a new era of peace, with a vanquished foe and predictably tighter budgets, French army leaders saw little reason to change their doctrine—and even less of a need to lobby for new equipment. The extensive numbers of rifles and helmets, machine guns and artillery pieces, and even new tanks left over from the victory in 1918 seemed to be more than sufficient to provision its army for years if not decades to come.

In stark contrast to the Germans, in the French army there was "no large-scale examination of the lessons of the last war by a significant portion of the officer corps."[13] Partly as a result, the lessons that the French army drew from World War I led to a warfighting doctrine that was nearly the polar opposite of that developed by the Germans. The French army assumed that the next war in Europe would largely resemble the last. The staggering number of French casualties during World War I led French leaders to conclude that an offensive doctrine would prove both indecisive and prohibitively costly.[14] They reasoned that a defensive doctrine would best preserve their fighting power and prevent the enemy from winning another major war through an offensive strike. As a result, nearly all French interwar thinking focused on leveraging defensive operations to prevail in any future war.[15]

French interwar doctrine inevitably also had to find ways to best utilize massive stockpiles of aging World War I equipment and materiel. A defensive doctrine enabled the army to place its large stocks of artillery and ammunition in fortified defensive positions that could pummel any attacker with heavy firepower. Furthermore, defensive tactics were much simpler to train for and could more readily be employed by the hundreds of thousands of new soldiers whom France envisaged training during a rapid mobilization. Defending static and fortified positions also required little physical prowess, which better suited the largely older cohort of reservists whose physical ability to undertake strenuous offensive operations was questionable.

By the early 1930s, concerns about the German rejection of key provisions of the Versailles Treaty, combined with French plans to withdraw from the Rhineland, caused the French government to fund perhaps the most famous defensive buildup of all time: the Maginot Line. Begun in 1929, the Maginot Line became the crown jewel of French defensive doctrine by the mid-1930s. Its 450 miles of fixed fortifications was a massive national investment, eventually extending from Luxembourg to Switzerland and protecting the easternmost flank of France against a direct invasion from Germany. Notably, the line did not extend north of Luxembourg along the lengthy border with Belgium, since Brussels was expected to be an ally in any future war with Germany.

French assumptions about the character of the next war were also very different from those of the Germans. The French looked back on World War I and saw the protracted horrors of relatively immobile trench warfare dominated by heavy firepower and sustained infantry combat. As a result, they projected that future wars would be prolonged conflicts giving the advantage to the side that could mobilize its population and industry for the long haul. The principles of mass, firepower, and human endurance therefore remained deeply embedded in French defensive doctrine. Speed seemed impossible to achieve in such an environment, so they believed that the next war would be slow and methodical. They also believed that no enemy could achieve a surprise victory, since the Maginot Line would adequately defend their borders and provide ample time for mobilization if necessary.

French interwar military leaders were battle-hardened veterans, but their hard-won victory of 1918 had dulled their senses to the potential of new thinking. Led by the aging icons of World War I, including national heroes Marshals Joffre and Petain, they doubled down on the lessons they chose to learn from their recent experience. These decorated veterans largely rejected the disruptive ideas emerging about mechanization, while innovative military thinkers in Germany, Britain, the United States, and the Soviet Union were all starting to grasp that offensive mechanized forces could play a potentially decisive role in the next war.[16] In France, forward-thinking officers who advocated for independent armored formations, including Colonel Charles de Gaulle, were ridiculed and marginalized.

By the mid-1930s, the French had allocated significant resources to develop first-rate new tanks and modern aircraft, but their doctrine for employing these new capabilities remained largely the same as it was at the end of World War I. Modern French tanks were relegated to a defensive role supporting groups of infantry, and scattered in small numbers across formations of slow-moving troops. The newest French tanks were capable of moving quickly, but French doctrine called for them to operate at the same speed as infantry forces marching on foot. French generals stonewalled de Gaulle's suggestions about creating experimental armored units until early 1939, but by then it was too late.

Every military risks choosing the wrong doctrine for its next war. But French army leaders compounded this problem by suppressing innovative thought between the wars, embracing a culture that demanded fealty to rigid doctrinal principles at the expense of debate. Even worse, their doctrine contained little if any capability for adaptation. French military exercises were designed to reinforce its doctrine of defense and methodical battle, and effectively buried any criticism of performance by leaders or units.[17] This made it virtually impossible for the French army to critically examine its own experiences and to be ready to adapt if its doctrine proved wrong in the opening battles of the next war.[18]

By the end of August 1939, the standing French army was one of the largest in the world, with 900,000 active troops and nearly 5 million reservists available for mobilization.[19] The nearly complete Maginot Line defended most of the French frontier, and France strongly believed that its ally Great Britain would come to its aid in the event of war. French aircraft and tanks were among the best in the world, outclassing most of their German opposites, and were deployed in impressive numbers. The French believed that time and their robust defensive capabilities placed the prospects for victory firmly on their side. The French army was prepared to slow any German offensive to a crawl through its fortified defenses. It intended to block any penetration of the Maginot Line and swing its more mobile forces forward into Belgium alongside British forces to defend against German thrusts there. Its well-developed defensive doctrine would enable France to fight a long-term war of attrition. The mobilization of the French population and industry, French allies in Europe, and the overseas French Empire would eventually exhaust and defeat the Germans.

The German invasion of Poland on September 1 plunged France and Britain together into war once again. As the Germans quickly and efficiently decimated the Polish army in just five weeks, Allied generals began to realize that the German army was employing an entirely new doctrine that used mechanized and air forces in innovative and unexpected ways. They also realized that a German attack in the west was inevitable. Yet while the victorious German army intently studied every aspect of its short fall campaign, the French and British armies largely ignored the conflict. They spent the winter of 1939–1940 improving their

defensive positions in the west while learning few new lessons about their dangerous enemy. This would prove a fatal oversight.

Reflecting a cultural bias and the military snobbery characteristic of western
European militaries, French military leaders blamed the speedy Polish defeat on
the ineptitude of the ill-equipped Polish army, and brushed aside any suggestions
of superior German doctrine and battlefield performance. It was simply inconceivable to them that the swift German victory had been caused by a completely
new doctrine applied with stunning proficiency. Thus, even though France had
received a detailed preview of German doctrine during the swift Polish campaign,
it almost entirely dismissed what should have been the very apparent dangers of
Germany's new offensive doctrine. As a result, France did not significantly adapt
any of its doctrine, training, or deployments after the Poles' rapid defeat. The
French remained supremely confident that their military was superior to the
German army, tank for tank and warplane for warplane. Their unshakable belief
in the supremacy of their defenses and the impregnability of the Maginot Line
led to a lethal complacency in the face of the looming German threat.

The Battle of France: May 10 to June 25, 1940

After a winter characterized by minimal Allied military activity, Germany
struck west on May 10, 1940. A large German force invaded the low countries
of Belgium and Holland, drawing large parts of the Allied armies forward into
Belgium in response. Unbeknownst to the Allies, a second German force with
the bulk of its armored units was simultaneously negotiating the supposedly
impassable Ardennes region in southern Belgium and Luxembourg. The German
armored columns moved slowly through the wooded Ardennes terrain largely
undetected by French intelligence. After scattering a screen of Belgian defenders,
the strung-out forces steadily pushed through the rough going and reached the
open terrain north of the Meuse River in less than two and a half days.[20] German
columns burst upon the surprised French defenders along the Meuse and quickly
forced several crossings with infantry and armored formations—which they
had rehearsed in detail on German rivers during the winter. The French High
Command had expected that it would take the Germans several days to cross
the Meuse, but they did so in less than 24 hours and started racing west toward
the English Channel. French forces along the Meuse retreated in chaos as the
German spearheads drove deep into their rear areas, threatening to cut them off
from their supply lines. French and British units in Belgium also began to fall
back, realizing that their rear was threatened by the advancing panzers.

This powerful surprise attack from the Ardennes threw Allied plans into dis-
array. The Germans were moving at a speed that was incomprehensible to the

FIGURE 2.2 Troops of the German First Panzer Regiment Cross the Meuse River near Sedan, France, May 14, 1940

Source: https://en.wikipedia.org/wiki/Battle_of_Sedan_(1940)#/media/File:Bundesarchiv_Bild_146-1978-062-24,_Floing,_Pontonbrücke_über_die_Maas.jpg.

French, who were still wedded to their plodding doctrine of methodical battle from World War I. The speed and shock of the German armored thrusts induced panic and near-paralysis in the outmaneuvered French High Command. French commanders distrusted their wireless capabilities, fearing enemy eavesdropping, and, unlike the Germans, they had no mobile wireless capability. Partly as a result, French senior commanders were located far in the rear, often tied down to large staffs in chateaus equipped with messengers to ferry written orders to the front in the event of an attack. Their ability to make quick battlefield decisions based upon changing conditions at the front was abysmal.[21] Moreover, once a decision had been reached, they could not rapidly convey new instructions to distant units under fire since they still relied on motorcycle couriers. French commanders were often unable to stay abreast of the battle raging around them and were often beset by confusion as orders from their superiors were given, countermanded, or arrived too late to be any longer relevant.[22]

The French inability to understand the battle and rapidly communicate revised orders crippled their ability to defend against their far more agile German adversaries. Bypassing French forces at unprecedented speed, the German armored spearheads reached the French coast by May 20, splitting the Allied armies. The

British Expeditionary Force retreated under intense enemy pressure to Dunkirk, where most of its forces were evacuated by June 4. The French surrendered less than three weeks later.

This sudden outcome was far from preordained. Six months earlier, the French army had been considered the best in the world, outnumbering the Germans in size and in the quality of most of its equipment, including its tanks.[23] It partnered with very capable British forces fielding substantial numbers of well-trained motorized and mechanized units. Yet the Germans were able to crush this potent Allied force in less than six weeks, which remains a monumental military achievement.

Much of the German army's tremendous success during the battle of France can be attributed to its doctrine. Its innovative, combined-arms doctrine accurately anticipated and shaped the characteristics of the emerging war, and forced the French and British to adapt to an entirely new way of fighting. Having far-sighted doctrine at the beginning of a war provides a significant advantage, but that advantage can be fleeting once enemies observe and effectively adapt. It was even more important that German doctrine was highly adaptable and could continually change and improve. Its adaptability in the face of ever-improving enemies served the Germans well throughout the war, much to the dismay of the Allies.

Conclusion: German Adaptability and French Rigidity in World War II, 1939–1940

The German army demonstrated all of the factors that promote doctrinal adaptability identified at the beginning of the chapter, while the powerful French army demonstrated none. The doctrine that Germany developed during the interwar period was both innovative and farsighted, and remained flexible enough to adapt to a range of unforeseen circumstances. The Germans repeatedly assessed their battlefield performance at all levels, including most senior levels, during interwar training exercises and even after their swift victory in Poland. They sought ideas and feedback throughout the chain of command and encouraged junior leaders to exercise a great deal of initiative. And they rapidly disseminated doctrinal changes throughout the force so that good ideas could be quickly applied.

French doctrine stood in stark contrast to German doctrine in almost every conceivable way. It was rigid and dogmatic, leaving little room for flexibility or adaptability. The French did not systematically assess their experiences in World War I or pay much attention to what should have been clear lessons from the German campaign in Poland. The French military stifled criticism of its performance, even during interwar training exercises, so as not to challenge its preferred defensive doctrine. And it did not disseminate and implement doctrinal changes, because it did not think that any changes were needed. French doctrine was so rigid and inflexible that the French army would have struggled mightily against any adversary that operated differently than what the French had predicted. But it stood absolutely no chance against the strikingly capable and adaptable Germans. The fall of France to German control after less than six weeks of fighting is a stark reminder of the potentially catastrophic costs when a military fails to adapt under fire.

Israel and Egypt 1973: Mistaken and Adaptable Doctrine Defeats Accurate and Rigid Doctrine

The 1973 Yom Kippur War between Israel and a coalition of Arab states was the first large-scale conflict between heavily mechanized adversaries since World War II. A striking example of the changing nature of modern war, it was characterized by speed, lethality, and unprecedented destructiveness. A critical portion of that war took place in the Sinai Peninsula, where elements of the Israeli Defense Forces (IDF)[24] deployed along the Suez Canal clashed with the Egyptian army. The war in the Sinai began when the Egyptian army conducted a surprise attack across the Suez that nearly overwhelmed the shocked Israeli defenders in the opening days of the assault.

The Egyptian military had dramatically revised its doctrine and capabilities following its abrupt defeat by the Israelis in the 1967 Six-Day War. In the years after their defeat, the Egyptians devised a novel offensive doctrine built upon new tactics and detailed planning as they plotted to eject the IDF from its defenses along the Suez Canal. This new doctrine strongly relied upon detailed planning and thorough rehearsals to ensure coordination among the forces responsible for executing complex attack plans. This approach initially proved advantageous, but ultimately its doctrinal rigidity became a lethal vulnerability. In contrast, the IDF began the 1973 war overconfident in its military prowess after its stunning victory in 1967. Key parts of its prewar doctrine were mistaken, having been shaped from erroneous lessons learned.[25] But the IDF's doctrine retained strong characteristics of adaptability and combat improvisation.[26] As the IDF withstood unexpectedly powerful blows at the beginning of the war, it rapidly adjusted its doctrine during intense combat. This rapid adaptation staved off a quick defeat and soon helped turn the tide in the IDF's favor and led to an eventual victory. While the Egyptians began the conflict with a well-planned doctrine for the war's first battles, their doctrinal rigidity meant that they could not adjust to the IDF's agile and rapid adaptations. That inflexibility coupled with the speed of Israeli adaptation led to Egypt's ultimate defeat.

Israeli Army Doctrine: Mistaken and Adaptable

The Israeli army defending the banks of the Suez in October 1973 was overwhelmingly confident that it could defeat the Arab nations arrayed around its borders. In the 1948 War of Independence, the Suez War of 1956, and the Six-Day War of 1967, the IDF had crushed its Arab foes, inflicting decisive defeats on the far more powerful Arab armies. Israeli military doctrine emphasized offensive operations characterized by speed and shock, which were empowered by a philosophy that emphasized decentralized command and control and improvisation by field commanders.

In June 1967, the IDF achieved an overwhelming victory against the combined militaries of Egypt, Jordan, Syria, and their allies in less than a week of fighting. The war began on June 5, when the IDF launched a preemptive attack against threatening Arab air and ground forces preparing to strike across Israel's vulnerable borders. In the west, IDF armored forces attacked deep into the Sinai Peninsula, destroying Egyptian mechanized and infantry units and leaving shattered formations in their wake. The war unfolded with lightning speed and ended on June 10 with Israeli combat forces on the banks of the Suez, on the Golan Heights, and occupying East Jerusalem. The final IDF positions at the end of the war became the de facto new borders of Israel, a major expansion of Israeli territory that had to be subsequently defended by the IDF. Israel's decision to hold onto these occupied territories ultimately helped set the stage for the 1973 war.

The 1967 war cemented the myth of Israeli military superiority not only in the minds of the IDF leadership, but in Western observers and the Arab states themselves. But the IDF's stunning battlefield successes meant that it now had to defend territorial limits that stretched far beyond the 1948 boundaries of Israel. To do so, the IDF shifted dramatically away from its historically successful offensive doctrine and instead adopted defensive doctrine to protect its newly seized lands. In the Sinai, the Israelis began to build what became known as the Bar Lev Line, a series of 31 strongpoints built to guard the length of the 175-kilometer-long Suez Canal. Israeli military leaders, unaccustomed to holding long-term, fixed positions, argued incessantly about how best to defend such a prominent terrain feature.[27] Eventually, they settled on a line of well-prepared bunkered defensive positions surrounded by mines and barbed wire. A massive sand escarpment built along the banks of the 200-meter-wide canal protected these fortifications. Although the defensive line appeared formidable, it was not always fully manned, given the IDF's limited manpower. Its defensive nature also clashed with traditional IDF doctrine, which long contended that only offensive operations were decisive in war.[28]

But the IDF had not entirely abandoned the offense. Further to the east, IDF commanders in the Sinai stationed strong armored forces in depth, poised to counterattack against any Egyptian penetration of the Bar Lev Line. But these armored forces looked very different from the ones that the IDF fielded during the 1967 war. At that time, armored units were combined-arms formations, including tanks, armored infantry in half tracks,[29] combat engineers, and mobile artillery. That war unfolded so rapidly, though, that Israeli tanks outran all of the other elements in their formations as the Egyptian forces collapsed.[30] After the war, IDF leaders erroneously concluded that their tanks should operate without these other supporting elements in order to maximize their speed. They therefore abandoned the combined-arms approach and built new armored units that consisted entirely of tanks.[31]

Thus, by the eve of the 1973 war, the IDF had two big doctrinal problems. For the first time in its history, it had largely abandoned its offensive doctrine in order to hold a static defensive line along the Suez Canal. At the same time, the doctrine for its armored forces had changed as well. After the IDF's quick and overwhelming victory in 1967, its leaders failed to undertake a rigorous examination of the war's lessons and instead focused exclusively on the remarkable success of their tank formations.[32] As a result, by 1973, IDF armored formations were organized without the critical mix of supporting arms upon which IDF offensive doctrine had long relied. The hasty doctrinal lessons that the IDF leadership drew from the 1967 war turned out to be the wrong ones, as the opening battles of the next war would soon painfully uncover.

Egyptian Army Doctrine: Accurate and Rigid

The Egyptian military was thoroughly defeated in the Six-Day War. Most Egyptian military equipment lay abandoned or in ruins in the now Israeli-occupied Sinai Peninsula. The war ended with Israeli armored formations on the banks of the Suez, where they remained for years. The Israelis soon decided to defend this new forward line by adopting a static defensive posture, and eventually built the heavily fortified Bar Lev Line to overlook the canal—itself a significant water obstacle that would confront any attacker. Any future Egyptian military strike against the IDF in the Sinai would have to deal with this formidable barrier in the first hours of any new war.

After the disaster of 1967, the Egyptian military set about rebuilding its decimated forces. While the IDF focused on defending its newly captured territories, the Egyptian military prioritized the development of offensive capabilities that could cross the Suez and retake some or all of the Sinai Peninsula.[33] To support this objective, the Soviet Union delivered massive infusions of advanced Soviet-made equipment to Cairo, reinforcing the importance of its long-standing ally. Moscow's largesse enabled the Egyptians to upgrade almost all of their weaponry between 1967 and 1973. Egyptian military leaders purposefully acquired a wide array of antitank weaponry and antiaircraft guns and missiles to help offset and ultimately defeat Israeli dominance in armor and aircraft, respectively.

Despite receiving substantial amounts of advanced Soviet weaponry, the Egyptian military did not fully adopt Soviet military doctrine or organizations. The Egyptians consciously tailored their doctrine to their own unique circumstances and operational requirements, which were quite different from what the Soviets would face fighting NATO in western Europe. Instead of adopting the Soviet focus on heavy armored formations with infantry in a secondary supporting role, the Egyptians organized and deployed infantry forces as a primary fighting arm. They also chose to deploy their newly acquired Sagger antitank wire-guided missiles (ATGMs) and shoulder-fired antitank rockets well forward with their front-line attacking units, even though the Soviets used these weapons primarily in a defensive role.[34]

Yet the Egyptian military did adopt key aspects of the relatively dogmatic Soviet approach to doctrine. The Soviet army emphasized limited flexibility and initiative at the tactical level in order to ensure the success of the overall operational plan. Junior Soviet commanders were not expected to exercise significant initiative lest their inventiveness put the carefully coordinated parts of the overall battle plan at risk. The Egyptians doubled down on this principle and made it even more inflexible in their new doctrine. As they crafted plans to cross the Suez and drive the Israelis back into the Sinai Peninsula, the high command in Cairo "scripted the entire operation down to the last detail."[35] Battlefield leaders and their troops were directed to "memorize a series of programmed steps" dictating their exact role in crossing the canal, and "junior officers were expressly forbidden from taking actions not specifically included in the General Staff plan."[36] Individual initiative, in other words, was not allowed, no matter what the battlefield exigencies.

Egyptian doctrine emphasized rote repetition of the complex maneuvers required to get a sizable force across the Suez before the IDF could counterattack. Thousands of Egyptian assault troops rehearsed the attack across the canal 35 times in its entirety, with one principal yardstick for success: total conformity to the plan.[37] This highly centralized approach ensured effective, detailed coordination between armor, infantry, artillery, engineers, and commandos for the very complex initial operation—the canal crossing and a limited follow-on advance. But subsequent offensive operations beyond the initial assault and expanded bridgeheads were not planned in any similar detail. The Egyptians' rigid adherence to detailed and carefully rehearsed plans was ill-suited for fast-moving operations or seizing emerging battlefield opportunities. Such an inflexible doctrine, which required strict obedience to highly detailed plans directed from higher headquarters, would be paralyzing when the fog and friction of war intervened. The Egyptians had devised their own version of the French army's doctrine of methodical battle—with the same dangers, as would soon become apparent.

The Yom Kippur War, October 1973

The Egyptian army launched its surprise attack across the Suez on the afternoon of October 6, 1973. A furious opening artillery bombardment from 2,000 guns pinned down Israeli defenders in their bunkers all along the Bar Lev Line. Egyptian aircraft, protected by an umbrella of air defense missiles, simultaneously bombed Israeli airfields, command posts, and surface-to-air missile sites deeper in the Sinai. Egyptian tanks in protected positions on the west bank of the Suez fired directly into Israeli fortifications on the Bar Lev Line, smashing pillboxes and creating further confusion among the shocked defenders.

Under the protection of this carefully orchestrated bombardment, five Egyptian divisions with 80,000 infantry troops began crossing the canal in nearly 1,000 rubber assault boats.[38] Specially trained engineers used water cannons to blast away the Israeli sand barriers on the east bank to create access paths for follow-on mechanized forces. Some of the assaulting infantry forces, equipped with Sagger antitank missiles and rocket-propelled grenades (RPGs), crossed the canal but bypassed the Bar Lev strongpoints to move deeper into the Sinai and block expected Israeli tank counterattacks. This carefully planned and rehearsed operation with its vast array of moving parts shocked the IDF defenders and unfolded nearly without a hitch. It showcased the advantages of a prewar Egyptian doctrine that emphasized detailed planning and strict compliance with exact orders from the high command. The IDF was taken completely by surprise, as some of the forward strongpoints were overrun and others were cut off and besieged. Virtually all Israeli strongpoints along the canal were abandoned or lost within 24 hours.

IDF commanders in the Sinai soon responded by launching their long-planned armored counterattacks to attack and repulse the rapidly expanding Egyptian bridgeheads. Israeli tank formations aggressively attacked from deep behind the Bar Lev Line into the teeth of the quickly growing Egyptian forces now across the canal. Unlike in 1967, however, these IDF armored units now consisted solely of tanks, without any accompanying infantry, engineers, or artillery support. The results were disastrous. The most forward detachments of Egyptian

FIGURE 2.3 An Israeli M60 Tank Destroyed during Fighting in the Sinai, 1973
Source: https://en.wikipedia.org/wiki/Yom_Kippur_War#/media/File:Destroyed_m60.jpg.

infantry, now armed with ATGMs and large numbers of RPGs, were already in hasty defensive positions behind the Bar Lev. These newly established strongholds rained destruction on the advancing Israeli tanks. From the first small-scale IDF armored counterattacks on October 6 to the division-sized attack on October 8, Israeli tanks attacking without infantry, artillery, or air support ran headlong into a blizzard of Egyptian antitank missiles and rockets. IDF tank units were repeatedly repulsed, losing as many as 200 tanks in a single day.[39] After the battle, junior Israeli commanders reported finding numerous Egyptian missile guidance wires strewn across their tanks' turrets, a testament to the ferocity of the ATGM attacks. Postwar analysis suggests that Egyptian RPGs may have knocked out nearly as many IDF tanks as did the new wire-guided missiles. As the Egyptians employed these two antitank systems with their forwardmost units using totally unanticipated doctrine, they severely disrupted the IDF's ability to employ its tanks in long-planned parries.

Egyptian successes in the first three days of the war forced the IDF to quickly reassess its prewar doctrine, or risk losing the war. The IDF's initial doctrine proved inadequate and mistaken for the changed battlefield environment of 1973. Its doctrinal assumptions about the effectiveness of tanks operating without supporting arms in the face of capable enemy forces proved outdated, as new Egyptian tactics and weapons rendered these prewar tactics obsolete. Fortunately for the Israelis, IDF doctrine retained its deep-seated characteristics of improvisation, adaptation, and individual battlefield initiative. These core principles helped save the IDF from complete battlefield defeat.

The IDF rapidly adapted its doctrine as combat continued to rage. Less than a week into the war, Israeli commanders decided to restore their combined-arms doctrine, and began to quickly reorganize their armored units into such formations. Mechanized infantry, engineers, and self-propelled artillery were integrated into formerly all-tank formations before the next series of IDF offensive operations began.[40] Tanks were also supplied with high-explosive shells to defeat enemy ATGMs and RPGs instead of carrying only armor-piercing ammunition designed to fight tanks.[41] New tactical doctrine was improvised to counter Egyptian antitank tactics. IDF tank commanders started firing volleys of these new high-explosive shells against the ATGM missile crews, who had to remain stationary for as long as 30 seconds after firing while guiding their missiles toward their targets. Attacking missile crews during this short window typically threw off their aim, causing their projectiles to miss their targets. Tank crews also began to better use the rolling desert terrain, sending a pair of tanks maneuvering forward only after another pair of tanks was in position to cover their advance with effective observation and fires.[42] The IDF's rapid battlefield adaptations soon proved effective.[43]

Meanwhile, the rigidity of Egyptian doctrine quickly became a profound liability. As the Egyptian army consolidated its lodgment on the east bank of the Suez, it began to prepare defenses to hold its newfound gains against the next series of fierce IDF counterattacks. However, on October 14, the Egyptian High Command unexpectedly ordered the army in the Sinai to abandon these defensive preparations and continue to attack.[44] This decision was a profound mistake given the inflexible doctrine that underpinned the initial Egyptian successes. Unlike the minutely detailed plans for crossing the Suez, no plans existed for this new operation. Now the scripted and inflexible doctrine that had initially succeeded became a severe drawback. Egyptian commanders lacked the intelligence and time to develop a detailed and scripted plan of fire to support the new offensive. As the hastily planned attack began, the massive but improvised Egyptian artillery bombardment largely missed its targets. Egyptian tanks then advanced in waves without infantry support directly into devastating IDF tank fire, sustaining huge losses in one of the largest tank battles since World War II. The attacking Egyptian army was hurled back in defeat, returning to the defensive line around its new lodgment just hours after the attack began.

Suddenly, the Israelis were in a position to reverse their earlier losses and shift to the offensive. Stung by its sudden setback and heavy losses, the Egyptian army now stood paralyzed in its hasty defensive positions miles deep in the Sinai. The Israelis, leveraging the inherent flexibility and initiative of their doctrine, quickly seized the initiative. Their new combined-arms armored units soon pierced the Egyptian defenses, as they employed their new tactical doctrine to neutralize ATGMs and handheld antitank weapons. IDF infantry, accompanying the tanks, quickly defeated enemy infantry missile threats to armored formations. These assault forces helped create openings in the Egyptian lines for Israeli mechanized formations to rush through, plunging deep behind the Egyptian defenses. By October 17, IDF paratroopers and tanks had moved west and crossed the Suez Canal on hastily erected assault bridges. Israeli armored divisions, now striking down the west side of the Suez, raced south toward Cairo and soon threatened to enter the outskirts of the city. This sudden menace to the Egyptian capital prompted the United States to intervene diplomatically to end the crisis, leading to a ceasefire on October 25. The October 1973 war, one of the most destructive conflicts since World War II, was now over.[45]

Conclusion: Israeli Adaptability and Egyptian Rigidity in the Yom Kippur War

The Israelis faced potential defeat on the Sinai battlefield before rapidly adapting to the unexpected challenges they encountered. The IDF's overconfidence following the stunning success of the 1967 war created a sense of hubris about Israeli military capabilities and led Israeli commanders to draw the wrong lessons about the dominance of the tank on the battlefield. They also failed to anticipate that the Egyptians would be able to rapidly cross the Suez and employ their new ATGMs and handheld antitank weapons in innovative and lethal ways. That forced the IDF to adapt quickly or risk outright defeat. Though its prewar doctrine was deeply flawed, it nevertheless remained flexible. As a result, the IDF rapidly assessed its battlefield failures in the disastrous opening days of the war and started to adapt its doctrine almost immediately. Senior leaders sought out a wide range of input from throughout their fighting units, up and down the chain of command. They also rediscovered key lessons about combined-arms warfare that the IDF had learned in the 1948 and 1956 wars. Finally, even though its units were engaged in desperate combat, the IDF was able to rapidly disseminate and implement key doctrinal changes throughout the force in a matter of days, an impressive performance under extreme battlefield pressure.[46]

The Egyptian military entered the 1973 war with a rigid doctrine that enabled it to prepare effectively for the set-piece battle of crossing the Suez and attacking the Bar Lev Line. But Egyptian doctrine in 1973 was even more dogmatic and inflexible than the Soviet military doctrine upon which it was based. It relied so heavily on detailed planning, centralized control, and extensive rehearsals that it could not adapt effectively to the inevitable fog and friction of war. The initially successful doctrine thus contained the seeds of its ultimate defeat. Once the Egyptian military encountered setbacks, its rigid and top-down system of command could not effectively garner feedback and ideas from throughout the ranks. All key decisions and changes to plans could only be made at the highest levels, at headquarters far from the battlefield. Even if the Egyptian High Command had been able to change its doctrine quickly, it would have been extraordinarily difficult for its rigid and hierarchical command structure to disseminate and implement those changes effectively while simultaneously engaged in fierce combat. Decentralized execution and initiative among junior leaders are essential in such circumstances—principles that were expressly forbidden by Egyptian doctrine.

The 1973 war demonstrates that getting doctrine right *before* a war can be far less important than having doctrine that strongly promotes adaptability *during* a war. The Egyptian military entered the war with relatively sound doctrine for its original battlefield purpose. But when battlefield conditions changed, Egyptian doctrine fell short—it simply could not adapt. As a result, the IDF was able to adapt before the Egyptians could deliver a knockout blow, and turned a seemingly certain defeat into an overwhelming victory.

The Israeli army entered the war as an overconfident force with mistaken doctrine. Its defensive doctrine failed in the first days of the Egyptian attack across the Suez, since it had learned the wrong tactical lessons from its swift victory over Egyptian forces in the Six-Day War. The IDF made an ill-founded shift away from its long-standing combined-arms doctrine toward one that relied solely on tanks for offensive strikes. This doctrine totally failed in October 1973, and caused the IDF to suffer devastating losses during the initial counterattacks against Egyptian infantry in the Sinai. Fortunately for the Israelis, their doctrine still encouraged adaptability, flexibility, and improvisation on the battlefield. This doctrinal flexibility allowed the IDF to rapidly adapt to its failures, and ultimately go on to defeat its inflexible adversary. The 1973 war powerfully demonstrates that effective adaptation during war can be more important to avoiding defeat and eventually achieving victory than entering a war with accurate prewar doctrine.

Conclusion: The Importance of Doctrinal Adaptability

Doctrine provides the blueprint from which militaries construct how they will fight the next war. Inevitably, it is only a best guess about what the nature of that next conflict will look like. Previous wars provide some guide to that picture, but the ever-changing dynamics of war ensure that no future war closely resembles any of its predecessors. Each war has its own character, formed by the unique convergence of time, location, plans, technology, leadership, and chance. All militaries enter wars with a series of doctrinal bets, but the best ones also realize that all of those prewar bets are off once the first shots are fired. From the very first engagements, the opponents duel back and forth to a finish in which only one will prevail. Adaptation, and reacting to enemy adaptation, becomes absolutely essential.

Militaries never know the extent to which their prewar doctrine will work until the shooting starts. Fortunate military leaders may discover that they generally guessed correctly, and that the war unfolds much as they had planned. The very lucky ones may win outright before their adversaries can respond effectively. This dynamic well describes the German army in 1939–1940 in its battles against Poland, France, and the Low Countries. The German military also had the additional substantial advantage of having inherently flexible and adaptable doctrine, which allowed it to continually improve and learn from each battle. In marked contrast, the French army was doubly doomed. Not only was its prewar doctrine largely mistaken, but it was completely incapable of adapting once the German offensive blows began to fall. French doctrine proved ill-suited for the war of 1940 and was brittle and inflexible in the face of battlefield setbacks, leading to a catastrophic French defeat in just six weeks.

Accurate prewar doctrine confers significant advantages, but is not necessarily sufficient for victory. The Egyptians launched the 1973 Yom Kippur War with effective doctrine for their opening battles and dominated the vaunted IDF during the first days of the war. By contrast, IDF prewar doctrine was largely mistaken but remained exceptionally adaptable. Despite enduring a series of jarring early setbacks and heavy initial battlefield losses, the Israelis swiftly began to adapt to the unexpected battlefield realities they faced. IDF leaders quickly assessed their flaws and rapidly implemented doctrinal changes to regain the offensive and ultimately defeat their surprisingly capable Egyptian foes. In doing so, they took full advantage of a flexible doctrine that was built upon long-standing principles of improvisation and small-unit initiative. The Egyptians were frozen in an inflexible doctrine that could not adapt to the fast-moving IDF operations, resulting in their total defeat just weeks after their breathtaking initial battlefield victories. In the 1973 war, the ability to rapidly adapt proved far more valuable than having accurate prewar doctrine.

No army can ever predict with confidence that its vision of the next war—and thus its doctrine—is entirely or even substantially accurate. But every army *can* confidently predict that adaptation will be necessary in the next war. The military whose doctrine prepares it best for rapid adaptation is far more likely to prevail in future conflicts than a military that has developed an accurate prewar doctrine but cannot adapt once the battle begins. Militaries that fail to deeply nurture the seeds of adaptation in their prewar doctrine take on the gravest of risks as they seek victory in the fog and friction of real war.

3

The Role of Technology

MILITARIES GO TO war with the military technology they have on the ramps of their motor pools and airfields. But surprises inevitably abound after the shooting starts, as weapons fail to perform as expected and long-planned battlefield systems do not function properly amid the fog and friction of war.[1] In this chaotic environment of opening battles, every military must find ways to adapt existing technologies and develop new ones that can help them prevail on the battlefield—or risk defeat.

Militaries face two types of technological adaptation, which are markedly different. *Tactical adaptation* is near-term, often bottom-up, and usually responds to immediate battlefield problems. It may involve adjusting or modifying an existing technology for an entirely new purpose, or developing some new invention on the battlefield itself. Tactical adaptations are typically driven by individual soldiers and junior leaders in the field, but they become more effective and consequential when they are adopted more broadly throughout the force. They are often the only form of adaptation when wars are short, lasting days or weeks rather than months or years. Institutions, and the bureaucratic rules that govern them, usually cannot adapt quickly enough during such brief wars.

Institutional adaptation falls at the other end of the spectrum. It is longer-term, often top-down, and usually involves major modifications to existing weapons systems or developing entirely new ones. It requires the headquarters of the military, far removed from the battlefield, to remain acutely attuned to technological shortcomings and to act rapidly once a problem has been discovered. But speedy wartime adaptation can be particularly difficult for large military organizations, whose peacetime processes for developing and fielding technology are often deliberate, bureaucratic, and risk averse. Institutional adaptation must be led by senior leaders—colonels and generals, and sometimes civilians—who drive supply and acquisition systems within the bureaucracy, and who can also reach into civilian industry to find a needed solution. Although far from the front lines, these leaders are the only ones positioned to drive these large institutions to make their balky bureaucracies respond with agility and speed to emerging battlefield demands.

Because tactical and institutional adaptation are so different, they require different attributes in different parts of the force. Successful technological adaptation at the tactical level requires combat units and their leaders, especially those closest to the front lines, to be able to

- **Approach unexpected problems with creativity and a willingness to experiment.** Front-line soldiers facing deadly shortfalls in their equipment have an extraordinarily strong incentive to fix them. Doing so effectively requires soldiers—especially in the junior ranks—to be problem-solvers and try new ideas. Militaries that encourage individual initiative and include plenty of resourceful troops are far better able to improvise on the fly, devising new ways to overcome problems with their weapons and other battlefield equipment.
- **Manufacture improvised solutions in or near the combat zone.** All armies have some level of field maintenance and repair capabilities. The most adaptable ones can repurpose their repair shops into ad hoc manufacturing plants that can rebuild or improve existing materiel, or even craft some limited new variations of weaponry and equipment. In some cases, they may also leverage government or civilian manufacturing facilities close to the combat zone to rapidly repair or modify equipment and return it quickly to nearby battlefields.[2]
- **Approve and disseminate solutions quickly throughout the operating force.** This requires far more than just manufacturing capacity; it requires senior field commanders to recognize problems, approve emerging ad hoc solutions, and then push hard to build and distribute those solutions during ongoing combat operations. Yet combat commanders are often wholly consumed with fighting and winning battles. They may not be aware of some of the technological problems facing their troops or may continue to believe that existing weaponry remains adequate even when presented with evidence to the contrary. But without their support, even the best field solutions will go nowhere. Senior commanders must not only embrace the need to adapt, but must also do everything they can to accelerate the adaptation process and disseminate successful adaptations across the battlefield.

Even if combat units and their leaders possess all of these characteristics, tactical adaptation may still fall short. As we will see later, the scope of the battlefield problem may simply exceed what can be fixed in the combat zone. In those cases, the problem requires an institutional adaptation—an entirely new tank, for example, or a munition that does not currently exist. Effective institutional adaptation requires different but equally formidable attributes than those required for effective tactical adaptation. Every delay in institutional adaptation to field a new or revamped system risks greater casualties and increases the chance of battlefield defeat. It is primarily the responsibility of senior staff officers far away from the front lines, since they are the ones responsible for pressuring their bureaucracies to change.

Successful technological adaptation at the institutional level requires militaries to

- **Maintain clear and candid communication with battlefield leaders to identify technology shortfalls early and regularly reevaluate shifting challenges.** First and foremost, effective institutional adaptation requires a clear recognition of a battlefield problem. If a military does not understand that its fighting forces have a serious deficiency that cannot be addressed in the combat zone, it cannot pursue an institutional solution. Senior leaders both in combat and at home share an obligation to communicate effectively, searching out and identifying dangerous technological shortcomings that have been exposed by the rigors of combat in order to ensure they can be addressed. Moreover, the challenges to battlefield technologies never remain static. New enemy weapons, changing front-line conditions, or even unexpected uses of existing technologies may all expose new weaknesses that must be recognized and corrected. Senior leaders must continuously reevaluate the performance of their own weaponry and equipment to avoid falling behind an adversary who is also constantly adapting.
- **Overcome internal turf battles and the peacetime acquisition bureaucracy to rapidly develop and scale new technologies.** Senior leaders and their staffs at home are responsible for developing and procuring new weapons systems, not units and their commanders on the battlefield. Yet these large bureaucracies often resist changes even when combat provides overwhelming evidence that existing systems are failing and that new ones are needed.[3] When higher headquarters do accept the need for change, they can become stuck over disagreements about how to solve the battlefield problem, especially when multiple proposals exist. And even when they do agree on a technical solution, they then must overcome peacetime processes for developing and testing new systems. These processes are inevitably cumbersome and slow, frequently prioritizing timeworn deliberations and careful resource management over innovation and imagination—exactly the opposite of what is needed in wartime. Effective institutional adaptation requires militaries to

overcome these bureaucratic challenges so they can devise, produce, and field new hardware rapidly enough to affect the outcome of the war.

This chapter examines some historical examples of technological adaptation during wartime. As Figure 3.1 shows, we look at two instances of tactical adaptation and two instances of institutional adaptation, with one success and one failure for each. They all involve armored warfare, to help facilitate the comparison, but the dynamics of technological adaptation we illustrate here apply to all types of combat operations. Both examples of tactical adaptation involve US Army armored operations in northwest Europe during World War II. In the successful one, junior army leaders preparing for the breakout from Normandy in the summer of 1944 devised an improvised cutting device for the front of their tanks. Called the Rhino plow, these ad hoc steel cutters allowed tanks to slice open hedgerows that were being used as fortifications by the defending Germans. The First US Army developed, tested, produced, and installed these devices on hundreds of tanks in just two weeks. This rapid adaptation was instrumental in the success of the subsequent offensive in Operation Cobra, since it enabled the Americans to surprise the Germans, break through their hedgerow defenses, and ultimately permit the Allies to rapidly advance across France. In the failed example, American tankers could not improvise any effective technological countermeasures to offset the lethally effective German high-velocity tank and anti-tank cannons that were devastating their formations. Despite numerous efforts, including layering their M4 Sherman tank hulls and turrets with sandbags and even concrete, US tankers remained fatally vulnerable to the penetrating power of high-velocity German guns until the end of the war. Thousands of M4 tanks were lost and many more thousands of crewmen killed and grievously wounded

	Success	Failure
Tactical Adaptation	US Army Rhinoceros tank plow, 1944	US Army supplemental tank armor, 1944–1945
Institutional Adaptation	French Army tank development, 1915–1918	US Army tank development, 1941–1945

FIGURE 3.1 Examples of Technological Adaptation during Wartime

as tank crews futilely struggled to adapt to this potent and widespread German threat.[4]

We also provide two examples of institutional adaptation. The French army's development of the tank during World War I is a remarkable story of successful adaptation. The French developed two medium tank models in 1916, but even before either one had been used in combat, they tested a third, lighter tank that proved to be far more successful. Between 1917 and the end of the war in 1918, France built and fielded more than 2,000 of these new light tanks—a truly remarkable accomplishment. By contrast, the lack of US Army tank development during World War II stands as a costly institutional failure. We identified the inability of US tankers to find ways to protect their tanks in 1944 and 1945 from German guns as a failure of tactical adaptation, but the fact that the army was still fielding M4s with few substantial improvements that late in the war was the far deeper problem. When the war began, the firepower, mobility, and protection of the M4 seemed adequate. But the Germans continued to develop and field better tanks and antitank guns throughout the war, while the US Army remained complacent about the performance of the M4. Starting as early as 1943, there was increasing evidence that the M4 could not stand up to the improving German tanks, yet the army failed to correct the problem before the war ended two years later. This striking failure of institutional adaptation resulted in staggering losses of US tank crews throughout the European campaign.[5]

Tactical Adaptation in Technology
Successful Tactical Adaptation: US Army Rhinoceros Tank Plow, 1944

In 1942, the US Army and its allies started preparing to invade Nazi-occupied Europe from bases being built up in England. Over the next two years, the growing Allied armies developed plans to break through the German beach defenses lining the English Channel in France and thrust deep inland. The invasion force prepared and rehearsed for thousands of hours, examining every imaginable problem that might threaten the massive and unprecedented amphibious landing.[6] Fortunately, the Allied landings in Normandy on June 6, 1944, were an overall success, but the Allied attack soon stalled. When the Americans stormed ashore on the western D-Day beaches, they discovered a dense patchwork of small farm fields starting about 10 miles inland in the path of their advance. The fields were bounded by earthen parapets that were four feet thick, up to 15 feet high, and topped by thick hedgerows that stretched for miles behind the beachheads.[7] The plots usually had only one tiny entrance, and that and their high, shrub-covered borders helped make each tract a natural defensive position for the Germans. This *bocage* terrain (from the French word for hedgerow) forced US soldiers to fight for one field after another against well-protected and concealed German defenders who adeptly used the hedgerow banks for cover. US tanks could not crawl over the top of the hedgerows without exposing their thinly armored underbellies to deadly German antitank fire. This patchwork of small fields was interlaced with sunken roads, creating a natural defense in depth that limited visibility and made prospects for rapid armored advance even more dismal.[8]

Senior commanders were aware of the *bocage* country before the invasion, but did not think it would be a major obstacle to the US advance. The assistant division commander of the 82nd Airborne Division, Brigadier General James Gavin, reflected the prevailing views of senior US commanders by noting, "Although there had been some talk in the U.K. before D-Day about the hedgerows, none of us had really appreciated how difficult they would turn out to be."[9] But junior American officers encountering the *bocage* were taken completely by surprise, and had received no specialized training in dealing with this warren of obstacles that extended for mile after mile behind the front.[10] As a result of this unexpected obstacle, General Omar Bradley's First Army, which had responsibility for rapidly breaking out of the beachhead after D-Day, instead ground to a halt. Bradley's tank units were confined in these natural mazes and began to suffer more and more losses, as did the US infantrymen vainly trying to assault these natural fortresses. They had a few bulldozer plows that could fit onto some tanks, but these proved too slow and far too few to break through the earthen walls and launch a major attack.[11] By late June, the entire US plan to break out of the Normandy beachhead was in serious jeopardy. In early July, First Army commanders hastily requisitioned nearly 300 additional dozer blades, but they could not arrive in time to support the breakout attack in Operation Cobra.[12]

In the face of this unexpected tactical setback, junior leaders of First Army tank units began to improvise. Recognizing that the hedgerows were the primary cause of such hard fighting and most of their casualties, they had strong incentives to find some sort of battlefield solution. In late June and early July, they experimented with a wide range of possible technological solutions to the hedgerow dilemma.[13] Some units, for example, developed ad hoc devices to breach the hedgerows that they mounted on the front of their Sherman tanks. But most of these improvised devices proved too slow to employ under fire and too fragile to withstand repeated use.[14]

But the widespread field experimentation eventually produced results. In the 102nd Cavalry Reconnaissance Squadron of the US Second Armored Division, a junior sergeant assisted by two majors and a captain invented an effective hedgerow cutting plow.[15] Sergeant Curtis Culin reworked scrap metal from German roadblocks into a cutting device of prongs and teeth that could be mounted on

FIGURE 3.2 A US Army M4 Sherman Tank with Rhino Plow Attached, 1944
Source: https://commons.wikimedia.org/wiki/File:Yanks_advance_into_a_Belgian_town.jpg.

the front of M4 Sherman and M5 Stuart tanks. Unlike the standard-issue tank plows, the soon-dubbed "Culin device" could slice directly through the hedgerows at their base, tearing a hole wide enough for a tank to push through in minutes. It proved both effective and relatively durable, and worked quickly enough for the units to achieve surprise against the waiting German defenders.

On July 14, the Culin device was demonstrated to Bradley, who immediately saw its promise and ordered the First Army Ordnance Section to put it into mass production.[16] With the preparations for the Cobra offensive accelerating, time was short.[17] Scrap metal was salvaged from German beach defenses, and ordnance shops in Normandy and back in England went into overdrive to produce enough devices—nicknamed the Rhinoceros or simply the Rhino plow—to outfit hundreds of First Army tanks for the coming attack. The distribution of Rhino plows to front-line units was amazingly fast. According to the 735th Tank Battalion's after-action report, its tanks were sent back for modification with the Rhino plow on July 17—only three days after they had been first demonstrated to Bradley.[18]

In the 11 days before the Cobra offensive, over 500 Rhino plows were produced and distributed to tank units. Since there were not enough Rhinos to give to all armored units, M4 Sherman and M5 Stuart tanks received the priority.[19] Other units continued to improvise on their own, and developed alternatives such as the Greendozer to ensure they had some capability if needed.[20] By late July, 60 percent of First Army's M4 Sherman tanks had been outfitted with some type of hedgerow cutter.[21] Installing the Rhino plows so quickly was a monumental logistics undertaking that had to be conducted secretly to preserve surprise while not weakening the front line as tanks were rotated back for modification. Executing such a complex ballet in just two weeks required great ingenuity to juggle the complex rotation of units in and out of the line without tipping off the Germans.

The Rhinos paid off handsomely during the Cobra offensive. Following a heavy Allied bombing of German front lines,[22] teams of infantrymen working with Rhino-equipped tanks once again attacked the formidable hedgerows, but this time surprised the bomb-shocked German defenders. The Rhinos quickly sliced through the hedgerow banks while infantry forces assaulted the German defenders from a new direction, altering the tactical balance in favor of the US attackers. This effort repeated itself in field after field, eventually routing the long-entrenched defenders. The ensuing successful breakout enabled the Allies to race across France to the German border in a mere six weeks—a stunning success that could not have happened without the Rhino plow. The unexpectedly rapid advance buttressed the morale of tank crews and the infantry, who faced many more months of hard fighting until the German army was finally defeated.[23]

The Rhino plow was a remarkably successful tactical adaptation. When First Army unexpectedly encountered the *bocage*, its junior leaders engaged in impressively novel and speedy efforts to improvise a technological solution while at the same time continuing to battle the enemy in daily combat. Bradley, the most senior field commander, immediately recognized the value of the Rhino prototype and energized all of First Army's ordnance support units to manufacture and disseminate it throughout as much of the force as possible in 11 short days. One author argues that the Rhino plow and its improvised cousins were "the only example where American forces in Europe institutionalized field expedient modifications across a large cross section of forces."[24] The speed and effectiveness of this battlefield improvisation still remains one of the most striking examples of effective technological tactical adaptation in modern combat.

Failed Tactical Adaptation: US Army Supplemental Tank Armor, 1944–1945

The US Army struggled in the European theater of World War II to address the threat from increasingly effective German tanks and antitank weapons. As we discuss in our example of failed institutional adaptation subsequently, army leaders repeatedly discounted the threats posed by these increasingly lethal capabilities, especially those posed by high-velocity German tank guns.[25] They continued to believe that the mechanically reliable M4 Sherman tank would be more than sufficient for the rest of the war, and ultimately built and employed tens of thousands of these easy-to-produce medium tanks. As a result, M4 tank commanders in the field were forced to figure out how to avoid destruction when fighting superior German tanks.

When the Allies invaded Normandy on June 6, 1944, they had been fighting the Germans on the ground in Italy and North Africa for over 18 months. German tanks had steadily continued to improve during this time, with new models arriving on the battlefield nearly every year. In Normandy, the Allies encountered significant numbers of the German Panzer V Panther and the Panzer VI Tiger I, both of which had previously seen battle against the Allies in the Mediterranean in 1943. Both the Panther and Tiger I were equipped with deadly high-velocity cannons as their main armament, while older German tanks continued to fight with less advanced (but still deadly) weaponry. But when the hundreds of M4 Shermans in the invading Allied armies encountered the sizable numbers of Panthers and Tigers in France, they quickly discovered that the cannons on these tanks could slice through the thickest parts of the M4's armor with frightening ease. Both German tanks also boasted angled armor plating that could readily defeat shots from the M4's low-velocity cannon.[26]

The extent of the M4's vulnerability to the newer German tanks shocked Allied tank crews. The M4's cannon could not penetrate the thick frontal armor of the Panthers and Tigers, and had a hard time penetrating their sloped but thinner side armor. Even worse, the German tanks could easily knock out the M4s from any direction, often destroying Shermans at long range with just a single shot. M4 tankers faced heavy losses as they fought against these improved German tanks. Since the M4 crews could not improve the capabilities of their guns, they tried to adapt in the only way they could: they began to improvise supplemental armor to shield their tanks from the lethal high-velocity fire. The most common approach was adding sandbags to the front of the tank—as many as 200 at a time.[27] This improved the confidence of the tank crews, but later tests indicated that it did little to defeat enemy tank rounds, and added so much weight that the tanks lost a substantial amount of mobility.

As the campaign in northwestern Europe progressed and losses mounted, units tried a wider range of possible solutions. The Seventh Army's 14th Armored Division created an elaborate kit of wooden frames to hold sandbags on the sides and turrets of their M4s.[28] The Ninth Army took the sandbag model to its pinnacle, welding spare steel tracks to the front slope of the tank and then covering them in an additional layer of sandbags draped with camouflage netting. The 12th Armored Division's ordnance shops applied steel-reinforced concrete to the front and sides of their tanks, hoping that the resulting loss of mobility caused by such massive additional weight would be balanced by better crew survival.[29]

The most effective approach was developed by Third Army, perhaps because its commander, General George S. Patton, had banned sandbag armor after deciding that the debilitating loss of mobility more than offset any benefits of increased protection. Instead, Third Army ordnance shops salvaged armored plates from destroyed Allied and German tanks and welded them directly onto M4 turrets and hulls. Although imperfect, this improvisation succeeded enough that Patton's superior, General Omar Bradley, recommended to Washington that this become the standard for all Sherman tanks in the European theater.[30] Some other tank units in Europe replicated this makeshift solution, but the impending end of the war precluded wider implementation or changing the tank production process.[31]

Despite the best efforts of these tactical innovators, ad hoc armor simply could not defeat German high-velocity tank rounds. While it could provide some protection against the growing numbers of handheld antitank weapons employed by German infantry in the last months of the war,[32] it largely failed to protect crews from German tank and antitank cannon fire.[33] Observers from the War Department reported that the primary advantage of supplemental armor for M4 tank crews was "psychological" rather than practical.[34] While tanks crews and armored units across the theater continually experimented to try to find ways to significantly improve their survivability, they ultimately fell short. Over 4,200 M4 tanks were lost to enemy action in western Europe from June 1944 to May 1945.[35] The M4 tank simply lacked enough armored protection to remain survivable against the superior German tanks and antitank guns. This was all too evident to the tanks' stalwart crews—but Allied leaders also should have been aware of this problem from earlier experiences in Italy and North Africa. Despite the great resourcefulness shown by M4 tank crews, tactical adaptation by soldiers in the field simply could not solve this fundamental problem.

Institutional Adaptation in Technology
Successful Institutional Adaptation: French Army Tank Development, 1915–1918

World War I ushered in an entirely new era of warfare. When the war began in August 1914, the armies of all of the belligerents were still largely built upon weapons and technology that had changed little since the mid-19th century. Conflicts such as the Second Boer War in South Africa (1899–1902) and the

Russo-Japanese War (1904–1905) each foreshadowed some emerging new weapons, but their broader lessons for future conflict were largely missed.[36] Machine guns, rapid-firing artillery, smokeless powder, motorized vehicles, and barbed wire were all used well before World War I began, but they were not used on a mass scale until then.

World War I began as a war of maneuver in August 1914, but soon turned into a frozen stalemate in western Europe. As offensive actions stalled, the opposing French, German, and British armies on the western front became locked into a static defense line stretching from Switzerland to the North Sea. This system of trench lines was soon protected by hundreds of kilometers of barbed wire and defended by infantry equipped with machine guns and backed by thousands of pieces of artillery. Mobility on the battlefield was paralyzed by the marked dominance of firepower and the unexpected effects of these new technologies. Many of the ensuing years of fighting were aimed at finding ways to break this stalemate, with little success.

The French led one of the most successful wartime efforts to restore mobility to the battlefield.[37] One of the most vexing obstacles to advancing troops was the dense defensive belts of barbed wire that effectively halted any advance by assaulting infantry. The French initially sought to physically clear a path through these belts using massed artillery fire, but that rarely worked.[38] To meet this and other newfound challenges on this unexpectedly complex battlefield, the French High Committee for Inventions actively sought out ideas for new systems from civilians and junior officers.[39]

By late 1914, a number of French officers and engineers were working on proposals for a tracked caterpillar-like vehicle that could break through barbed wire obstacles.[40] In 1915, engineers of the Schneider firm in France began to build a prototype based upon the American Holt tractor design.[41] A French artillery colonel named Jean Estienne also noted the Holt tractor's potential, and in December 1915, he forwarded his own proposal for an "armored landship" to Army General Headquarters (GHQ).[42] Estienne is often credited as the foremost proponent of the French tank concept, and soon became a tireless advocate for its adoption. Two weeks after his proposal was submitted, the French chief of the General Staff and commander-in-chief, General Joseph Joffre, placed Estienne in charge of the army portion of the tank project, directing him to "pursue its creation with all the required speed and secrecy."[43] Joffre and his successor, Philippe Petain, were both instrumental in moving the unconventional project forward. Petain was a particularly strong supporter of Estienne, who had once served as the chief of divisional artillery in his division.[44]

By early 1916, the French had developed two medium-sized tank prototypes: the Schneider CA-1, which weighed about 12 tons, and the St. Chamond, which weighed about 20.[45] Both tanks mounted a 75mm cannon, making them the most heavily armed tanks that would see combat in the war. Yet they were large, heavy, cumbersome in design and performance, and prone to mechanical breakdown. These inherent flaws were compounded by the inefficiencies of producing two different designs simultaneously. Each model required completely different parts and independent production lines, wasting scarce resources and complicating subsequent logistical support.[46] This unfortunate decision resulted from an unresolved turf battle between Estienne at GHQ, who supported the CA-1, and the Ministry of Armaments (and the Automobile Department), which advocated for the St. Chamond. Wartime resource shortfalls, especially in steel, further slowed medium tank production. In November 1916, for example, the CA-1 factory was only producing one medium tank per day.[47]

Even before these first-generation French tanks saw action, however, Estienne had already started designing an entirely new light tank, which would become the most successful and widely produced tank of the war. The Renault FT (commonly known as the FT-17 for its year of introduction, and shown in Figure 3.3) was the first truly modern tank ever produced. Weighing only 6.6 tons, it was crewed by two men and mounted a revolving turret armed with either a 37mm cannon or a machine gun.[48] The FT-17 entered trials on April 9, 1917, one week before the CA-1 and St. Chamond were first employed in combat during the Nivelle Offensive.

Neither of these first-generation medium tanks performed particularly well in the battles of April and May 1917. Between extensive mechanical failures and unexpectedly effective German antitank defenses, both CA-1 and St. Chamond tank units suffered heavy losses. The other shortcomings of these tanks also became evident quickly: inadequate top armor, poor visibility for the crew, and an inability to communicate with nearby infantry.[49] Subsequent actions, including the October battle at La Malmaison, produced mixed results but confirmed that the two new medium tanks had serious problems with mobility, visibility, and mechanical reliability. The need for a lighter, more agile, and more reliable tank had become abundantly clear.[50]

Fortunately for the French, the FT-17 trials went very well. In June 1917, Petain expanded the initial production of the new light tank from 1,150 to 3,500.[51] The French planned to build up their force of new FT-17s before employing them in a major offensive, but the surprise German Ludendorff Offensive that began in March 1918 demanded their immediate use. The FT-17s were quickly thrown into combat as part of a broad French defensive battle, and immediately proved their worth. In their first action at Chaudun in April 1918, a force of FT-17s "crippled" two German divisions in a six-hour action.[52] In another engagement, five FT-17s "reduced a German division to complete disorder in less than three hours."[53]

FIGURE 3.3 US Army Troops Advancing into Battle with FT-17 Tanks, 1918
Source: https://commons.wikimedia.org/wiki/File:FT-17-argonne-1918.gif.

Shifting from the medium tanks to the FT-17 not only helped solve the seemingly intractable problem of battlefield mobility, but also simplified the production challenges that had hobbled delivery of medium tanks. Splitting French production efforts between two very similar medium tank models immensely complicated their manufacturing and maintenance, especially since they shared few if any common parts. Only 400 CA-1s and 400 St. Chamonds were produced during the entire war.[54] Moving to a single, simpler light tank that required far fewer resources to manufacture facilitated rapid and efficient industry production. By November 1918, 3,187 FT-17s had been produced, a remarkable accomplishment that reflected both concerted French leadership efforts to speed their delivery and their relative ease of production.[55]

Producing such a stunning number of tanks in the midst of a war required extraordinary institutional flexibility and adaptability in the French army and industry. Between 1916 and 1918, French manufacturing firms introduced and produced an entirely new vehicle of a type they had never seen before. Although the effort was marred initially by bureaucratic infighting and slow deliveries, French industry was able to deliver initial batches of the CA-1 and the St.

Chamond to the French army by late 1916. Yet even though the French had two new tanks entering production, they wisely continued to experiment. They ended up developing a third vehicle, the lightweight FT-17, that better met their needs, even as the first two medium tanks were just starting to be used in combat. The French leaders quickly grasped the value of the FT-17 and pivoted rapidly, ordering full-scale production of this new light tank. They also subsequently catalyzed French industry to produce prodigious numbers of these tanks in the last year of the war, with dramatic battlefield results.

The French army's speedy development, record-breaking production, and large-scale delivery of the FT-17 tank ranks as one of the most impressive examples of successful institutional technological adaptation in any war. The fact that France, Germany, and Britain entered the war without having any concept for a tank makes this achievement all the more remarkable. The French government and military incentivized creative thinking to address intractable battlefield problems, which was institutionalized in the French High Committee for Inventions. Moreover, senior French military leaders such as Joffre and Petain empowered and supported creative subordinates like Estienne to boldly conceive of innovative solutions to battlefield problems. When these innovators found success, the full weight of the military and industry was marshaled behind them to rapidly deliver the product on a heretofore unimaginable scale, even as combat operations continued without interruption. This tremendous example of institutional adaptation played a major role in enabling the final Allied offensives in 1918 that ultimately ended the war.

Failed Institutional Adaptation: US Army Tank Development, 1941–1945

Earlier in the chapter, we identified US efforts to develop supplemental tank armor between 1944 and 1945 as an example of failed tactical adaptation. But we also noted that the challenge of protecting the M4 Sherman tank on the battlefield from high-velocity tank and antitank gunfire was simply too hard. Despite valiant efforts by US tank crews to improvise some form of armored protection, they could not overcome this fundamental problem with the M4's design. Yet this is not simply a case of the army having chosen a piece of technology that did not pan out on the battlefield. Between 1942 and 1944, there was growing evidence that the M4s were not performing well against the best new enemy tanks, which were superior in firepower as well as protection.[56] The army's failure to recognize the M4's growing inadequacy and accelerate the fielding of a better US tank in time for the campaign across northern Europe makes this a clear case of failed institutional adaptation.

When the United States entered World War II, it was still developing the technology needed for mobile armored warfare. In 1939, its armored forces were equipped with light tanks for reconnaissance, tank destroyers to fight enemy tanks,[57] and medium tanks to exploit breakthroughs deep into the enemy's rear. In other words, tanks would not fight other tanks; tank destroyers, with their heavier guns, would take on that role. Instead, US tanks would exploit breakthroughs made by infantry attacks, penetrating the broken enemy lines to attack vulnerable rear areas and destroy softer targets such as supply depots and headquarters. The M4 medium tank was partly designed with this task in mind. It was built for mechanical reliability, speed, and mobility, and it was relatively easy to mass produce. Given its intended mission, firepower and crew protection were effectively secondary concerns.

Allied battlefield experiences in North Africa, Italy, and central Europe between 1942 and 1944 should have shattered these prewar assumptions. M4s were first used in combat by British forces operating in North Africa in November 1942 at El Alamein. The British M4s were immediately thrown into battle against German tanks, since the British did not employ tank destroyers or embrace US tank destroyer doctrine. The M4s performed fairly well in this role, since their capabilities were at least as good as (if not better than) the opposing German tanks of the time.[58] But the Germans continued to aggressively improve their tanks to meet the new challenges of fighting the Allies in North Africa and against the Red Army on the Russian steppes. Starting in 1943, despite a devastating Allied bombing campaign, Germany managed to develop and field a range of new medium and heavy tanks of increasing quality and capability before the end of the war.[59] The Germans started deploying their new Tiger I tanks in North Africa in early 1943, against both British and American forces. The Tiger I was a heavy tank with a high-velocity 88mm gun, and its impressive design gave it vastly better firepower and armored protection than the M4. However, few were available for the campaign in North Africa, which masked their battlefield importance to the Allies.

By July 1943, the North African campaign was over and the US and British invasion of Sicily was underway. The Germans employed Tigers in Sicily, but these units took such heavy damage from naval gunfire and air attacks that the vulnerabilities of the M4 in tank-on-tank fighting were again largely obscured. The latter phases of the Italian campaign, from late 1943 into early 1944, involved relatively few tank battles, which further lulled the US Army into a sense of complacency about the M4's effectiveness.[60] But the US Army also should have absorbed important lessons from the eastern front during this time, as the Red Army engaged the Germans in some of the largest tank battles in history. The Soviets began encountering large numbers of highly capable new German tanks there by mid-1943, including the Panzer V Panther and Panzer VI Tiger I in large numbers. After the largest tank battle of the war at Kursk in July 1943, the Russians rapidly upgraded their previously successful T34/76 medium tank with an entirely new turret and high-velocity 85mm gun in response to these new German tank capabilities. This speedy adaptation proved highly successful, and Soviet tank capabilities remained roughly equal to German tank capabilities until the end of the war. Even though the US Army missed important lessons in Italy and North Africa, the Soviet experience should have revealed the rapidly improving German tank capabilities and alerted army leaders to the M4's growing vulnerability.

Institutional biases and turf battles within the army staff in Washington further hindered the army from fielding a better medium tank before the invasion of France. Generals in the Ordnance Branch, who were responsible for developing new weaponry, consistently advocated for an improved medium tank that could take on the newest panzers, and even developed several prototypes. Major General Jacob L. Devers, who commanded the Armored Force (responsible for training tank crews and developing tank doctrine) also supported this effort. But generals in the Infantry Branch continued to view the tank's role as supporting the infantry and exploiting breakthroughs, and argued that the M4 was adequate for this mission. The chief of Army Ground Forces, General Lesley J. McNair, agreed with the advocates in the Infantry Branch and strongly opposed producing any new prototypes.[61] Armor chief Devers was so concerned about McNair's decision that he went around his superiors and brought the issue directly to Army Chief of Staff General George Marshall. Marshall overruled McNair, and in December 1943, he ordered production to begin on the T26. This tank, which was later known as the M26, was the most promising of the new prototypes at the time,[62] but it was already too late to get replacement tanks to Europe before D-Day. Most M4s that landed in Normandy in June 1944 and began the final battles across northwest Europe were only marginally better than the M4s that were first used by the British in the North African desert in late 1942.[63]

As discussed earlier, M4 crews became acutely aware of how poorly their tanks performed against their German counterparts shortly after D-Day. By 1944, many German tanks in France were armed with high-velocity 75mm and 88mm cannons that could devastate M4s from long range, often killing them with a single frontal shot. For US tank crews, using this same tactic was out of the question; even the M4s that had improved 76mm guns could not stand up to most German tanks in direct engagements. M4 crews learned to attack the German tanks from the sides and rear, but the geometry of many combat engagements often made that approach either impossible or suicidal. Losses of M4s mounted swiftly after the invasion. For example, a tank battalion of the US Second Armored Division fighting in France lost 70 percent of its tanks and 51 percent of its men in two weeks of hard fighting after D-Day.[64] As scholar David Johnson notes, "American soldiers began to lose faith in their tanks—with good reason."[65]

In the months after D-Day, US generals in Europe slowly became aware of the deadly disparity between the M4s and their German counterparts. The US tanks could not effectively kill German tanks with their low-velocity cannon, and also lacked enough armored protection and mobility to avoid being destroyed. As previously noted, as the fighting went on, some M4 crews tried to protect themselves from tank and antitank assaults by welding extra steel onto the turrets and frontal armor plates of their tanks, while others layered hundreds of sandbags along turrets to provide yet another layer of protection. Even though many army leaders continued to believe that the M4 was more mechanically reliable and agile across varied terrain than its German counterparts, the advance across western Europe into Germany soon disproved that notion as well. German tanks could move much more effectively across soft ground, because they had much wider tracks with lower ground pressure. The M4s became bogged down in terrain that German tanks could easily maneuver across, dispelling yet another of the Sherman's ostensible advantages.[66]

Between June and December 1944, the 12th US Army Group in Europe lost over 2,000 medium tanks.[67] Such staggering losses should have prompted a widespread reappraisal throughout the army, but the impending end of the European war dampened any efforts to do so. In December, US troops faced a surprise onslaught of massed German armor in the Battle of the Bulge, a large and entirely unexpected enemy attack through the Ardennes. The army lost hundreds of tanks during the bloody and desperate battle, prompting a media outcry. Supreme Allied commander General Dwight D. Eisenhower was taken aback by the press reports and apparently learned about the problems with the M4 only by reading critical press accounts. In January 1945, Eisenhower demanded more information about these problems from his armored division commanders.[68] The responses he received largely confirmed the deficiencies of the medium tank and were illustrated by a series of grim anecdotes from unhappy tank crews.[69] But by then, it was simply too late to fix the problem. The M26 replacement that Marshall ordered over a year earlier had not even entered production until November 1944.[70] That meant that it would be too little, too late; it could not reach Europe in enough time or numbers to have any significant effect. Only 310 of these more capable M26 tanks arrived in Europe by the end of the European war in May 1945, and only about 20 actually participated in combat operations.[71]

The US Army's continued reliance on the M4 Sherman throughout World War II was an unmistakable failure of institutional adaptation. The tank's major shortcomings in firepower, protection, and mobility were evident to tankers in the field as early as 1943, and its relative inadequacies only grew as the Germans fielded ever more capable tanks year after year. Army leaders in the United States and overseas failed to give serious attention to the continued German tank improvements, despite plenty of available information on their ever-increasing capabilities.[72] Senior US battlefield commanders repeatedly failed to recognize the M4's growing technological shortfalls, objectively reevaluate its performance, and demand an institutional solution. Equal blame accrues to the institutional leaders on the home front who failed to grasp the M4's problems even as they grew increasingly evident, and wasted precious time squabbling over bureaucratic turf and flawed prewar concepts of armored battle that had already become obsolete. US M4 tank crews in Europe paid a heavy price in blood for this enormous institutional failure.

Conclusion: The Importance of Technological Adaptability

Tactical adaptation can deliver short-term fixes to respond to some of the most immediate technological challenges on the battlefield. Militaries that encourage their front-line soldiers to approach these types of problems creatively and experiment with unconventional solutions have a strong advantage, but that alone is not sufficient. At the tactical level, even the most innovative technological solutions matter little if they cannot be manufactured and disseminated beyond the units that developed them. If Sergeant Culin's improvised hedgerow cutting tool had never been shown to General Bradley, or if Bradley had not immediately ordered it into mass production, First Army's successful Cobra offensive might have turned out very differently. Senior battlefield leaders must recognize the value of these technological adaptations and energize their command to ensure that they are rapidly produced and disseminated throughout the force.

Sometimes, however, even the best efforts at tactical adaptation fail. No matter how hard they tried, the M4 tank crews fighting in Europe after D-Day could not improvise adequate supplemental armor to protect themselves from superior German tank and antitank firepower. At the point of tactical failure, institutional adaptation is required to overhaul a weapons system or develop an entirely new one. This requires higher headquarters at home to continuously reevaluate the combat performance of their systems. Yet that is an enormously difficult challenge in practice, since the leaders of those organizations must work tirelessly to overcome the inevitable turf battles and entrenched bureaucratic interests that stubbornly resist change—even when soldiers are dying in battle. The French army did this relatively well in World War I, establishing the French High

Committee for Inventions and developing a new and lighter tank even before two earlier models had entered combat. The US Army in World War II, by contrast, stubbornly clung to its prewar views about how to employ tanks long after they should have been demolished by the experience of combat. Furthermore, infighting among different parts of the army's bureaucracy effectively stonewalled any new tank development during the entire war. As a result, M4 tank crews were stuck fighting superior German tanks with a vehicle that simply could not stand up to ever-improving German capabilities. The calcified army bureaucracy failed to recognize the threat, despite the ever-increasing number of casualties, and battled internally for years instead of solving the problem. The war might not have been won any faster with a better tank, but it certainly would have reduced many of the horrific losses in tanks and men that American forces sustained during the last year of the European campaign. As we will see again in Chapter 6, which describes an eerily similar problem that US troops faced in the first years of the Iraq War, senior leaders who fail to overcome bureaucratic stonewalling in the face of urgent battlefield requirements fail in their duty to the American soldier.

4

The Role of Leadership

THE TWO PREVIOUS chapters examined how doctrine and technology affect a military's ability to adapt. But the third factor in our framework—the role of leadership—may be even more important for effective adaptation. Battles and wars are frequently decided not by the doctrines that armies bring to war, nor by the technology that equips military forces, but by the human beings charged with making the crucial battlefield decisions that will lead to either victory or defeat. The effectiveness of a nation's military leaders ultimately may be the most important factor in determining whether its forces prevail or succumb in battle.

Human beings are inherently involved in virtually all aspects of military operations. They conceptualize, refine, and promulgate the doctrines that guide armies in battle. They invent, develop, and distribute the diverse types of technology that are employed in war. They fill the ranks of the soldiers who fight a nation's wars, are wounded and die in battle, and often are scarred by their wartime experience. These activities are vitally important, but they differ from military leadership.

The US Army officially defines military leadership as "the process of influencing people by providing purpose, direction, and motivation to accomplish the mission and improve the organization."[1] That definition says little about *how* leaders are expected to do those things, but army doctrine consistently stresses that its leaders must be adaptable. Its primary manual on leadership explicitly notes, "Military history is replete with accounts of adaptation, hinging on a leader's ability to have uncanny insights into the situation, to be keenly self-aware, and to have a mindset and knowledge that promotes adaptation."[2]

Why is adaptability such an important characteristic of military leadership? As we discussed in Chapter 1, the fog and friction of war make the battlefield inherently unpredictable. Adaptability, according to the US Army, serves as "insurance against the ambiguity, adversity, and uncertainty found on every battlefield,"[3] and it enables leaders "to achieve mission accomplishment in dynamic, unstable, and complex environments."[4] Adaptable leaders need to be comfortable with uncertainty and ambiguity and "see each change thrust upon them as an opportunity rather than a liability."[5] That requires open-mindedness, cognitive flexibility, and creativity, so military leaders can figure out how to use their knowledge and experience in new and different ways.[6]

Adaptability is important for leaders of every rank, but requires different characteristics at different levels. *Tactical leadership* is carried out by combat leaders, from sergeants through colonels, who are responsible for fighting and winning on the battlefield.[7] *Theater leadership* is carried out by generals who craft and implement campaigns designed to win wars. These senior commanders sit atop the chain of command at the operational level, and, as the name implies, they are usually physically located within the theater of war. They often report to more senior generals who operate at the strategic level back in national capitals, but theater commanders are the ones who determine how to use tactical forces to achieve strategic objectives.[8] Yet they may also have broader strategic and political responsibilities, such as maintaining unity within an alliance or coalition, or managing relations with host nation or other friendly local forces.

Militaries are closed organizations that grow their leaders entirely from within. In the civilian world, rising leaders and even the most senior executives are routinely hired away from other organizations. In the military, virtually the only way to become a multistar general is to begin your career as a second lieutenant and slowly work your way up, one step at a time, through the rank structure. Moreover, each rank cohort plays a significant role in developing the cohorts coming up behind it. Generals mentor colonels, colonels develop captains, and captains help grow lieutenants and sergeants. This means that wartime leaders at every level have been selected and developed though years of peacetime training and education. They inevitably reflect whatever qualities were valued by the military before the war and are typically rewarded with success and promotion for adherence to those peacetime attributes.

Unfortunately, many of the qualities required for successful advancement in peacetime do not encourage—and sometimes even undermine—the attributes necessary for adaptive leadership in war. This affects leaders of all ranks, but is most pronounced at the highest levels among the generals and senior colonels. As noted in Chapter 1, militaries are large and complex organizations, requiring leaders to manage them in ways that are often similar to managing big civilian organizations. In between wars, generals craft budgets, conserve resources, administer pay and personnel actions, develop new weapons and tactics, oversee vast and far-flung facilities, and maintain order and discipline. Yet these critical activities may have little in common with the demands of combat and the skills required to win in battle.

Moreover, the profusion of bureaucratic requirements that inevitably grow during peacetime affects even the most junior military leaders. Petty rules and regulations and leaders who prioritize compliance over boldness can rob rising leaders of their initiative and audacity. The abject risk-aversion often found in peacetime militaries can produce leaders entirely ill-suited for the hazardous and risk-filled decision-making that war requires. At times, ambitious military leaders in peacetime may seek to ensure future promotions by doing everything possible to prevent their subordinates from making mistakes. Yet such micromanagement strips young leaders of the vital trial-and-error experiences in training needed to develop the confidence and independent judgment that they need to succeed in battle. A well-known military saying notes, "Good judgment comes from experience, and experience comes from bad judgment."

The best militaries are able to overcome the bureaucratic necessities of peacetime service. They find ways to instill characteristics in their rising leaders that allow them to adapt and prevail when the disruption and chaos of war finally arrive. And they reward, select, and promote such leaders and work hard to ensure they can lead the army effectively in the next war.

Adaptable tactical and theater leaders share some similar characteristics, but they also differ because they operate at two distinct levels of war. At the tactical level, adaptable leaders must

- **Rapidly assess the battlefield and identify a need for change.** One of the first tasks of a combat leader is to understand the nature of the battle he or she is fighting. This requires leaders to overcome the many systematic and unconscious biases that characterize how people make decisions under uncertainty,[9] though that can happen quickly amid the immediate and lethal feedback of the battlefield. Tactical leaders must also physically move to where they can best see and absorb the situation unfolding around them, even if doing so exposes them to grave personal danger. They must be able to calmly assess ongoing combat results amid incoming fire and the chaos surrounding them, and identify what must be done next in order to prevail.
- **Abandon accepted procedures when needed and rapidly improvise new approaches to address battlefield setbacks.** Successful adaptation requires that leaders be able to seize new tactical approaches quickly and effectively. These could include directing a bold counterattack, reorganizing disrupted and ineffective units, grasping an unexpected battlefield opportunity, or employing unanticipated resources (such as captured enemy weapons) in unorthodox ways to accomplish the assigned mission.

- **Candidly advocate for organizational change when needed.** Tactical leaders are often the first to realize that the tactics and strategies of their higher commands are not working. When they face recurring battlefield setbacks that require fixes that lie beyond their authority to impose, they must unhesitatingly flag the problem to higher-level commanders and press for change. Their current combat experience and fresh ideas are invaluable in helping higher commanders identify what is going wrong and figure out the right solutions.

All theater leaders have already been successful leaders at the tactical level, or they would not have been promoted to such high ranks. But the skills they need to succeed at the operational level can be very different from the skills that they have already mastered at the tactical level. Effecting change throughout very large, complex organizations in the midst of combat is very different from issuing orders to a platoon or company and directly supervising their execution. Senior leaders also face longer time horizons than their more junior counterparts, since military campaigns may last weeks, months, or even years. At the theater level, adaptive commanders must

- **Understand the situation and identify problems.** Like tactical leaders, theater leaders must understand the nature of the fight in which they are engaged, and dispassionately determine whether their approach is succeeding or failing. Yet this challenge is even more complex for them, because of the large size of their commands, the geographic and bureaucratic distance between them and their front-line soldiers, and the fact that they must often rely upon second- or even thirdhand reports when assessing the battlefield. It is also harder for them to make decisions under uncertainty than it is for tactical leaders, since they receive less immediate and direct feedback about the effects of their decisions. This makes it even more important for them to regularly observe front-line units and to speak personally with combat soldiers and their leaders, in order to gain their perspectives and feedback.
- **Modify or abandon established procedures when they are not working.** Senior commanders are expected to have a high degree of self-confidence and skill. But it is a short step from there to being overconfident in one's own abilities, even when faced with clear evidence that current approaches are failing. Hubris is a constant threat to their effectiveness.[10] Successful theater leaders constantly challenge their assumptions about the enemy, the environment, and their own forces and plans. Open-mindedness and cognitive flexibility are essential attributes. They recognize and accept evidence of failure, even failures directly caused by their own decisions, and rapidly develop new and improved alternative approaches.

- **Seek out and incorporate good ideas throughout the entire chain of command.** Senior leaders must work continuously to break down the barriers of rank and deference in military organizations so that they can hear the unvarnished truth from those in their chain of command.[11] They must establish a climate of trust and confidence that encourages and rewards subordinates for reporting unfavorable news, challenging conventional wisdom, and offering new or unconventional ways to solve intractable battlefield problems. They must seek out and nurture candor and creativity among their subordinates. This includes finding ways to promote and absorb feedback from even the lowest levels on the effectiveness of current operations, and seeking out new and creative ways to accomplish the assigned mission from throughout their organization if current approaches are faltering.
- **Systematically organize and lead rapid battlefield change within an army at war.** Once a theater commander recognizes the changes that need to be made, he or she must aggressively lead the organization through making the necessary (and sometimes unpopular or dangerous) adaptation. Leaders must catalyze change throughout their command, personally making the case for the changes to subordinate leaders and units. Supervising and monitoring the day-to-day implementation of a new approach is as important as devising the approach itself. Commanders must put their full weight into overcoming the natural bureaucratic inertia inherent in any large organization faced with the disruptive force of change. Only the full-throated support of the theater commander can energize the entire organization to shift gears and embrace a new approach, especially in the midst of combat.

As Figure 4.1 shows, we look at two instances of leadership adaptation at the tactical level and two instances of leadership adaptation at the theater level, with one success and failure for each. At the tactical level, we examine how Captain John Abizaid[12] successfully adapted to his rapidly changing mission and unexpected battlefield developments during the US invasion of Grenada in 1983, which helped contribute to the swift success of the operation and certainly saved American lives. By contrast, in Vietnam in November 1965, Lieutenant Colonel Robert McDade proved unable to adapt to a new enemy, a new terrain, and a new war—despite having a clear and compelling preview of each in the 72 hours before his first clash. His unit lost the ensuing battle and suffered over 300 casualties in 24 hours, many of whom might have survived if McDade had proven more adaptable.

At the theater level, we discuss how Field Marshal Viscount William Slim completely transformed an army that was utterly broken by Japanese forces in Burma in 1942 into a well-disciplined and highly effective unit that defeated

	Success	Failure
Tactical Adaptation	Captain John Abizaid, Grenada, October 1983	Lt. Colonel Robert McDade, Vietnam, November 1965
Theater Adaptation	Field Marshal William Slim, Burma, 1942–1945	General William Westmoreland, Vietnam, 1964–1968

FIGURE 4.1 Examples of Leadership Adaptation during Wartime

the Japanese on the exact same terrain three years later. We contrast that little-known example of successful theater adaptability with what may be one of the best-known cases of failure to adapt in a wartime theater. During his four years commanding the war in Vietnam, General William Westmoreland never adapted to the nature of the war he was fighting, despite substantial available evidence showing that his approach was failing. The flawed approach that he championed set the overall course for the most consequential period of the Vietnam War, and ultimately led to a humiliating withdrawal in 1973 that cast a long shadow over the US military for over a generation.

Tactical Adaptation in Leadership
Successful Tactical Adaptation: Captain John Abizaid, Grenada, October 1983

Captain John Abizaid exemplified successful tactical adaptation during the US invasion of Grenada in October 1983.[13] Grenada is a tiny island in the southeastern Caribbean and was wracked with bloodshed in the fall of 1983 after a bloody coup brought a Marxist regime to power. Cuban paramilitary construction workers were building a 9,500-foot runway on the southwest tip of the island that could accommodate the largest Soviet military aircraft. After the coup, the Soviet-armed Grenadian army declared a 24-hour shoot-on-sight curfew, posing a threat to the hundreds of US citizens on the island. Facing worries about the fate of these Americans, President Ronald Reagan launched an extremely short-notice military operation to safeguard them and restore order on the island. Two US Army Ranger battalions would seize the huge unfinished airfield, while special operations forces and a marine task force would secure other key locations.

Abizaid commanded A Company of the US Army's First Ranger Battalion, which included 150 of the nearly 900 rangers who participated in the attack. Neither he nor any but a handful of his men had seen combat before.

Abizaid's original mission was to lead his company to secure the unfinished airfield at Point Salines, clearing obstructions from the runway to allow reinforcing troops aboard US Air Force cargo aircraft to begin arriving immediately thereafter. He originally planned to parachute about 40 men of his company onto the runway at night and clear it of any obstacles, so that the remainder of the 500-man ranger force could quickly land in blacked-out MC-130 transports.[14] Rangers would then fan out to secure the area around the airfield and rescue the hundreds of nearby American medical students.

The mission was fraught with problems from the very start. On the seven-and-a-half-hour flight to the island, Abizaid's battalion commander, Lieutenant Colonel Wesley B. Taylor, worried about the still unknown size of enemy forces guarding the runway. He ordered Abizaid to parachute his entire company onto the runway to ensure he had enough men to overcome any resistance he found there.[15] Abizaid now had to abandon his original plan of attack. He was forced to activate contingency plans to drop all of his men in by parachute, without the jeeps and other heavy weaponry they had planned to roll off the aircraft when

FIGURE 4.2 Captain John Abizaid (right) receiving an award from Lieutenant Colonel Wesley B. Taylor, late 1983.

Source: John Abizaid.

they landed. In the darkened aircraft, over 100 A Company rangers began donning parachutes they would now need to parachute onto the island.

Unexpected changes continued to cause Abizaid to keep adapting his plans. When the aircraft took off, the invasion time was set for 2:00 in the morning on October 25. But during the long flight to the island, Abizaid found out his drop time had been changed to 5:00 in the morning for seemingly inexplicable reasons.[16] His company's parachute drop was now pushing perilously close to dawn, putting both jumpers and aircraft at increased risk. The rangers' bad luck continued. Abizaid's MC-130 was originally the lead aircraft for the drop, with the two other jump aircraft trailing close behind. Now on the way to the island, his transport lost its navigation system, forcing it to drop back to the number-three position in the flight. The new leading aircraft now contained only one platoon of 40 rangers from another company, as well as Taylor and his small combat headquarters.[17]

As the fleet of invasion aircraft approached the drop zone, problems quickly turned from bad to worse. Alerted defenders sent up a barrage of fire from anti-aircraft cannons located all around the runway, which were detected after it was too late for the invasion force to change plans. The first MC-130 jump aircraft bored in and avoided the worst of the fire by dropping its rangers from an altitude of only 500 feet. But the two remaining MC-130s, carrying Abizaid and his men, turned away in the face of a sky full of flaming tracers and flew back out to sea.[18] The 40-man ranger platoon, along with the battalion commander and his assault command post, were now on the ground alone in the midst of a much larger enemy force.[19] Taken aback by this unexpected turn of events and circling in the sky off the island, Abizaid knew he had to find a way to reinforce this small isolated force on the ground before it was overwhelmed by the enemy. He took immediate action. Shedding his parachute so that he could more easily move forward through his heavily laden rangers up to the MC-130's flight deck, Abizaid confronted his unnerved flight crew. Told it was too dangerous to return in the face of such intense fire, Abizaid erupted at the pilots. After a heated debate, he finally convinced them to turn around to make another pass through the airfield's gauntlet of fire. The two MC-130s carrying Abizaid and the rest of his company now lined up in the brightening dawn skies for a second attempt at the parachute drop.

Moments later, the two aircraft dropped Abizaid and his men from 500 feet into a hail of enemy fire in nearly complete daylight.[20] Most of his men landed on the south side of the runway, away from enemy positions located among the low hills and building to the north. It was miraculous that no rangers had drifted in the brisk wind into the nearby surf, but many abandoned parachutes near the runway were marked by inflated life preservers activated by worried jumpers.

Recognizing that enemy resistance was much stronger than anticipated, Abizaid reorganized his men south of the runway and began to attack enemy positions overlooking the airfield from low hills that dominated the northern side.

Without the gun jeeps outfitted with heavier weaponry that should have arrived once the MC-130s landed, Abizaid had to improvise. His men were lightly equipped with small arms and machine guns, but had no heavy firepower beyond the 60mm mortars that had been airdropped. Rushing two platoons of his company north across the runway under fire, he was relieved to find no rangers had been hit. He dispatched another platoon on foot to race to the east end of the runway and secure the American medical students, something his original plans had assigned to his fast-moving jeep teams. Soon after, his company suffered its first fatality, a machine-gunner killed as he lay next to a Cuban vehicle while returning fire.[21] Under increasing enemy fire from the high ground north of the runway, Abizaid needed something to give his men an advantage as they advanced against the dug-in Cuban and Grenadian troops.

As shock troops, ranger companies were lightly equipped and often planned to use whatever equipment they could commandeer on the battlefield. Unlike virtually any other army units, rangers had practiced how to hot-wire captured enemy and civilian equipment and employ all sorts of enemy weaponry as part of their routine training. His troops used the hot-wire kits they had brought with them to start up and move Cuban construction equipment off the runway to enable follow-on aircraft to land. Abizaid soon uncovered another opportunity to put these skills to remarkably innovative use.

As A Company continued to come under fire from dug-in enemy troops, Abizaid spotted a possible solution. Directing one of his sergeants to start up a nearby Cuban bulldozer, Abizaid directed its use as an ad hoc tank to attack the Cubans raining fire on his men. A staff sergeant got the bulldozer started, raised the blade as a bulletproof shield, and began to maneuver the slow-moving vehicle up the nearby hills toward the dug-in Cuban and Grenadian positions.[22] Other rangers moved up behind the ersatz tank, using it for cover and firing their weapons as they advanced. By employing this unconventional armored attack, Abizaid was able to destroy or force the withdrawal of all the enemy positions that were threatening his company and denying use of the eastern end of the airfield. Other rangers soon turned captured Cuban antiaircraft weapons against the defenders, pummeling them with withering cannon and heavy machine-gun fire. The remaining Cuban and Grenadian forces abandoned their positions near the airfield, which made it possible to open the runway and for US aircraft to fly in reinforcements. The toughest fighting for Point Salines was now mostly over, although a strong enemy counterattack with armored vehicles was still to come in the afternoon.

Abizaid epitomized the most important attributes of an adaptive tactical leader during this chaotic mission. During the course of the operation, his environment, his mission, and his resources all changed with dizzying frequency. Yet Abizaid continually reassessed this ever-changing situation and calmly and repeatedly altered his plans despite sustained and potentially debilitating confusion. He refused to accept the judgment of his aircrew that the aircraft drop run was simply too dangerous to repeat, and convinced them to return to the fight. He reorganized his force several times while in the air, and retasked his men once again on the ground, reacting to rapid changes on the battlefield. In the face of a resilient and unexpectedly stubborn enemy, he brilliantly improvised a way to use a bulldozer as an ad hoc tank to help win a tough fight for his company. Abizaid's adaptability made him a highly effective leader under unique and challenging battlefield conditions and significantly contributed to the success of this little-known US military operation.

Failed Tactical Adaptation: Lieutenant Colonel Robert McDade, Vietnam, November 1965

The first battles of the Vietnam War proved a severe test of US forces. Long accustomed to conventional wars against similarly equipped adversaries, the demands of fighting an unconventional enemy in the jungles, rice paddies, and villages of South Vietnam presented a significant challenge to the US Army. Many of the army's midgrade tactical leaders had served in combat in the Korean War, and some were also veterans of World War II. The colonels who led battalions and brigades in the initial encounters in Vietnam were shaped by wars that were vastly different from the one they would fight in Southeast Asia. The army's opening battles with the North Vietnamese Army in late 1965 drove home those realities in bloody casualty counts.

Lieutenant Colonel Robert A. McDade commanded the Second Battalion of the Seventh Cavalry (2/7 Cav) in November 1965, an element of the army's new helicopter-borne First Cavalry Division, which had just arrived in South Vietnam. McDade had served as an infantry platoon leader in the Pacific in World War II and as an infantry company commander in the Korean War. When he took command of his battalion less than three weeks before the Ia Drang operation, he had not commanded troops in almost 10 years.[23] McDade and his battalion played a key role in the opening battles of the Ia Drang Valley campaign. These first battles with the North Vietnamese Army (NVA) would be among the most ferocious that the US Army faced during the entire war.

On the morning of November 17, McDade and his battalion were located at Landing Zone (LZ) X-Ray in the Ia Drang Valley, the site of a savage three-day

FIGURE 4.3 US Army Lieutenant Colonel Robert McDade, Commander of 2/7 Cavalry at Landing Zone Albany in 1965
Source: https://en.wikipedia.org/wiki/File:Mcdade_photo.jpg.

battle that had just concluded between the Americans and North Vietnamese. Only one of McDade's companies had been involved in the successful battle to defend LZ X-Ray, and now 2/7 Cav was to move away from the recent battlefield to secure its own landing zone. McDade and his battalion were ordered to trek two miles north on foot to establish another landing zone called Albany.[24] Once they arrived at LZ Albany and set up a hasty defense overnight, helicopters from the First Cavalry Division would arrive the following day to extricate them from the still dangerous area.[25]

After the fierce battle at LZ X-Ray, McDade knew that there were still large numbers of enemy forces in the area. The terrain near his route north was dominated by a huge massif called the Chu Phong. In the fight for LZ X-Ray, thousands of NVA troops had stormed down off the mountain in three days of withering assaults, nearly overwhelming the US defenders. Yet despite the obvious threat posed by the still-potent NVA on the massif, McDade took few precautions during his move north.[26] His men were exhausted from being awake on full alert the entire previous night, as they waited for an NVA attack that never occurred.[27] As his battalion slowly trudged north, the terrain grew more difficult as knee-high elephant grass gave way to shoulder-high vegetation, six-foot termite hills, and an increasing canopy of dense jungle.[28] Other battalion commanders in the valley chose to use supporting artillery fire to clear enemy forces that might be lurking along their routes, but McDade did not, apparently attempting to conceal his unit's movement from the enemy.[29] Even more inexplicably, some of his troops set fire to a straw building they encountered on their march, clearly marking their advance to any enemy observers.[30]

McDade's battalion moved in one long column stretching over 500 yards through the high grass, with his troops spread out at the front and back to provide some measure of security.[31] Most of the unit, however, moved in a ragged single file in the heat, plopping down at rest breaks with no security and little alertness, given their fatigue.[32] As the battalion approached the clearing that was designated as LZ Albany, its lead elements captured two enemy prisoners. McDade immediately moved to the head of the column so he could personally interrogate them. Unbeknownst to McDade, at least one other enemy soldier got away and reported the American force to nearby North Vietnamese troops.[33] Two North Vietnamese battalions in the area began to maneuver into hasty ambush positions all along the US battalion's long and exposed northern flank.[34]

After interrogating the prisoners, McDade realized that enemy troops were nearby, but he took no additional precautions. Despite the clear danger of imminent contact with the enemy, McDade called all of his company commanders forward to the head of the column for a meeting to explain his plans for the next day, including the occupation of the landing zone and the subsequent helicopter extraction scheme. All of his company commanders traipsed forward along the overextended column with their radio operators in tow.[35] McDade's inexplicable decision to bring all of his commanders to the head of the column stripped all of his companies of their senior leadership as well as a vital radio link. Combined with the column's utter lack of security, this tactical error by McDade produced deadly consequences.

The NVA troops struck just as the luckless company commanders arrived at the head of the column. NVA mortars, machine guns, and small arms rained down on the ragged file of Americans resting unsuspecting in the elephant grass. Enemy soldiers assaulted all parts of McDade's column, splitting his strung-out units into small clusters of men who were forced to fight for their lives.[36] McDade's commanders attempted to race back to their companies to take command of the fight, but most were trapped away from their men, and forced to fight where they stood. Chaos reigned as troops fought desperately without orders from their commanders. In the confusion, companies had difficulty calling for fire support as coordination between units broke down. The NVA troops decimated McDade's companies with deadly accurate fires followed by close assaults.

Despite the overwhelming fury of the sudden attack, McDade failed to alert his superiors or call for reinforcements. His second in command called in artillery and close air support from the front of the column, but his efforts were complicated by the wholesale intermingling of US and NVA troops in close combat. McDade called upon his units to cease fire at one juncture, fearing the Americans were firing into one another. NVA snipers climbed into the trees to shoot down on the Americans who were pinned down in exposed positions.[37] With little effective communication with his subordinates, McDade failed to grasp the immensity of the catastrophe that was erupting all around him in the high grass. By late afternoon, the brigade commander sent one company to reinforce McDade. But neither McDade nor his brigade commander accurately communicated the scale and intensity of the attack to the division headquarters, masking the true scope of the disaster. This unconscionable lapse prevented the division commander from dispatching substantial reinforcements to help 2/7 Cav escape this blazing inferno. The division commander later noted: "We had ample resources to reinforce Albany . . . but no one asked."[38]

As night fell, the battalion remained split apart, isolated from reinforcements, and in deep trouble. North Vietnamese soldiers roamed at will up and down the devastated line of Americans, wiping out small outposts and slaughtering wounded cavalry troopers in the dark.[39] McDade's 2/7 Cav was no longer a cohesive fighting force; it was now a collection of fragmented and isolated small bands fighting desperately for their own survival. At daylight, the enemy withdrew, and McDade took stock of his losses. In his lead unit, Alpha Company, 50 men were killed and two platoons overrun in the first minutes of the fight; in Charlie Company, around the middle of the column, 20 men were killed in the opening fusillade. In all, 2/7 Cav suffered staggering losses, with 155 killed, 125 wounded, and at least four missing.[40] The losses were epic in a war that ultimately produced tens of thousands of American casualties. McDade's catastrophic encounter with the North Vietnamese Army in 1965 remained the most costly single day for the United States in the entire Vietnam War.

Lieutenant Colonel McDade exemplified failed tactical leadership in several ways. Despite his previous combat experience in two wars, he was fundamentally unable to adapt to a new war fought on the very different battlefield he found in the Ia Drang. But his failure to adapt is all the more damning since his sister battalion successfully fought a furious and heroic battle against similar North Vietnamese units at LZ X-Ray only a few days earlier. Even though McDade observed that action, he apparently drew few lessons about the formidable fighting power and newfound tactics employed by the North Vietnamese. He completely failed to absorb the nature of this new enemy and the very different type of battlefield upon which he found himself and his battalion. And he fundamentally failed to grasp what needed to be changed to succeed against this manifestly capable new foe.

Most inexplicably, McDade utterly failed to ensure his unit maintained adequate security in the face of a dangerous and skillful enemy force that he knew was nearby. His most deadly decision was calling his subordinate commanders forward to join him at the head of his extended column for a conference. That decision might have been permissible under peacetime conditions, but was incomparably risky during wartime, especially given the bloodiness of the previous days' fighting. When the North Vietnamese onslaught occurred, McDade failed to improvise and adapt in the face of imminent battlefield disaster for his battalion. He organized no counterattacks to break the death grip the NVA had on his exposed column, and did not even attempt to pull his scattered units into a defensible perimeter.

Once the severity of the attack became clear, McDade failed in yet another fundamental responsibility of command: he did not communicate the grave peril facing 2/7 Cav to the First Cavalry Division headquarters and ask for immediate reinforcements.[41] During the previous days' bloody battle at LZ X-Ray, McDade's fellow battalion commander, Lieutenant Colonel Hal Moore, reached a point in his fight when he approved a Broken Arrow call to alert higher headquarters to his desperate situation.[42] Broken Arrow was a well-understood code word indicating that an American unit was in deep peril, its troops were intermingled with the enemy, and it was about to be overrun. It would mobilize all available US air power and ground troops across Vietnam to rally to the aid of the beleaguered unit. The Broken Arrow call may have been what ultimately saved Moore's unit from defeat.[43] McDade never made such a call, despite being in a situation that was even more desperate than Moore's. The results were disastrous for 2/7 Cav, which suffered the worst losses in a single engagement of any battalion in the Vietnam War. Tragically, McDade failed to adapt effectively to a new kind of war, despite the clear lessons of previous days' combat all around him.

Theater Adaptation in Leadership
Successful Theater Adaptation: Field Marshal Viscount William Slim, Burma, 1942–1945

Field Marshal William "Bill" Slim led large elements of British and Commonwealth forces in bitter fighting against the Japanese in Burma and India from 1942 to 1945. Slim's command began in the depths of failure as British forces were decisively defeated and driven out of Burma in 1942, and ended at the pinnacle of their successful return in 1944–1945. The remarkable adaptation Slim showed first as a corps and then as an army commander is a striking example of successful theater adaptation under very difficult military circumstances. As a commander with ever-growing responsibilities, he found himself battling an experienced and victorious Japanese army with largely unprepared Allied forces. At the same time, the China-Burma-India theater of operations in which Slim fought was a wartime backwater compared to better-known and higher-priority Allied efforts to retake Europe and directly battle Japanese forces in the central and Southwest Pacific. Slim's 14th Army was so sparsely supported during the war that it earned the title "The Forgotten Army." Slim would have to fashion the ability to fight and win on a shoestring, using whatever supplies and troops trickled in from a very long and constrained global pipeline.

Slim was an unlikely commander to ultimately lead the British Commonwealth forces in these battles. A decorated World War I veteran, he had seen combat in Mesopotamia and was badly wounded at Gallipoli, both battlefronts characterized by fighting much different from the trench warfare in western Europe. Slim's army service was also not in the regular British army, but in the British-officered Indian Army. These uncommon command experiences would provide him useful skills in dealing with both diverse troop units and limited resources, common experiences in secondary theaters of war. At the outset of World War

FIGURE 4.4 Field Marshal Sir William Slim
Source: https://collection.nam.ac.uk/detail.php?acc=1951-02-10-2.

II, Slim initially fought with Commonwealth formations in Sudan, Eritrea, and Ethiopia, and by 1940 was commanding an Indian division in offensive operations in Iraq. Building upon his World War I experience, his forte was fighting in the desert, where he approvingly noted, "You can see your man."[44] His later wartime battles against the Japanese proved almost wholly unanticipated and bore virtually no resemblance to any of the terrain, enemy, and tactics he had previously experienced.

In March 1942, Slim took command of the First Burma Corps, which consisted of the 17th Indian and First Burma Divisions. The "Burcorps" was reeling at the time under a powerful Japanese offensive aimed at seizing Burma from the British and ultimately invading India. Slim had little useful guidance from senior British commanders for his defensive campaign; it was not clear whether his objective was to fight to destroy the Japanese army, retain part of Burma at all costs, or slowly trade space for time as the defense of India was readied.[45] Although Slim was a subordinate commander in a larger effort, the lack of a clear objective for his difficult mission had a debilitating impact on his operational planning and the morale of his troops. It made a strong impression on him that greatly shaped his later approach to command.

Slim's corps was badly mauled by the Japanese. Trained and equipped for the Middle East rather than the Burmese jungles, the divisions of the Burcorps were outclassed and outfought by a well-led Japanese foe who was well prepared for a jungle environment.[46] Slim's forces were largely road-bound, laden with many vehicles strung out along lengthy supply lines upon which his fighting units wholly depended for reinforcement and resupply. His soldiers were intimidated by the inhospitable jungle, viewing the rugged western Burman hills and thick forests as hostile and impenetrable terrain.[47] Japanese troops, by contrast, moved easily through the jungle without relying heavily on roads or motor transport. They hooked around Slim's road-bound forces that were defending key roadblocks, cutting them off from the rear while pressing them on all sides.[48] Slim's troops were repeatedly forced to fight to regain their supply lines, prompting a series of perpetual retreats. These continuing tactical defeats decimated the morale of Slim's troops, and they began to view Japanese troops as invincible in the jungle. Defense, encirclement, and withdrawal became his corps' single operating principle.

Slim's leadership of Burcorps during this difficult period was sorely tested. He struggled to overcome the deficiencies of his own forces, while desperately attempting to stem the tide of the Japanese advance. During this trying period, he recognized that his troops' battlefield approaches were failing against effective Japanese tactics, but he was powerless to provide an effective counter other than his physical presence on the front lines and personal inspiration. Slim often found himself standing alone in mountain passes during the long retreats, encouraging his troops as they glumly marched by in one withdrawal after another.[49] In one desperate battle, his forward headquarters was nearly overrun, with one of his senior staff officers driven to defend their position at close quarters with a submachine gun.[50]

By late May 1942, the British and Commonwealth forces in Burma were soundly defeated by the Japanese and had retreated into India.[51] The defeat was a bitter one for Slim, but his calm and effective battlefield leadership under intense enemy pressure earned him a promotion. Slim was chosen to command the newly forming 14th Army, comprising the bulk of British Commonwealth forces assembling in India for the return to Burma. His command included British, Indian, Sikh, Gurkha, and East and West African soldiers, a remarkably diverse force spanning a dizzying array of nationalities, cultures, and military experiences.[52] Slim's challenge would be to mold this disparate collection into an effective fighting machine that could return to Burma and defeat a powerful, experienced, and confident Japanese army—and do so with fewer resources and support than any other Allied army experienced during the war.

Slim recognized that in an economy-of-force theater, he would be unable to leverage the massive resources available in higher-priority theaters. The Burma campaign would never muster the numbers of troops, thousands of landing craft, and fleets of warplanes devoted to achieving victory in Europe and the Pacific. But Slim recognized that with enough time to train, he could revamp his defeated forces into a capable and confident army despite his inevitable shortages of people and materiel.

Slim carefully analyzed the lessons of the defeat in Burma and tirelessly drove the 14th Army to address them. His personal assessment of the causes of the earlier failure was dispassionate, objective, and unsparing of his own generalship.[53] He identified fighting leaders below him who could help his army adapt to the jungle war, and removed those who could not.[54] Listening to subordinate commanders and sorting out good ideas from bad, he crafted his own detailed operational guidance to effectively adapt his army's approach to the environment and the enemy. Slim instilled a fighting spirit in his units by restoring their confidence in their ability to conduct jungle warfare. He began an intensive training program to teach *all* of his troops, from supply clerks to infantrymen, how to operate and patrol in the jungle.[55] He insisted that all of his formations conduct offensive patrols, regardless of their location or mission on the battlefield. This offensive spirit instilled confidence and a growing willingness among his troops to seek out and attack enemy forces aggressively.[56] Slim demanded that his units train for night operations, to dispel their natural fears of moving and fighting at night in the jungle. He also entirely revised the tactics with which he planned to fight his next battles, adapting to now well-known Japanese tactics and to the nature of the terrain he knew he would find.[57]

Despite the chronic shortages of men and materiel that plagued his command throughout the war, Slim was remarkably effective in finding new and creative ways to maximize his modest capabilities. He and his supporting air force commanders found ingenious ways to use air power to offset their shortages in ground power. Slim's airmen pioneered the use of transport aircraft to resupply isolated ground units, permitting far bolder independent offensive operations that no longer relied upon vulnerable roads for supply or even reinforcements.[58] Responding to one of his top theater priorities, Slim's air force commanders also devised innovative ways to quickly evacuate casualties by small aircraft from forward units to hospitals in the rear, which greatly improved troop morale.[59] Slim was also a forceful and determined advocate to his higher headquarters in India. He constantly lobbied for more of the critical resources he needed, especially aircraft and for always-short amphibious landing craft.[60] Yet he also became the master of capitalizing upon what he had, devising unorthodox means of overcoming the unyielding obstacles of weather and terrain when he could not rely upon getting more materiel.[61]

Slim personally visited large numbers of his dispersed units, and spoke quietly and confidently to his troops about the challenges they faced and the importance of their mission.[62] He placed an equal emphasis in his visits upon each of the types of troops responsible for making his command function effectively, from mail clerks to supply sergeants to ammunition handlers. His unobtrusive visits and warm words for his men drew unusual acclaim, establishing a powerful reputation among his soldiers of all nationalities.[63] Slim worked hard to understand and speak to the ethnic and cultural differences among his Indian, Gurkha, African, and British troops, even learning phrases from different languages to speak with them in their native tongue.[64] According to Williamson Murray and Allan Millett, "Slim found a way to get everyone into the fight"—which ultimately extended to native guerrilla fighters as well.[65] The troops and their leaders responded well to this quiet, unassuming leadership, and particularly appreciated Slim's candor in explaining what was to be done, why it was important, and how difficult the tasks ahead would be.[66]

Slim relentlessly held his commanders to account for his new changes in discipline, standards, and training. The China-Burma-India theater was rife with disease, and illnesses such as malaria put far more troops out of action than did enemy fire in the battles of 1942. Slim ruthlessly demanded that commanders enforce the use of malaria prophylaxis to drive down these unnecessary drains upon his combat power, relieving several commanders whose units failed to reduce their sick counts. Word soon spread, and malaria rates plummeted across his command, ensuring he was able to maximize his already limited manpower for the battles ahead.[67]

The 14th Army that started fighting the Japanese in 1944 was wholly transformed from the ill-trained and poorly equipped British Commonwealth forces in Burma that had been defeated in 1942. When a surprise Japanese offensive aimed toward India in March 1944 disrupted Slim's plans for a major attack of his own, he found his forces in desperate defensive battles in northeast India at Kohima and Imphal. The Japanese achieved unexpected initial success, but Slim employed innovative adaptations with his own forces. For the first time in history, two entire infantry divisions were airlifted into the battles, and large formations surrounded for months were solely provisioned by airdrop.[68] In brutal fighting, the 14th Army held its own, and the Japanese attack foundered in the face of the fierce resistance from Slim's troops. By July 1944, the 14th Army had crushed the large and capable Japanese force, ending the threat to northern India. Beginning in early 1945, Slim's forces went on the offensive in a strikingly innovative campaign to retake Burma despite strong Japanese resistance.[69] By early May, Slim's 14th Army joined other British forces in occupying Burma's capital, Rangoon. The Japanese occupation of Burma was largely over, and Japanese troops were in full retreat.

Slim represents an exceptional example of how adaptive theater leadership can transform armies. He clearly understood that the Burcorps was failing at jungle warfare when he took command in March 1942 during a full-blown retreat, but he had neither the time nor the resources necessary to prevent the Japanese victory in Burma a few months later. In 1943 and 1944, he systematically incorporated the lessons of that earlier defeat into the training of the new 14th Army and wholly revolutionized its approach to fighting in the jungle. He developed new training and unconventional tactics, especially finding creative ways to utilize air power, and ensured that these changes were rapidly and effectively implemented throughout his large force. He also actively sought out the perspectives and suggestions of those under his command, down to and including front-line troops, and incorporated them into his operational plans. His powerful leadership skills enabled him to reverse the dismal morale and broken fighting strength of his defeated army and build a new, confident, and capable force that went on to defeat the Japanese in every encounter throughout 1944 and 1945.

Failed Theater Adaptation: General William Westmoreland, Vietnam, 1964–1968

General William C. Westmoreland served as the commander of the US Military Assistance Command-Vietnam (MACV) from June 1964 to July 1968. He led the entire US effort during the Vietnam War's most significant period of buildup and its heaviest fighting. Following his departure, the United States began a prolonged withdrawal from the conflict that ultimately ended with a North Vietnamese victory in 1975. Westmoreland's failure to adapt to the distinctive character of the Vietnam War and devise an effective strategy for its conduct contributed substantially to the eventual US defeat.

Westmoreland was widely viewed as among the brightest and most talented of his generation of US Army officers. Westmoreland graduated from West Point in 1936, and commanded a field artillery battalion during the early stages of World War II. Later in the war, he became the chief of staff of an infantry division, and participated in battles from the North African campaign to the fall of Berlin in 1945.[70] After the war, he attended the army's parachute school and took command of a paratroop regiment, quickly becoming well known as a rising star in the army's influential "airborne mafia."[71] In the Korean War, "Westy," as he was known, commanded an airborne regimental combat team in combat, furthering his reputation as a fighting leader. As the army found itself in the uneasy global stand-off of the Cold War, Westmoreland proved equally adept at Pentagon assignments, moving rapidly up the ranks. He attended the prestigious Harvard

FIGURE 4.5 General William Westmoreland in Vietnam, 1967

Source: https://commons.wikimedia.org/wiki/William_Westmoreland#/media/File:William_Westmoreland.jpg.

senior executive program, served as superintendent of West Point, and commanded the army's famed 101st Airborne Division and later the 18th Airborne Corps. Everything in Westmoreland's career pointed to exceptional potential for the highest commands.

The growing war in Vietnam provided that opportunity. Handpicked by President Kennedy before his death to take over the faltering effort there, Westmoreland arrived in early 1964 to find himself in a conflict that bore no resemblance to his battlefield experience in World War II or Korea.[72] At the time of Westmoreland's arrival, US forces were limited to about 16,000 troops, including advisers to the South Vietnamese military and other supporting forces that were operating from bases hosting American air power. A nationwide network of Viet Cong guerillas, supported by the communist North Vietnamese regime in Hanoi, was putting immense direct and indirect pressure on the US-backed Saigon government and its military forces. As the enemy ramped up its offensive actions, American airbases came under assault, and North Vietnamese patrol boats and US warships seemed to have clashed in the Gulf of Tonkin. Despite confusion about the facts of the incident, President Johnson responded by seeking wide-ranging congressional authority to engage in combat operations, which was approved nearly unanimously by Congress. Westmoreland soon asked for more US forces, beginning a series of major escalations of the US presence that would eventually grow to more than 543,000 US troops.[73] The United States was now on the path to a much wider war, with Westmoreland commanding the rapidly growing US force.

Military strategist Carl von Clausewitz argued that "the first, the supreme, the most far-reaching act of judgment that the statesman and commander have to make is to establish . . . the kind of war on which they are embarking; neither mistaking it for, nor trying to turn it into, something that is alien to its nature."[74] Tragically, neither Johnson nor Westmoreland achieved this objective in Vietnam. As the theater commander, Westmoreland had a unique responsibility to understand the character of the conflict he was charged with winning, and to explain it to his superiors and subordinates alike. In 1964, the Vietnam War was primarily unconventional—a Viet Cong insurgency rooted in winning over the South Vietnamese population and undermining the government in Saigon. Unfortunately, Westmoreland was ill-suited by his outlook and experience to either understand such a threat or take on the myriad of complex political-military tasks required to defeat it. His extensive conventional war experience in World War II and Korea combined with a deep-seated army model of warfare that emphasized firepower and attrition left him ill-equipped to devise and lead a complex counterinsurgency effort.[75] Rather than focus the efforts of his quickly growing military forces upon protecting the population and building

South Vietnamese fighting capability, Westmoreland quickly deployed arriving US forces into direct combat against the enemy.

As the US effort shifted from a limited advisory and support mission into a massive intervention deploying hundreds of thousands of conventional troops, Westmoreland led a dramatic change in the war. With overwhelming public and congressional support for his effort in 1964, he had the opportunity as the senior US commander in Vietnam to shape the war in a variety of alternative directions. He could have directed his newly arriving forces to partner with and improve South Vietnamese units; he could have insisted on using his much larger resources to protect the vulnerable population; he could have demanded far larger numbers of better-trained advisers from the United States. But over the next four years, Westmoreland continued to focus on the enemy, not the Vietnamese population, in a war of attrition that relied heavily on superior US firepower. Westmoreland believed that if a "crossover point" could be reached, where US and South Vietnamese forces killed or captured more of the enemy than could be effectively replaced, the United States would win the war.[76] He directed his forces to conduct large-scale search-and-destroy operations against the Viet Cong and North Vietnamese infiltrators, convinced that killing large numbers of the enemy would force Hanoi to capitulate and end the conflict.[77] Building capable South Vietnamese forces and shielding the population from insurgents were much lower priorities, and received substantially fewer resources and attention.[78] The army's firepower-intensive force structure and deeply ingrained model of warfare helped make the "war of the big battalions"[79] Westmoreland's first choice, reinforced by his previous wartime experience. As one army officer observed, Westmoreland "never thought about doing it any other way."[80] Attrition—killing more and more of the enemy until Hanoi could no longer sustain its effort—became the principal American strategy.

Westmoreland did not alter this approach in any substantial way over the next four years, despite mounting US casualties and little objective progress. At the end of 1965, Westmoreland commanded 35 maneuver battalions, the equivalent of nearly four divisions, and had requested reinforcements to double that number.[81] By the end of 1966, he commanded seven divisions and five separate brigades, more than 2,400 helicopters, and over 500 other aircraft.[82] He used these forces to conduct a series of major offensive search-and-destroy operations, which involved units ranging from several battalions to several divisions and often lasted for months at a time. By 1967, alternative views about how to conduct the war were circulating both inside and outside his command.[83] But after so many years of forcefully advocating an attrition approach focused completely on killing more enemy troops, Westmoreland seemed simply unable to consider a different option.

The attrition strategy inevitably meant that the enemy body count became the most important measure of success for American units and their commanders.[84] This unfortunate metric led to wildly inflated numbers of reported enemy deaths, little distinction between dead civilian and enemy bodies, and eventually the deep erosion of US military integrity and professionalism.[85] Killing as many enemy troops (or "suspected" troops) as possible overrode any impetus toward safeguarding villages, strengthening the South Vietnamese government, or reinvigorating the South Vietnamese military. This proved to be self-defeating, since the North Vietnamese and Viet Cong repeatedly proved that they could withstand far higher losses, and for much longer, than could the Americans. Yet Westmoreland continually regaled his superiors in Washington and the American public with an unremitting series of highly optimistic assessments of the war's progress.[86]

By the end of 1967, there was ample evidence from the field and from Westmoreland's commanders that the attrition strategy was failing.[87] Brigade commanders expressed concerns to visiting officials about the effectiveness of the US firepower-based approach. At the same time, many officials in the Johnson administration and members of the press became convinced that the United States was losing the war.[88] US casualties continued to rise, with no evidence that the North Vietnamese felt any pressure to come to the negotiating table. By virtually any objective assessment, there was no apparent path to win the war. Nevertheless, Westmoreland doggedly pressed on with his strategy, becoming even more optimistic in his press briefings and interviews.

These dynamics set the American people up for a massive psychological shock. In January 1968, Viet Cong and North Vietnamese forces launched a stunning nationwide surprise attack during the Tet holiday. Large portions of major southern cities came under siege, and the attacks were only repulsed after heavy losses on both sides. Although the North Vietnamese and Viet Cong assaults were rebuffed, this surprise enemy offensive and the record US losses it incurred proved the ultimate undoing of the war at home. After Tet, the US population and government had lost the will to continue a seeming never-ending war of attrition in Vietnam. Westmoreland's requests for major new troop increases after the Tet offensive were denied, and by June he had been replaced as MACV commander and promoted to army chief of staff. His four years in command of the Vietnam War resulted in an unprecedented strategic reversal for the United States. The largest, most powerful, and most technologically advanced army in the world had been battled to a standstill by lightly armed Viet Cong guerrillas and dedicated, well-trained North Vietnamese regulars. From mid-1968 onward, the United States was on a path of drawdown and ultimate departure from Vietnam—an outcome that resulted mostly from Westmoreland's flawed leadership during the four most critical years of the war.

Westmoreland fundamentally failed the first and most important test of a theater commander. He lacked the ability to understand the nature of the war he was fighting, and to recognize that his approach was failing. Despite a growing chorus of disapproval by civilian observers and military leaders, even by mid-1968 he remained unable to see the evidence of his failures all around him. After four years of deep personal commitment to his strategy, Westmoreland seemed unable to consider the possibility that he needed to adapt to the war as it was, not how he would like it to be. As a result, he remained closed to change and to new ideas from his subordinates, and he failed to create or lead any fundamental organizational change. Under his leadership, the US Army in Vietnam grew ever more efficient at delivering overwhelming American firepower at the tactical level, but failed at the strategic level by misunderstanding the nature of the war it was fighting.[89] Westmoreland's inability to grasp that his strategy had failed and produced an unending bloody stalemate is in many ways the central story of the US loss in Vietnam. As theater commander at the critical juncture in the war, Westmoreland bears a major responsibility for the ultimate US defeat.

Conclusion: The Importance of Leadership Adaptability

Leadership may be even more important than doctrine and technology in determining the extent to which a military force is adaptable. Successful wartime leadership requires leaders at all levels of command to constantly and dispassionately assess their operating environments, and determine when new approaches are needed. This is much easier said than done, however, because it requires a tremendous amount of focus and discipline amid the fog and friction of war. It also requires leaders to overcome the challenges of decision-making under uncertainty, and especially what behavioral economists call the availability heuristic: the more easily an example comes to mind, the more likely we are to think it will represent the future.[90] We discuss this idea further in Chapter 9, but the leadership failures of McDade and Westmoreland demonstrate this dynamic all too clearly. Both men were profoundly shaped by their combat experiences in World War II and the Korean War, and neither seemed to shed the blinders caused by those experiences to truly understand the very different character of the Vietnam War.

Adaptive leaders must also be able to develop creative approaches to the problems they face. At the tactical level, this frequently requires exploiting unexpected opportunities and improvising with unanticipated resources. Abizaid, for example, demonstrated great leadership by convincing his aircrew not to abort the flight into Grenada, which enabled him to finally parachute into battle with his company and continue their mission. But what enabled his lightly equipped rangers to eventually prevail against their well-equipped enemy was the spark of his personal inspiration. Abizaid was able to look at an abandoned bulldozer and reimagine it as a tank that his rangers could advance behind.

At the theater level, adaptive leaders must not only develop innovative ways to solve operational problems, but also be able to transform large and unwieldy organizations to embrace their solutions. Upon taking command in Burma, Slim quickly grasped that his forces were completely unprepared for jungle warfare, as evidenced by their repeated defeats at the hands of the jungle-savvy Japanese. But over time, he emerged as a transformative theater commander because he revolutionized the ways in which his forces were trained, rebuilding their skills and fighting spirit. By 1945, he had developed new tactics that used his limited resources to such effect that his "Forgotten Army" in Burma defeated the powerful Japanese forces that had beaten them three years earlier.

Finally, the examples in this chapter demonstrate the importance of clear and open communication throughout the chain of command, but especially flowing upward. Subordinates are almost always closer to the fighting than their superiors. As we saw earlier in this chapter and in Chapter 3, they also often know far more about what is going right and wrong on the battlefield than their superiors do. Subordinates are keenly aware what information their superiors favor and what they dislike, and often tend to provide them with the former even if the latter reveals key problems that should be addressed. Superiors must actively seek out all types of information from their subordinates, including bad or unwelcome news, by building a climate of openness, trust, and candor. This is challenging at the tactical level, but it becomes extraordinarily difficult at the theater level, because the military's deference to rank and seniority reaches its zenith when generals are involved. Adaptive theater commanders find ways to overcome these challenges, as Slim did to such great success in Burma. Westmoreland, by contrast, created an environment that encouraged only good news to be delivered to the top. The contrast between the two men could not have been greater, nor the results more starkly different.

PART II

The Recent Wars

5

Doctrinal Adaptation in Iraq and Afghanistan

US ARMY DOCTRINE before the war in Iraq remained profoundly shaped by the lessons that the organization chose to learn after the Vietnam War. Militaries frequently use the term "lessons learned" as if there is only a single, definitive set of lessons to be drawn from a situation, but it is rarely that straightforward. As historian Edward Drea has noted, "The way an Army *interprets* defeat in relation to its military tradition, and not the defeat itself, will determine, in large measure, the impact an unsuccessful military campaign will have on that institution."[1] After Vietnam, the principal lesson that the US Army chose to learn was simple. It should avoid counterinsurgency (COIN) warfare at all costs and should instead focus primarily on preparing for the large conventional operations that would be required during any confrontation with the Soviet Union. The army soon actively eliminated counterinsurgency operations from its doctrine and its educational institutions, which meant that an entire new generation of officers had no understanding of what they required and how they differed from conventional war. As a result, the army as a whole, and especially its senior leaders, remained disastrously unprepared for the insurgencies it encountered after 2001, particularly the virulent one that emerged in Iraq after Saddam Hussein was removed from power in 2003.[2]

In December 2006, the US Army finally published a new counterinsurgency manual that was its biggest doctrinal adaptation in four decades.[3] This remarkable document was one of several factors that helped rescue the US military from the brink of defeat in Iraq,[4] and has been rightly acclaimed as a tremendous success by a wide range of scholars and pundits. Yet that success masks tremendous failures in the army's doctrinal adaptability. The new counterinsurgency manual was not issued until almost four years after the war began—longer than the United States spent fighting all of World War II. And even then, it only managed to happen

because of a remarkable confluence of events and individuals who completely cir-
cumvented the army's normal processes for developing doctrine. As a result, the
new manual was both a pivotal doctrinal adaptation and one that paradoxically
demonstrated the poor doctrinal adaptability of the US Army.

In Afghanistan, the key doctrinal adaptation to the growing insurgency
involved the gradual development of provincial reconstruction teams (PRTs).
The US Army rewrote its civil affairs doctrine during the 1990s to incorporate
many lessons about how to work with local populations during peacekeeping
and stability operations. Yet US military planners began to adapt that doctrine
even before the first civil affairs teams arrived in Afghanistan, marking the start
of over a year of continuous adaptation as security and reconstruction became
increasingly important elements of the US strategy. The first PRTs were established in 2003, and their numbers increased rapidly throughout 2004 and into
2005 as US operations shifted from intermittent counterterrorist operations to a
more comprehensive counterinsurgency approach. PRTs resulted from a remarkable process of doctrinal adaptation, and significantly helped maintain support
among the Afghan people for the US and coalition military presence throughout
the country.

Figure 5.1 summarizes these examples, which we examine in the rest of this
chapter. We should note, however, that both the COIN manual and the PRTs
have weathered a great deal of criticism. The new counterinsurgency doctrine has
been accused of focusing on tactics over strategy, among other criticisms,[5] and
the PRTs have become exemplars of the continuing US failure to pursue a whole-
of-government approach in its conflicts.[6] These criticisms are important ones
and, over the long term, may be accurate assessments of their effectiveness and

	Iraq	Afghanistan
Prewar Doctrine	AirLand Battle and rejecting counterinsurgency (mistaken)	Civil Affairs for peacekeeping (mistaken)
Wartime Adaptation	Relearning counterinsurgency (failure followed by success)	The evolution of Provincial Reconstruction Teams (success)

FIGURE 5.1 Doctrinal Adaptation in Iraq and Afghanistan

impact. Yet both nevertheless remain examples of successful military adaptation during wartime, because they required those in uniform to recognize that existing approaches were failing and to develop new and creative solutions to their battlefield challenges.

Doctrinal Adaptation in Iraq:
Relearning Counterinsurgency
Mistaken Prewar Doctrine: AirLand Battle
and Rejecting Counterinsurgency

For most of its history, the US military has fought its wars through a strategy of annihilation. It has used firepower and mass, and where possible overwhelming force, to wear down and ultimately destroy the enemy's armed forces and ability to fight. It has relied on that strategy so frequently that historian Russell Weigley has called it "the American way of war," noting that the US military has had little patience for indirect methods and other approaches that lead to less decisive outcomes.[7] The US Army, as the oldest and the largest of the four military services, has deeply incorporated the strategy of annihilation and decisive victory into every element of its doctrine, training, and operations.

In his landmark work on the US Army in Vietnam, Andrew Krepinevich argues that the army's vision of war, which he calls the "Army Concept," has always had two key elements. It includes a focus "on mid-intensity, or conventional, war and a reliance on high volumes of firepower to minimize casualties—in effect, the substitution of material costs at every available opportunity to avoid payment in blood."[8] This approach has left the service poorly suited for counterinsurgency operations,[9] but the Army Concept continued to prevail because it had been so successful in conventional wars. The US Army had never lost a major war before Vietnam, so it never had to question whether its approach was correct. The Army Concept was further reinforced by the tremendous victories in both world wars, and even the lessons of the inconclusive Korean War.[10]

The Army Concept proved disastrous in Vietnam, since, as Krepinevich argues, it was a "gross mismatch with the needs of a counterinsurgency strategy."[11] As discussed in Chapter 4, General William Westmoreland and other US Army leaders continued to focus on destroying enemy forces with large amounts of firepower throughout the war despite numerous signs that the strategy was failing.[12] Although the army did conduct some more traditional counterinsurgency efforts in the later years of the war, its strategy—even after the 1968 Tet Offensive—remained focused on attriting enemy forces rather than providing population security to strengthen support for the South Vietnamese government.[13]

The US Army that emerged from Vietnam was a defeated and broken force. It had been deeply riven by the war, and was plagued by drugs and racial tensions as well as a shattered NCO corps and undisciplined troops.[14] Army leaders at that time faced the daunting challenge of rebuilding their service into a competent fighting force once again. They did so in many different ways, including discharging substandard soldiers from the army, improving standards and discipline, revamping training, and developing new weapons and fighting systems.[15] But perhaps one of the most consequential rebuilding efforts occurred in the area of doctrine. Here the army leadership decided to turn its focus almost exclusively toward preparing the army for conventional warfare against the Soviet Union in Europe. Fighting the Soviets was a vital, real-world mission for the army at the time, of course, but that key decision also had institutional benefits. As one of us has previously written, "Refocusing on a major war in Europe offered to heal the painful scars of failed counterinsurgency."[16] The army set out to prepare itself for the "anti-Vietnam"[17]—a traditional conventional war against a clear adversary where victory and defeat would be definitively decided on the battlefield.

General William DePuy took command of the newly created US Army Training and Doctrine Command (TRADOC) on July 1, 1973—only a few months after the last US combat troops withdrew from Vietnam. DePuy believed that the army needed to prepare itself for the large tank battles that would characterize any war with the Soviets in Europe—a conviction that was further strengthened by the October 1973 Yom Kippur War. Though Israel ultimately prevailed in that brief conflict, the Soviet-trained and Soviet-equipped Arab forces performed far better than expected.[18] The three-week war destroyed more tanks and artillery pieces than the US military had in its entire inventory, and captured Arab equipment revealed that Soviet vehicle technology was more advanced than American vehicle technology at the time. DePuy concluded that the army would have to find a way to defeat well-equipped Soviet forces in the opening days of any future conflict.[19]

Over the next several years, DePuy oversaw a major revision of the capstone army doctrine for operations. The 1976 version of Army Field Manual 100-5, *Operations*, was based on the concept of active defense, which focused on defending Europe against a numerically superior and increasingly lethal Soviet adversary. It reaffirmed the Army Concept's focus on destroying the enemy through firepower to such an extent that, as one army officer later wrote, it marked "the apogee of attrition theory's influence over US Army doctrine after the Second World War."[20] The new field manual was also deliberately designed to expunge the painful experience of Vietnam, with no mention of counterinsurgency at all.[21] It formalized in doctrine what Army Chief of Staff General Creighton

Abrams stated unequivocally in 1973: that the army was leaving counterinsurgency behind forever.[22]

The army also sought to erase counterinsurgency from its institutional memory in three other ways. First, it did not update its counterinsurgency doctrine to incorporate any lessons learned from Vietnam.[23] Field Manual 100-5 was the capstone operations doctrine for the army, but there were also dozens of other doctrinal publications on more specific topics. The army had revised its counterinsurgency doctrine in 1963, but then elected not to revise it again for more than three decades.[24] Army doctrine on low-intensity conflict was revised in 1981, but it basically restated the principles contained in a 1967 publication about counter-guerrilla operations, and it too remained unrevised for many years thereafter.[25] Other army doctrinal publications also contained little or no mention of counterinsurgency operations.[26]

Second, the army essentially eliminated counterinsurgency as a subject of military instruction. As retired army colonel and history professor Peter Mansoor has noted, "For three decades the professional military education system all but ignored counterinsurgency operations."[27] In 1973, the Army War College stopped teaching its five-week course in unconventional warfare, and a few years later, it stopped teaching counterinsurgency and foreign internal defense.[28] The army did continue teaching the broader concept of low-intensity conflict, but even that was severely curtailed. In 1977, the yearlong course at the US Army Command and General Staff College (CGSC) included only 40 hours on low-intensity conflict, and in 1979, that was slashed to only eight hours.[29] Preparing for high-intensity war in Europe dominated the curricula at virtually all army schools.

Third, the army ignored counterinsurgency to such an extent that such missions were almost exclusively relegated to special forces. As one of this book's authors has written, based on personal experience in the army of the time, "Small-scale contingencies were the purview of airborne troops and a growing force of special operators, but 'real Soldiers' rode to battle in armored vehicles."[30] The army did create two new light-infantry divisions after the 1983 invasion of Grenada to improve its capabilities for low-intensity conflict.[31] Yet these still constituted a small fraction of the army's force structure, and they enabled the rest of the army to prepare for what it viewed as real combat.[32] Investments in special forces also grew in the 1980s as they were deployed to counter some of the insurgencies that erupted in Central and South America, most notably in El Salvador. Even then, however, Army Special Forces focused on what one analyst called "resistance forces in support of conventional operations"—unconventional warfare designed to lead insurgencies, instead of conducting counterinsurgency operations themselves.[33]

The army deliberately removed counterinsurgency from its institutional memory despite the fact that many of its own studies, conducted both during and after the Vietnam War, recommended significant changes to its counterinsurgency doctrine.[34] Instead of choosing to examine the war's painful lessons, the army instead gravitated to the far more positive conclusions drawn by Colonel Harry Summers in his 1981 book *On Strategy*.[35] Summers argued that the army's attrition strategy in Vietnam had actually been correct, but that US civilian leaders prevented the army from being able to execute it properly. Historian Conrad Crane argues that the book "quickly became the prism through which the Army viewed Vietnam," because it effectively absolved the army from its humiliating defeat and reaffirmed the Army Concept.[36] *On Strategy* begins with an epigram attributed to a US colonel telling a North Vietnamese colonel, "You know you never defeated us on the battlefield." The army failed to absorb the importance of the North Vietnamese colonel's response: "That may be true, but it is also irrelevant."[37] The book soon became required reading at CGSC and the Army War College, as well as at many other institutions involved with professional military education.[38]

The 1976 version of Field Manual 100-5 was not well accepted within the army, but not because of its silence on counterinsurgency.[39] The army revised the field manual again in 1982, now incorporating some of the key ideas from *On Strategy*.[40] The new doctrine of AirLand Battle was deliberately designed to "fight outnumbered and win"[41] against the Soviets in Europe, by attacking all echelons of their force simultaneously.[42] AirLand Battle, with its offensive focus and highly conventional approach to warfighting, was readily embraced across the army. Its influence permeated virtually every aspect of the institution and continued to shape major procurement, training, and leader development efforts until long after the 9/11 attacks.[43]

AirLand Battle was never tested against the Soviets, but its approach to high-intensity conventional war was seemingly vindicated by the swift and decisive US defeat of Iraqi forces in the 1991 Gulf War. The ground campaign of Operation Desert Storm lasted only 100 hours. Inflicting a crushing defeat on a sizable and well-equipped Iraqi army in such a short time was a tremendous accomplishment for a US military that had been floundering less than two decades earlier.[44] It also further reinforced the lesson that the army had chosen to learn after Vietnam: that it should focus solely on high-intensity conventional warfare. General Colin Powell, who served as the chairman of the Joint Chiefs of Staff during the 1991 Gulf War, later wrote that he and the other officers of his generation who had been shaped by Vietnam "vowed that when our turn came to call the shots, we would not quietly acquiesce in halfhearted warfare for half-baked reasons that the American people could not understand or support."[45]

Though all army leaders took tremendous pride in the post-Vietnam rebuilding efforts that led to victory in the Gulf, retired General Barry McCaffrey, the commander of the 24th Infantry Division during the Gulf War, sounded a particularly prescient warning. He later said, "I fear the majors of Desert Storm," because now a whole new generation of soldiers believed that the overwhelming success of the Gulf War proved once and for all that the army had learned the right lessons from Vietnam.[46] As we will see in Chapter 8, most of the senior military leaders who later led the wars in Afghanistan and Iraq were deeply influenced by their experiences as captains, majors, or lieutenant colonels in the early 1990s, and especially by the lopsided victory against the Iraqi army in Operation Desert Storm.

After the Soviet Union collapsed in December 1991, the army no longer faced the imminent threat of a major conventional war. Instead, throughout the 1990s it found itself conducting numerous peacekeeping and stability operations—including in Somalia, Haiti, Bosnia, and Kosovo. But even though these missions demanded skills different from those required for conventional war, the army's experience in the 1990s did little to prepare it for the counterinsurgency operations that emerged after 9/11. The 1993 revision of Field Manual 100-5 did acknowledge that the army needed to be ready for the full spectrum of conflict and devoted a chapter to what it called "operations other than war."[47] But despite its extensive experience with these types of operations in the 1990s, the army continued to view missions at the lower end of the conflict spectrum as lesser cases of bigger wars and a distraction from its real mission of high-intensity conventional combat.[48] The 2003 revision of *Operations*, now called Field Manual 3-0, mentioned counterinsurgency on only one page, and even then only in the context of providing support for host nations.[49]

The US Army therefore had entirely the wrong type of doctrine for the wars it would soon fight in Afghanistan and postinvasion Iraq. Army doctrine after Vietnam explicitly and consistently rejected counterinsurgency and low-intensity conflict, and instead focused exclusively on preparing for a major conventional war against the Soviets in Europe. That doctrine proved to be accurate for the 1991 Gulf War, and that success reinforced its stature so thoroughly that it outlasted the Soviet Union and persisted throughout a decade of peacekeeping and stability operations afterward. But the army's overarching doctrine was fundamentally mistaken: it deliberately excluded an entire form of warfare that would come to dominate the wars of the early 21st century.

Adapting Doctrine in Iraq:
The Counterinsurgency Manual

Early Counterinsurgency Doctrine, 2003–2004

The planning for the invasion of Iraq began on November 27, 2001, and assumed from the very beginning that the war would end once Saddam Hussein was removed from power.[50] This assumption was widely shared by both senior civilian officials and those senior military leaders who had been the Desert Storm majors and other field-grade officers whom McCaffrey had warned about after the 1991 Gulf War.[51] As discussed in Chapter 8, most of them were entirely unprepared to counter the insurgency that quickly emerged and had engulfed most of Iraq by the summer of 2004. Some commanders experimented with counterinsurgency principles, emphasizing the need to provide security for the local population. Yet most fell back on the doctrine that they had been trained in throughout their careers: killing or capturing the enemy through firepower, raids, and other traditional methods of warfare.[52]

In early 2004, Lieutenant General William Wallace recognized that the army needed to update its counterinsurgency doctrine. As the commander of the US Army Combined Arms Center at Fort Leavenworth, Kansas, Wallace oversaw a wide range of training and educational programs and also had primary responsibility for developing army doctrine.[53] A group of officers had just come through Leavenworth to conduct their last major exercise before deploying to Iraq, and Wallace observed that the exercise bore little resemblance to the types of operations that would be required to counter the growing Iraqi insurgency. The fundamental problem, he realized, was that the army had no current counterinsurgency doctrine.[54]

In February, Wallace ordered Colonel Clinton Ancker, the director of the doctrine division, to develop a new counterinsurgency field manual within the next six months. This was a significant challenge, since field manuals usually took more than two years to produce. But Wallace believed that the army needed some sort of counterinsurgency doctrine quickly, even if imperfect, in order to adapt to the challenges in Iraq.[55] Ancker passed the task to Lieutenant Colonel Jan Horvath, who sought to absorb as many historical lessons learned as he could before starting to write. In a telling anecdote about the post-Vietnam period, Horvath visited the army's special forces headquarters at Fort Bragg to read through whatever old counterinsurgency materials its library contained. He found nothing there and was told it had all been thrown out years before.[56]

Horvath finished the draft in August and asked various army commands and colleagues to comment upon it. One of those colleagues was Kalev Sepp, who taught at the Naval Postgraduate School. Sepp told several of his students, young officers just returned from Iraq, to read and comment on it. The officers were all extremely critical, identifying many different ways in which the manual did not address the fundamentals of counterinsurgency. Horvath agreed with many of their comments and tried to incorporate much of their feedback. But Wallace thought that the draft was sufficient, despite its many problems, since it would at least provide some improved guidance to soldiers on the battlefield.[57]

In October 2004, the Combined Arms Center published Field Manual-Interim 3-07.22, *Counterinsurgency Operations.* The "interim" designation was very unusual; the cover explicitly noted that the doctrine would expire in two years when, presumably, it would be replaced by more developed and permanent doctrine.[58]

The Rise of the COINdinistas, 2004–2005

While the interim field manual was being drafted, the level of violence throughout Iraq continued to escalate. By the summer of 2004, a small group of military officers and civilian experts had concluded that the army's traditional approach of killing or capturing enemy leaders was proving counterproductive, and began to investigate the precepts of irregular warfare and counterinsurgency. Gradually, their shared concerns about the trajectory of the war led them to cross paths. They formed a nascent intellectual community dedicated to promoting counterinsurgency as an alternative approach. Their critics began deriding them as "COINdinistas," combining the military abbreviation for counterinsurgency with the Sandinistas, an insurgent group that took control of the Nicaraguan government in the 1970s. Yet some members of the group took the name as a badge of honor, since they knew that they were essentially mounting an insurgency against the type of doctrine that had become deeply embedded in the army ever since Vietnam.[59]

The first major conference that drew some of these disparate thinkers together was held on October 6, 2004. Brigadier General Robert Schmidle, a marine who was then serving as the director of the Expeditionary Force Development Center at Quantico, was concerned that US forces in Iraq did not know how to fight an urban insurgency. The marines had participated in more small wars and counterinsurgencies in their history than the army had, but even marine commanders in Iraq were relying heavily on firepower to try to defeat the burgeoning insurgency. Schmidle called it the Irregular Warfare Conference, deliberately avoiding the still-controversial term "insurgency." Jim Thomas, then the deputy assistant secretary of defense for resources and plans, offered to cosponsor the conference because of his own concerns about the direction of the war.[60] Other speakers included an Australian lieutenant colonel named David Kilcullen, who had written a paper in 2003 warning that an insurgency was erupting in Iraq. Although his prescient report had drawn little attention, Kilcullen's remarks on counterinsurgency at the conference soon landed him a job in the Pentagon working for Thomas.[61]

Another speaker was a then-obscure army lieutenant colonel named John Nagl. Nagl was working in the Pentagon for Deputy Secretary of Defense Paul Wolfowitz, an outspoken proponent of the 2003 US invasion of Iraq. Nagl had served as a lieutenant leading a tank platoon during the 1991 Gulf War, and had concluded that the utter dominance of the US military in that operation meant that few future US adversaries would choose to fight that way. He decided to study counterinsurgency lessons in graduate school, and his dissertation on the lessons from Malaya and Vietnam had been published in 2002.[62] He served with an armored battalion in Iraq's restive Anbar province during 2003, and became the subject of a cover story in the *New York Times Magazine*.[63] Nagl passionately believed that the US military needed to quickly adopt counterinsurgency principles in Iraq or risk losing the war. He soon became one of the subterranean leaders of the growing COINdinista movement.

By this time, a growing number of articles about counterinsurgency were being published in military journals. The most prominent was *Military Review*, published by the US Army Command and General Staff College, which was part of the Combined Arms Center that Wallace commanded. In March 2004, shortly after ordering the drafting of the interim field manual, Wallace hired a new editor for the journal and told him to make it directly relevant to the war in Iraq.[64] Between March 2003 and March 2004, *Military Review* published nine articles that had some connection to counterinsurgency. Between March 2004 and March 2005, that number skyrocketed to 29, with many authors writing about their own operational experience.[65]

The most important of these articles was written by Kalev Sepp, the Naval Postgraduate School professor who had required his students to critique the interim field manual. Sepp, a former special forces officer, was friends with Colonel William Hix, the officer in charge of strategy working on General George Casey's staff in Iraq. As discussed in Chapter 8, Hix asked Sepp to help review Casey's campaign plan in the fall of 2004. As part of his work, Sepp wrote a paper called "Best Practices in Counterinsurgency." Hix thought it was so good that he wanted it to become an unclassified appendix to the staff's next campaign plan; in the meantime he encouraged Sepp to publish it in *Military Review*.[66] It appeared in the journal's May–June 2005 issue, alongside several other articles that addressed counterinsurgency themes.[67]

Another key workshop was held in early June 2005 at Basin Harbor in Vermont. It was hosted by Eliot Cohen, a professor of strategic studies at the

Johns Hopkins School of Advanced International Studies in Washington. Cohen had spoken at Schmidle's Irregular Warfare Conference and remained very concerned about the trajectory of the Iraq War. The conference included some now-familiar faces, including Kilcullen, Nagl, and Sepp. The intellectual debates at Basin Harbor were deep and sometimes controversial and helped sharpen the thinking of all of the participants. Journalist Fred Kaplan later observed,

> Basin Harbor's larger impact flowed not so much from the substance of the discussions as from the fact of the meeting itself . . . [and] the dawning realization among these scholars and officers and officials, many of whom were meeting one another for the first time, that they formed a community; that their ideas and interests had common grounding and intellectual heft; and that, in the context of the country's two floundering wars, which the current political and military leaders seemed to have no idea how to handle, they—this insurgency of counterinsurgency thinkers— might be of use, might have impact.[68]

Although they did not know it at the time, the next major conference they attended would mark the beginning of the effort to develop an entirely new counterinsurgency doctrine for the US Army and Marine Corps.

A New Counterinsurgency Manual

In October 2005, Lieutenant General David Petraeus succeeded Wallace as the commander of the Combined Arms Center at Fort Leavenworth. Unlike most of his peers, Petraeus had studied irregular warfare from the beginning of his career. He spent the summer of 1986 working at US Southern Command for his mentor, General John Galvin. There he observed the regional effects of the insurgencies in El Salvador and Colombia and ghostwrote an article for Galvin arguing that winning those types of wars required gaining popular support rather than applying firepower.[69] He earned a PhD at Princeton, writing his dissertation on US Army counterinsurgency efforts in Vietnam. And as the commander of the 101st Airborne Division in Iraq from 2003 to early 2004, he successfully implemented many aspects of a counterinsurgency strategy in the northern part of the country, especially in the city of Mosul.[70]

The Leavenworth assignment was part of the army's training and doctrine establishment and was not an especially high-profile job, especially since it did not involve commanding troops in the field. But Petraeus soon saw the job as an opportunity to make big changes, especially in army doctrine.[71] Jan Horvath had been revising the interim field manual ever since it had been published the previous year, but Petraeus thought a new effort was needed that started completely from scratch.[72] He decided to bring together more experienced officers and academic experts to completely overhaul the doctrine. He later described their efforts as "an attempt to fill a genuine doctrinal void,"[73] and hoped it would have as much impact on the army as AirLand Battle had had more than two decades earlier,[74] an ambitious but unlikely aspiration.

The counterinsurgency community reconvened at another conference held in Washington in November 2005. Unusually, it was cosponsored by the US Army War College and the Carr Center for Human Rights Policy at Harvard University, which was led by Sarah Sewell. The 90 attendees included Nagl, Sepp, and Hix; Conrad Crane, the director of the Military History Institute at the US Army War College; several participants from the Basin Harbor workshop; and numerous other scholars and practitioners who were working on counterinsurgency issues.[75] At an evening social event, Nagl sketched out his ideas for an outline for the draft manual. His restaurant outline would soon become a key blueprint for the nascent publication.[76]

Petraeus soon recruited Crane, a West Point classmate, to lead the writing team and signed on Nagl to assist with the effort. Nagl was only able to get permission to work on the project in his personal time, given his sensitive position working for Wolfowitz. Petraeus told Crane that he wanted the manual to be done within a few months and that he planned to host a two-day conference in late January or early February 2006 where outside experts would comment on an initial draft and offer help with the revisions.[77] These plans caused some discomfort in the doctrine development office, since it usually took several years for new doctrine to wend its way through the bureaucratic approval process—a drawn-out practice that was precisely what Petraeus wanted to avoid.[78] In December, Wallace—now Petraeus's boss as the four-star head of Training and Doctrine Command—also expressed concern about the planned conference. If the experts failed to reach a consensus, Wallace said, it might lead to a "doctrinal publication watered down to the point of uselessness."[79] Crane reassured both Wallace and Petraeus that the text would produce "the exact opposite of the pabulum joint doctrine normally becomes."[80]

Crane and Nagl attended an unrelated conference at Fort Leavenworth in mid-December and used it as an opportunity to meet with Horvath and other members of the doctrine office.[81] By that time, they knew that the manual would stress two main principles: that successful counterinsurgency operations required protecting the population, and learning and adapting more rapidly than the insurgents.[82] Crane came to the meeting with a list he had prepared called "COIN Principles, Imperatives and Paradoxes," and started the meeting by writing them on a whiteboard.[83] By the end of their daylong meeting, the group had turned it into a list of key paradoxes that, as discussed later, became one of the most important parts of the final manual.[84] They then decided that they should publish this list of paradoxes in *Military Review* as a preview of the forthcoming doctrine, enabling soldiers in the field to read it as soon as possible and offer their suggestions for revision. They asked Eliot Cohen to coauthor the article with them, which he agreed to after some persuasion, and it was published in the March–April 2006 issue.[85]

During that meeting, they also developed a more detailed outline of the manual, which, after an introductory overview of insurgency and counterinsurgency, included chapters on integrating civil and military activities; intelligence; operations; supporting host nation forces; combat support and combat service support; and leadership and ethics. The outline also included a series of appendices, on specific topics like cultural assessment, the role of interpreters, and legal issues, as well as a bibliography.[86] They also assigned lead authors for each of the chapters, and Crane asked them to submit their drafts by January 16—only four weeks away.[87]

Most of the authors worked over the holidays to meet the very short deadline. Crane later described January 2006 as "a month of frenzied writing," as the entire writing team busily edited, revised, and improved each of the chapters. The integrated draft was sent to the conference attendees on February 6, a remarkable achievement, and the authors continued to incorporate more changes as they received additional feedback. The lead authors of each chapter met on February 22, the day before the counterinsurgency conference began, and many were surprised by how extensively the entire manual had been revised in the previous couple of weeks.[88]

In the meantime, the number of people attending the February conference had exploded. Though Petraeus initially envisaged a meeting of about 25 experts, he and his team kept extending more invitations, and the final participant list included 106 people. They more than filled the main room at the conference, and since many of them had brought aides and other assistants, many of the 250 seats in the large overflow room were taken as well.[89] Petraeus saw the conference as an opportunity to introduce the new doctrine to influential people both inside and outside the US military. He hoped that by being as open and transparent as possible, he would gain participants' support for the new doctrine and their help in promoting it as an alternative strategic approach to Iraq. This was particularly important since the levels of violence were continuing to spiral out of control and an increasing number of Americans supported withdrawing from Iraq altogether.[90]

The final participant list included most of the COINdinistas who had now been attending conferences and working together for more than a year, but the list extended far beyond that expert circle. Petraeus invited his counterpart in the Marine Corps, Lieutenant General Jim Mattis, to participate, and also convinced him to cosign the final manual so it would be officially issued by both services. Sarah Sewall said that the Carr Center would cosponsor the conference if she could invite several human rights activists and leaders of international aid organizations, an offer that Petraeus readily accepted. Bill Hix and Kalev Sepp attended, bringing invaluable perspectives from their continuing work for General Casey in Iraq. The meeting also included officials from the State Department, the US Agency for International Development, and the CIA; veterans of the counterinsurgency operations in El Salvador and Vietnam; people who had published interesting articles on counterinsurgency or Iraq in *Military Review*; and several journalists who had covered the war.[91]

The now-sizable conference began on February 23, with opening comments from Petraeus and a few others. Crane then took the stage and reminded the participants what was at stake. He opened by showing a graduation photo of First Lieutenant David Bernstein, whom Crane had known at West Point, with a caption that read, "Why Are We Here?" He then showed the next slide—an image of Bernstein's gravestone, showing that he was killed in action in Iraq on October 18, 2003, and declared, "To minimize this."[92] Bernstein was one of more than 2,200 US soldiers who had been killed in Iraq by that time.[93] Crane later wrote:

> I wanted to emphasize that this was not an academic exercise, to make sure, by putting a face on the price of failure, that everyone understood the main reason that we were there. There were many justifications for the presence of so many important people at Fort Leavenworth that day, but the primary one was that the nation's best and brightest were dying in faraway places trying to achieve its policy objectives, and we had a responsibility to lower those costs and improve the results.[94]

The rest of the conference agenda explored the draft manual one chapter at a time, incorporating sharp feedback as well as praise for each. The discussions were wide-ranging and lively, with particularly intense debates over the proper roles and responsibilities of training host nation forces, what the doctrine would say about detainees, and whether large-scale counterinsurgency efforts were even possible.[95] These discussions, as well as the tremendous amount of feedback that poured in after the conference, provided the writing team with invaluable new ideas and suggestions for the way forward.[96]

As soon as the conference ended, the writing team went back to work. They produced a new draft that incorporated the extensive feedback from the conference in about three months.[97] The team met again in mid-May, and after another round of revisions, released the final coordination draft on June 20.[98] It was sent to every army command and was posted on Army Knowledge Online, an online portal that all military personnel and army civilians can access. The official deadline for feedback was July 17, but the deluge of comments kept flowing in for months thereafter. Military personnel in the field submitted more than four thousand comments—three thousand from army troops and the rest from marines—which ensured that the final revisions included the latest operational experience.[99] The draft was also leaked to the press and reprinted on several websites, which led to an even wider range of comments and feedback.[100] As Nagl later observed, "No previous doctrinal manual had undergone such a public review process before publication or provided so many opportunities to comment to both those inside and outside the Army/Marine Corps tent."[101] The draft doctrine also started influencing troops in Iraq very quickly. The commander of the Fourth Infantry Division, whose troops were operating in Baghdad, sent the final coordinating draft to all of his battalion and brigade commanders even as the writing team continued its revisions.[102]

The new manual, simply titled *Counterinsurgency*, was officially published on December 15, 2006. Because Mattis had agreed to sign the document, it was simultaneously issued as Army Field Manual 3-24 and Marine Corps Warfighting Publication No. 3-33.5.[103] The release of what would ordinarily have been a soporific new doctrinal manual drew extraordinary attention in the press.[104] Petraeus, who was well known for cultivating relationships with the press throughout his career, cleverly engineered most of this tremendous publicity. Popular interest soared as well. The manual was downloaded more than 1.5 million times in the first week after its release, and the version that was later reprinted by the University of Chicago Press drew a positive review in the *New York Times Book Review*.[105]

Some of this tremendous interest resulted from the fact that the violence in Iraq had escalated so badly throughout 2006 that the war seemed unwinnable, and public pressure was steadily building for a US withdrawal. The Iraq Study Group, a congressionally appointed bipartisan panel, issued its grim report in early December. It started with the blunt conclusion: "The situation in Iraq is grave and deteriorating," and its recommendations included considering a phased withdrawal of US forces.[106] Published only a few days later, the counterinsurgency manual offered a completely new approach for US forces that might help transform the entire course of the war. Less than a month later, President George W. Bush announced his plans for what became known as "the surge," and, as discussed in Chapter 8, selected Petraeus to succeed General George Casey as the commander of all US forces in Iraq. After taking command in February 2007, Petraeus ensured that the new counterinsurgency doctrine was disseminated and implemented throughout the force—and that US troops knew that their primary mission was now to provide security for the local population instead of killing or capturing those conducting the violence. The military's rapid adaptation of this new and wholly different wartime doctrine was one of the key factors that contributed to the remarkable decline in violence that began in early 2007.[107]

Conclusion: The Successful Adaptation That Demonstrates Poor Adaptability

The US Army unquestionably had the wrong doctrine for the counterinsurgency that erupted soon after Saddam Hussein's ouster. The Cold War doctrine of AirLand Battle, which had seemed vindicated by the 1991 Gulf War, enabled US forces to defeat conventional Iraqi armed forces during the first three weeks of major combat operations. But by refusing to acknowledge that the military should play any substantive role in postconflict reconstruction efforts, that same doctrine also meant that the US military did little to stop the violence that spread throughout Iraq in 2003. Even more disastrously, it proved entirely inadequate once the insurgency took root by the summer of 2004.[108] US Army doctrine in this case proved rigid and unadaptable.

Wallace deserves great credit for recognizing in early 2004 that the army needed new doctrine to counter the growing insurgency. He pushed his staff to publish the interim counterinsurgency manual in October 2004, realizing that US troops in Iraq needed some sort of guidance. Yet his limited staff did not have the expertise or capacity needed to develop a new approach that was antithetical to the doctrine that had been deeply inculcated into the force for more than two decades. Even his interim manual, designed to fill a gaping wartime hole in army doctrine, was not widely disseminated or adopted. In fact, one of us commanded all US forces in Afghanistan when the interim manual was issued, but had no idea that it existed until he conducted the research for this book. When the interim manual was released in late 2004, he had already developed and implemented his own counterinsurgency strategy for Afghanistan (as discussed later in this chapter).

The 2006 counterinsurgency manual stands as an extraordinary example of doctrinal adaptability in two key ways. First, Petraeus and his team succeeded where Wallace and his team had made a valiant attempt but fallen short. In just over a year, Petraeus's disciples resurrected the key elements of the counterinsurgency doctrine that the army had deliberately expunged after Vietnam and updated it to reflect the modern battlefield and the specific challenges facing US forces in Iraq. Second, and even more importantly, the doctrine itself stressed the need for continual adaptation and flexibility in its implementation, especially regarding providing security for the population. Nowhere is this seen more clearly than in the section in Chapter 1 called "Paradoxes of Counterinsurgency Operations."[109] These paradoxes were designed to demonstrate how deeply counterinsurgency differs from the types of war that the US military was accustomed to fighting, and how much constant learning and adaptation it required.[110] The paradoxes include "Sometimes Doing Nothing Is the Best Reaction," and "If a Tactic Works This Week, It Might Not Work Next Week; If It Works in This Province, It Might Not Work in the Next."[111]

The doctrine contained in the 2006 counterinsurgency manual meets all the criteria for successful doctrinal adaptability identified in Chapter 2. It was flexible and not dogmatic. It resulted from a clear understanding that existing doctrine was failing. It incorporated suggestions not just from the entire chain of command, but from a wide variety of outside experts. And it was rapidly disseminated—through the media as well as through traditional army channels—and implemented throughout the force within months of its publication.

Yet, as noted at the outset of the chapter, the new counterinsurgency manual may well be the exception that proves that army doctrine on the whole is *not* adaptable. This doctrinal adaptation succeeded because Petraeus bypassed the normal process for developing doctrine—by hand-selecting a team of outsiders to write the document; by seeking input from an unprecedented range of military, government, civilian, and international experts; and by enabling thousands of soldiers and marines, from youngest privates to oldest generals, to provide feedback and comments so that the final version would incorporate the most recent battlefield experience.

Yet it also took nearly four years of repeated battlefield setbacks and thousands of US casualties before this manual got to the troops who needed it most. Even then, it is far from clear that the doctrine would have been disseminated and implemented if President Bush had not decided to completely change the US strategy in Iraq at essentially the same time that the new manual was complete, *and* if he had not decided to put Petraeus in charge of the US war effort shortly thereafter. In other words, the 2006 counterinsurgency manual demonstrates that rapid and effective doctrinal adaptation is possible, but it also suggests that the normal doctrine review process is nowhere near up to the task. Instead, success in this case required several different stars to align almost perfectly—and there is no guarantee that they will do so the next time that the US military needs to adapt its doctrine in the middle of a failing war.

Doctrinal Adaptation in Afghanistan: The Evolution of Provincial Reconstruction Teams

Mistaken Prewar Doctrine: Lessons from the Peacekeeping Era

During the 1990s, the US military thoroughly rewrote its civil affairs doctrine. Civil affairs units are responsible for working with local populations and helping fill some of the functions that are usually the responsibility of civil government.[112] As we discussed earlier, army doctrine after Vietnam focused almost exclusively on major conventional combat operations, which prepared it well for the 1991 Gulf War. Civil affairs units conducted a wide range of largely unpublicized missions in Operations Desert Shield and Desert Storm, including managing dislocated civilians, securing local resources for combat service support, and helping the Kuwaiti government-in-exile plan to restore services after liberation.[113] Moreover, shortly after Saddam Hussein signed the cease-fire, the US military began a humanitarian relief mission to help the Kurdish population in northern Iraq. That began a series of US humanitarian and peacekeeping operations that extended throughout the following decade, in which US civil affairs capabilities were suddenly central rather than secondary skills. During the 1990s, US civil affairs units operated in Somalia, Rwanda, Haiti, Bosnia, and Kosovo. By the middle of the decade, US joint doctrine for civil affairs was revised to incorporate the many ways in which civil affairs units had adapted to their burgeoning new roles and missions. In February 2000, the army issued a new civil affairs manual that built upon and expanded that joint doctrine, and in February 2001, joint doctrine was updated once again.[114]

The September 2001 attacks on the United States ushered in a new era for US civil affairs operations. The following two decades would find civil affairs units committed to missions all across the greater Middle East and Africa. Afghanistan was the first of those missions, as US forces started arriving in October 2001. As military civil affairs units began to arrive in the following months, civil affairs doctrine began to adapt to this unexpected and challenging new environment. The newly revised army and joint doctrine for civil affairs contained two important changes that were particularly relevant for Afghanistan. The first outlined a new task force structure, which would report directly to the overall commander and thereby elevate the importance of civil affairs. The second was a civil-military center that facilitated military coordination with nongovernmental organizations (NGOs), international organizations (IOs), and civilian elements of the US government. Both were useful ideas, but each required significant adaptation in Afghanistan.

The Joint Civil-Military Operations Task Force

The Joint Civil-Military Operations Task Force (JCMOTF) was described in the 1995 revision to joint civil affairs doctrine as a way for a theater commander to oversee a wide range of joint civil-military operations.[115] Previously, the US Army's senior civil affairs commander would serve as the main adviser to the theater commander, even though all of the services provided some civil affairs capabilities.[116] This new joint structure, however, would include units or personnel from multiple services and facilitate planning, coordinating, and conducting all civil-military operations throughout the entire area of operations. It would also report directly to the theater commander, just like a joint task force of conventional forces or a joint special operations task force.[117] That would make it far easier for the theater commander to integrate civil affairs into the campaign plan and to ensure effective coordination with other military forces.

The 2000 version of army civil affairs doctrine and the 2001 version of joint civil affairs doctrine both included many details on the new JCMOTF structure, including extensive lists of possible responsibilities and even a notional organization chart.[118] But even though the JCMOTF was described in doctrine as early as 1995, none of the commanders of US military operations throughout the 1990s felt it necessary to establish one, likely because their operations were small enough for civil affairs to be effectively integrated without it.[119] By the time the war in Afghanistan began, the JCMOTF was an untested doctrinal concept that had so far only existed on paper—and one that would inevitably require adaptation when first put into practice.

Civil-Military Operations Centers

Unlike the JCMOTF, civil-military operations centers (CMOCs) began as ad hoc adaptations to the complex demands of humanitarian operations in the early 1990s. CMOCs were popular and had been used several times before they were incorporated into army and joint doctrine. They represent a positive story of doctrinal adaptation that came from the bottom up and was quickly embraced by the civil affairs senior leadership.

Immediately after the end of the 1991 Gulf War, the US military began Operation Provide Comfort to address the overwhelming humanitarian needs of the Kurds who had fled their homes in and around northern Iraq. US Army Special Forces began operating in the area on the same day that Saddam Hussein's forces capitulated. They soon started organizing local relief efforts, since few NGOs or IOs were operating in the area. Those organizations started to arrive soon afterward, though, as it became safe to do so. Despite their initial mutual mistrust, the relief organizations and special forces soldiers quickly realized that they needed to coordinate their efforts in order to alleviate the most suffering.[120]

In one refugee camp, for example, special forces and NGO workers began meeting every morning at 10:00, to get to know each other and share information about where they were operating. The meetings gradually became more structured, and the NGOs realized that the special forces had a more complete picture of needs and likely risks throughout the area. They cautiously began to share their information with the military, and the synthesized results enabled everyone to establish priorities and assign tasks for the day. Most importantly, the military forces soon started providing logistical support and vital air and ground transportation for the NGOs. This synergy enabled the military forces and the relief workers to assist many more people than either could on their own.[121] Similar coordination meetings soon emerged at higher levels as well.

A similar daily coordination meeting developed independently in the Turkish city where most of the relief agencies were headquartered. In mid-April, two US officers arrived in the city to liaise with these agencies, which had not been formally coordinating their activities. One of the officers started organizing a voluntary nightly meeting where all of the NGOs and IOs could share information, prioritize tasks for the next day, and gain access to military support. The meetings grew extremely popular, and were soon chaired by the US embassy team in the city. These meetings and continuous informal coordination, both in the refugee camps and in Turkey, were the first CMOCs, though they were not yet known by that name.[122]

At the end of April 1991, a tropical cyclone devastated the coast of Bangladesh, killing more than 140,000 people and leaving hundreds of thousands more without food or shelter. In early May, US Marines began conducting Operation Sea Angel to help provide humanitarian relief and restore infrastructure throughout the affected area. This was a very different situation than in northern Iraq, since there was a functioning government and US embassy. Yet once again, a CMOC-like structure naturally emerged to coordinate civilian and relief efforts in the coastal town where the relief agencies and military forces were located. The US command hosted the morning meeting of what was called the Military Coordination Center (MCC), which set priorities and tried to plan up to two days ahead.[123] Military forces helped distribute supplies throughout the country, and 17 US Marine communications teams ensured that the MCC, the aid distribution points, and the supply warehouses could constantly coordinate with each other.[124]

CMOCs proved so useful that they were employed by the US military in most of its subsequent relief operations. On December 9, 1992, US Marines landed ashore in Somalia to conduct humanitarian operations as part of Operation Restore Hope. Two days later, they established a CMOC in Mogadishu to coordinate with the United Nations and NGOs; regional CMOCs were also later established.[125] In late July 1994, US military forces in Rwanda established four CMOCs as part of Operation Support Hope to provide humanitarian relief for hundreds of thousands of Hutus fleeing from the genocide raging across the country.[126] And in September 1994, US military forces set up CMOCs in Haiti as part of Operation Uphold Democracy.[127]

CMOCs were officially codified into doctrine in June 1995, in the same revision to joint civil affairs doctrine that included the JCMOTF. They were defined as the "nerve center" of civil-military operations, and designed to promote effective coordination with civilian agencies and NGOs. CMOCs would be the "primary coordination agency" when the Department of Defense was leading US operations, and play a "supporting role" when other US agencies or international organizations were in charge.[128] In October 1996, the first edition of joint doctrine for interagency coordination provided more information on how CMOCs should operate, as well as helpful descriptions of their roles and organization.[129] The 2000 revision to army civil affairs doctrine significantly expanded the joint doctrine by incorporating many recent operational lessons learned. It included a detailed appendix about CMOC functions, organization, and staffing.[130] In February 2001, the Joint Staff published a revision of the June 1995 doctrine, which aligned it with the new army doctrine.[131]

By the fall of 2001, then, CMOCs had become a regular part of US military operations. They remained fairly controversial within the NGO and IO communities. Many aid workers believed that working with the military undermined their neutrality, which they held as one of the fundamental principles of their work. Nevertheless, these organizations had become accustomed to working with military forces through the CMOCs. Much like the US military, they now expected CMOCs to be a part of any future military humanitarian operation.[132]

Civil-Military Operations, September 2001–December 2002

Within hours of the September 11 attacks, the US military began planning military operations to eliminate al-Qaeda in Afghanistan and defeat the Taliban regime that had allowed the terrorist group to operate from its territory. Though the Pentagon regularly develops and refines a variety of plans for possible crisis scenarios around the world, few US military or civilian leaders believed prior to September 11 that Afghanistan posed enough of a strategic threat to warrant such planning.[133] As a result, military planners had to quickly develop a campaign plan entirely from scratch, without the time or detailed wargaming that characterize peacetime planning efforts. Bush administration officials believed that the Afghan people would not tolerate a massive or prolonged US military operation. They well remembered that the Afghans had fought ferociously against Soviet occupation forces for most of the 1980s and had forced their withdrawal only a dozen years earlier. As a result, the Pentagon's top leadership directed the military to design a campaign that prioritized immediate offensive operations against al-Qaeda and the Taliban, but that involved as few US troops as possible. Given the immense pressure for a quick response, they deferred thornier issues like who would take power after the fall of the Taliban.[134] But the White House did make clear that the US military would not be involved in nation-building efforts like those in Bosnia and Kosovo—something that President Bush had opposed throughout his presidential campaign.[135]

Civil-military operations evolved through three distinct phases in the first 15 months of the war, as US forces adapted to unanticipated circumstances and a constantly changing environment.[136] During the first phase, the military planned to support IOs and NGOs that would provide humanitarian relief and other civil assistance. The second phase started only two months later, immediately after the fall of the Taliban in mid-November, when the US military started directly providing such assistance. The third phase began in the summer of 2002, after most initial combat operations had ended and stability and reconstruction tasks became a much higher priority.

September–November 2001: "Wholesale" Support

The initial campaign plan that US Central Command (CENTCOM) developed in September was based on what was then the standard four-phase planning construct.[137] The first three phases involved preparing forces for the operation, conducting air strikes and deploying special operations forces, and then decisively defeating the Taliban and eliminating al-Qaeda in Afghanistan.[138] Phase 4 planning, however, was more limited than usual, because none of the questions of postwar governance had been settled.[139] Instead, Phase 4 emphasized the strategic importance of providing support for humanitarian assistance efforts. By helping to provide food, water, and other basic necessities, military planners hoped to gain the support of the Afghan people and to show them (and, ideally, Muslim communities around the world) that the United States was not conducting a war against Islam.[140] They also hoped that such assistance would provide enough stability for NGOs and IOs to establish a long-term presence, thus enabling US forces to leave Afghanistan. Anything that could remotely be construed as nation-building was notably absent.[141]

Operation Enduring Freedom (OEF) began on the evening of October 7, 2001, with US and British air strikes against al-Qaeda and Taliban facilities in Afghanistan.[142] Four days later, an officer from the 96th Civil Affairs Battalion flew to Islamabad, Pakistan, to begin setting up a CMOC.[143] The strategic importance of humanitarian aid in Afghanistan meant that civil affairs units needed to begin work right away. Their initial efforts focused on sharing information about security conditions and available routes in and out of the country, as well as coordinating aid operations and deconflicting them with ongoing combat operations.[144]

Two important civil-military adaptations occurred during this early phase of the war, one at the headquarters level and the other in the field. First, CENTCOM established the first-ever Joint Civil-Military Operations Task Force. This decision, which enabled the theater commander to directly oversee civil-military operations, reflected the fact that aiding the population was a crucial strategic objective. Because the task force could include coalition partners, the word "Combined" was added at the beginning of its name, leading to the unwieldy acronym CJCMOTF (pronounced see-jick-mo-tiff). Since a CJCMOTF had never been built before, there was extensive debate about how it should be organized—especially about whether assistance efforts should focus on providing logistical support for others or providing direct support to local populations.[145]

The pressure to maintain a small footprint led planners to choose the logistics support option, in what author William Flavin called the "wholesale approach." The CJCMOTF would facilitate the assistance efforts of the aid community (including NGOs, IOs, and coalition partners). But wherever possible, they would use existing civilian infrastructure and provide transportation and coordination only until those organizations could deliver relief supplies on their own.[146] This would involve a relatively small number of troops and would help prevent US forces from becoming too deeply involved in nation-building or other direct reconstruction efforts.[147] Since most army civil affairs capabilities were in the reserves and not yet mobilized, however, the initial decision to provide logistics support for other actors may have been driven as much by practical necessity as the need to minimize the number of troops involved.[148]

Brigadier General David Kratzer, the deputy commander of the army's 377th Theater Support Command, became the commander of the CJCMOTF.[149] In September, his unit and the 122nd Rear Operations Cell had participated in an exercise that included establishing several remote logistics bases.[150] Those two units became the center of the CJCMOTF, though it also included personnel from other civil affairs and military police units.[151] Kratzer reported directly to Lieutenant General Paul Mikolashek, who commanded all land forces in

Operation Enduring Freedom. He initially set up the CJCMOTF at Mikolashek's headquarters in Atlanta, but soon established a presence at CENTCOM in Tampa, which was conducting most of the coordination efforts with coalition partners and aid agencies. Kratzer deployed to Kabul, the recently liberated Afghan capital, in November. The rest of his headquarters soon followed, and the CJCMOTF was officially established there on December 13, 2001.[152] Once in theater, Kratzer became dual-hatted as the deputy chief of the newly established Office of Military Cooperation (OMC) at the US embassy in Kabul. The location in Kabul was partly to facilitate coordination with the various civilian aid agencies located there, but partly because the CJCMOTF's mission was expected to be temporary. Mikolashek's staff had already planned to shrink the CJCMOTF by the spring of 2002 and transfer its functions to the OMC.[153]

The second important adaptation during this period of time was the development of smaller units called coalition humanitarian liaison cells—whose acronym, CHLCs, led them to be nicknamed "chicklets."[154] The CJCMOTF commander originally established two subordinate organizations that were dubbed CMOCs, one with responsibility for the north and the other with responsibility for the south. But these organizations did not perform any of the traditional functions of a CMOC described previously. Instead, they provided logistics and command and control for the small teams of personnel that formed mini-CMOCs throughout the country. Although many NGOs and IOs had become used to working with the US military in the postconflict humanitarian operations of the 1990s, they were very uncomfortable about working with military organizations during ongoing combat operations. To mitigate these concerns, the mini-CMOCs were renamed CHLCs in order to emphasize their nonmilitary nature.[155]

Yet even though their name omitted any reference to the military, CHLCs nevertheless provided critical support for ongoing military operations. Their initial mission focused on coordinating civilian aid efforts, but they were deliberately designed and deployed in ways to generate Afghan support for the US military presence. These small teams provided a way to extend the reach of the military beyond major cities like Kabul and Kandahar, and they were purposely sent to areas where US commanders believed they could help build relationships and trust with local community members. CHLCs therefore provided a way for the US military to achieve the strategic benefits of supporting humanitarian relief efforts.[156]

November 2001–June 2002: "Retail" Support

By the time that CHLCs began deploying around Afghanistan, it was already clear that the wholesale support model would be inadequate. Many civilian aid agencies had fled Afghanistan during the Taliban era, and the remaining ones left when combat operations began in October 2001. Even after the fall of the Taliban in mid-November, the resulting instability and ongoing US military counter-terrorism operations meant that these agencies did not yet feel safe enough to return.[157] With no civilian assistance efforts to coordinate, and increasing pressure to build local support for the US military operation, CHLCs quickly began to conduct their own relief and assistance missions. Once they started directly providing assistance, they became something far more than just mini-CMOCs. They essentially moved from "wholesale" support, coordinating the assistance efforts of others, to "retail" support, directly delivering aid and assistance, especially in the least secure areas of Afghanistan.[158]

The mission of the CHLCs soon expanded to include executing quick-impact projects to rapidly build local support. These tended to be small infrastructure efforts, such as rebuilding schools, hospitals, and wells, in areas where local communities could see tangible results in a short period of time. These projects generally succeeded at building goodwill, but were far from politically neutral. CHLCs depended on local leaders to provide information and access to the population, as well as some degree of security. That meant that many of their initial aid and contracts went to those who supported the local officials and warlords who gave them access and protection.[159] Some CHLC members built such close relationships with these leaders that they lived in their homes or compounds.[160] This quick-impact approach inevitably meant that local projects were chosen according to their ability to rapidly meet military objectives, rather than through a more thorough needs assessment that considered long-term sustainability, among other factors.[161]

By early 2002, CHLCs were devoting more effort to these quick-impact reconstruction projects than on relief and aid delivery. But because the Bush administration remained so adamantly opposed to any nation-building efforts, all CHLC activities continued to be described as humanitarian assistance.[162] Because CHLCs were deliberately operating in the least stable parts of Afghanistan, these small teams needed US military protection as they conducted these projects. They were therefore colocated with US special forces and, foreshadowing the provincial reconstruction teams to come, the two began conducting operations together.[163] The CHLCs would do the civil affairs part of the mission, while a special forces A-team provided protection and conducted more combat-related tasks. Both communities agreed that this was a win-win situation that helped each of them be more effective.[164]

Despite their challenges, CHLCs became one of the most, if not the most, effective providers of humanitarian assistance and quick-impact projects throughout Afghanistan.[165] This caused considerable friction with the

NGOs and IOs that started gradually returning to the country. Based on their experiences working with the military throughout the 1990s, they expected that the US military would provide them with a wide range of support, such as engineering and transportation capabilities. Yet none of these were available, because the pressure to keep the operation small meant that these finite capabilities were all committed to supporting combat operations.[166] They also expected that CMOCs would help coordinate their support, but as noted earlier, the CMOC coordination function essentially disappeared after the CHLCs were formed. As the CHLCs expanded, NGOs and IOs grew increasingly frustrated that these organizations had both the resources and force protection necessary to operate throughout the country, while they did not.[167] They were also very upset that CHLC members often wore civilian clothes instead of uniforms while carrying their weapons, since they were afraid that the Afghan populace would be unable to distinguish their aid workers from troops dressed as civilians.[168] By April, they had convinced the military chain of command to require any troops involved in humanitarian or reconstruction assistance to wear clothing that clearly distinguished them from civilian aid workers.[169]

By May 2002, CJCMOTF and the CHLCs had spent more than $2.5 million to support a wide range of quick-impact projects throughout the country. Most of the funds were used to restore and build schools, hospitals, and medical facilities, and for water projects, though small amounts were also used for agriculture projects and bridge and road infrastructure.[170] Yet just as the CHLCs were up and running, their mission changed again.

June–December 2002: Stability Operations

Civil-military operations became an even more critical part of Operation Enduring Freedom during the summer of 2002, soon after the arrival of a new US military headquarters dubbed Combined Joint Task Force (CJTF)-180. Built around the headquarters staff of the 18th Airborne Corps, CJTF-180 had responsibility for overseeing the ever-expanding US mission. Its units included a growing number of conventional forces hunting Taliban and al-Qaeda remnants, others training the Afghan security forces, and still others working with the ministries of the interim Afghan government.[171] Because of continuing fears that the Afghans would reject the US military presence if it became seen as an occupation, CJTF-180 had to figure out how to improve stability and security while still relying on a relatively small number of troops. Its commander, Lieutenant General Dan McNeill, emphasized civil-military operations as a crucial element of his full-spectrum approach.[172]

At the headquarters level, McNeill put all the forces in theater—including the CJCMOTF, conventional forces, and most special forces—under his operational control to better coordinate their activities.[173] He also ordered the CJCMOTF to create more CHLCs and deploy them in more heavily contested areas. In June, six CHLCs were operating in or near Herat, Kandahar, Bamian, Mazar-e Sharif, Konduz, and Kabul. By August, three additional CHLCs had been established near Khost, Gardez, and Jalalabad—sites chosen not only because the provincial governments had requested them, but also because they had historical and cultural connections to the Taliban.[174]

By December 2002, the CJCMOTF and the CHLCs had spent more than $8 million, the majority of which was spent in the latter half of the year.[175] The CJCMOTF had also received 492 proposed projects and had approved 305 by that time, which were worth a total of more than $14 million. These projects still spanned a wide range of needs, but many were substantially larger than earlier projects had been. One particularly large project, for example, was designed to reconstruct a pharmaceutical plant so that it would be able to employ hundreds of Afghans and make more medicine available to the population.[176]

Yet even as their work expanded, it became increasingly clear that the CHLCs were simply too small and limited for the more extensive stability and reconstruction operations that were needed. As purely military teams designed for quick impact, they depended on NGOs and IOs for longer-term projects, but relations with these organizations remained poor, and many refused to coordinate with or support any military-led reconstruction efforts. This problem was exacerbated by the fact that some of the commanders and staff within the CJCMOTF had deliberately fostered a sense of competition with these civilian organizations. For these leaders, success was measured in the number of projects and dollars spent, so transitioning longer-term projects to NGOs and IOs was a sign of failure instead of success.[177] The CJCMOTF and the CHLCs, which had started as effective adaptations in reaction to an unforeseen war with a rapidly shifting operational environment, were no longer sufficient. During the summer and fall of 2002, military planners dusted off an idea that had been floating around since the spring and developed a new way to conduct civil-military operations. It proved so successful that it was soon incorporated into army and joint doctrine.

Provincial Reconstruction Teams, 2003–2006
The First PRTs

As early as the spring of 2002, CJCMOTF leaders had started discussing new ways to improve their operations. They came up with the idea of a new organization called a joint reconstruction team (JRT), which would bring together representatives from DOD, the State Department, the US Agency for International Development (USAID), and other partners to review proposed projects together and thereby shorten approval timelines.[178] Support for this idea grew throughout the summer, especially as the limitations of the CHLCs became increasingly obvious. The multinational International Security Assistance Force (ISAF) was promoting stability and security in Kabul, but the Bonn Agreement had limited its mandate to that city only. During the summer of 2002, JRTs were increasingly seen as a way to distribute what some US officials called "the ISAF effect" throughout the country.[179] They would be a "super CMOC on steroids," according to one officer who helped develop the concept, enabling a wider range of civil affairs efforts beyond quick-impact projects.[180] Even more importantly, the JRTs would help the new interim Afghan government to extend its reach and credibility as they worked together to improve security and reconstruction.[181]

In September, CJTF-180 directed Lieutenant Colonel Michael Stout to develop this concept into an actual program.[182] He was told to establish three pilot JRTs in the cities of Gardez, Bamian, and Konduz, in order to further develop the concept and identify lessons for improvement.[183] Stout had few resources at his disposal when he began putting together the first JRTs. He had to convince other parts of the command to give him what he needed, which was no easy task. McNeill eventually ordered all units within CJTF-180 to provide support for the JRTs, but resources still remained in short supply.[184]

In late November 2002, as CJTF-180 was getting ready to deploy the first JRT to Gardez, its staff held two press conferences in Kabul—one at the US embassy, and the other at the offices of the UN Development Programme. These public rollouts were designed to share details of the program with interested representatives from NGOs and IOs, the diplomatic community, and aid donors. Both briefings emphasized that the US military operation had now reached the reconstruction phase (even though that term had not been used in the initial campaign plan). Since most organized active terrorist activities had been eliminated, the briefers stressed that the JRTs had an entirely new mission. They would form a key part of the US military's effort to improve security and help expand the authority and influence of the central Afghan government before the presidential and parliamentary elections scheduled for the summer of 2004.[185]

According to the briefers, the JRTs would help identify reconstruction projects, conduct needs assessments in villages, and coordinate the activities of NGOs and IOs, the Afghan government, and the US military.[186] At this point the proposed organization chart included approximately 50–100 military personnel; representatives from other US agencies and international organizations were added later.[187] In perhaps the most important adaptation, the JRTs would include representatives of the Afghan government, which enabled the government to participate in decisions about which projects would be pursued. It also made the JRTs the cornerstone of the effort to transition to the reconstruction phase of military operations, and ultimately to the end of military operations altogether.[188]

Hamid Karzai, the president of the Afghan Transitional Administration, strongly supported the JRT concept but did not like its name. He thought that "regional" implied that the teams would work for regional leaders and warlords, which could empower them at the expense of the fledgling central government. Karzai requested that they be renamed provincial reconstruction teams, to emphasize their substantive contributions and to emphasize that his administration was providing benefits to Afghanistan's 34 provinces. They were called PRTs from that time forward.[189] NGOs and IOs were far less enthusiastic about PRTs, however, as were some State Department and USAID officials. These traditional purveyors of humanitarian and assistance remained very concerned about cooperating with military forces in the middle of an active conflict. They also strongly opposed what they saw as the increasing militarization of humanitarian and reconstruction efforts.[190]

The first PRT officially opened in January 2003 in Gardez—a site chosen by Karzai to help extend the influence of his government into a particularly unstable area.[191] It included 30 soldiers from the 450th Civil Affairs Battalion, as well as six representatives from international organizations. Five of those six civilians were initially required by their organizations to travel back to Kabul every night—a trip that lasted three hours—because of the frequent rocket attacks on the PRT compound. CJTF-180 assigned the PRT an infantry platoon and a special forces A-team to improve security, as well as 12 additional civil affairs soldiers. By April, six additional civilians from various IOs had joined the team, and all were now allowed to stay in Gardez. Yet security still remained a challenge while members of the PRT traveled around the region. They had to travel in large groups for protection, which limited the number of projects they could manage.[192]

The next two PRTs, which opened in Bamian on March 2 and in Konduz on April 10, demonstrated the notable flexibility of the concept. Although all PRTs shared the common mission of enhancing security, promoting reconstruction, and strengthening the Afghan government, each one was tailored to meet the specific needs and conditions of its area. The Bamian PRT, for example, had a much smaller security force than the Gardez PRT because the security environment was far more favorable. The PRT in Konduz, where the environment was far more dangerous, grew to somewhere between 200 and 300 personnel.[193] Even in the more permissive areas, however, security remained a major constraint on PRT efforts, since civilian personnel still had to rely on the relatively limited military resources available within each PRT for security during their travels.[194] PRTs also faced serious coordination and deconfliction challenges when they operated in the same areas as conventional or special operations forces, since all three of those elements reported to CJTF-180 through different chains of command.[195]

The early PRTs suffered from a great deal of tension between the military members and the civilian representatives of other US government agencies. Part of this was due to the fact that the PRTs continued to pursue quick-impact projects, but most of this tension resulted from unrealistic expectations on both sides. This was the first time that most military personnel had ever worked so closely with civilians in a combat environment. The soldiers greatly overestimated what the civilians could deliver and quickly grew frustrated about their small numbers and lack of resources and authority. The PRTs often included only one civilian, who was typically fairly junior, untrained, and limited to 90 days in theater. The civilians on the teams—including ones who had military experience—were equally frustrated with their military counterparts. They faulted the military for being reluctant to support them, treating them as outsiders, and not being better organized. These relationships improved over time, however. By the end of 2003, most of the four PRTs included civilian representatives from the State Department, USAID, and the Department of Agriculture, many of whom served for a full year. CJTF-180 did a better job of coordinating combat and PRT activities in the same areas, so the teams could accomplish more. Perhaps most critically, the PRTs gained access to Economic Support Funds through their State Department civilians, which finally enabled them to support longer-term reconstruction projects.[196]

From their inception, PRTs were envisaged as a way to promote burden sharing among the coalition partners. The first international PRT, led by the United Kingdom, opened in Mazar-e Sharif on July 10, 2003. New Zealand and Germany followed by the end of the year in Bamian and Konduz, respectively, and several other countries established new PRTs in 2004 and 2005. The US military also planned to transfer leadership of many US PRTs to other nations once they had become established in an area. The different national models for PRTs demonstrated their adaptability, but the profusion of national agendas meant that the PRTs often pursued very different and sometimes contradictory missions.[197] While this greatly complicated coherent reconstruction efforts, these international adaptations lie outside the scope of this book.[198]

PRT Expansion, 2004–2005

In mid-2003, General John Abizaid, who had succeeded Tommy Franks as the commander of CENTCOM, decided to add a theater-level headquarters above CJTF-180 to focus on synchronizing the political and military elements of the war.[199] In September, Abizaid asked Major General David Barno, with whom he had served in Grenada,[200] to establish the new headquarters and assume command of the war.[201] Barno flew to Afghanistan in early October to start setting up the new organization. He was promoted to lieutenant general and assumed command responsibility in November, and officially established Combined Forces Command-Afghanistan (CFC-A) in February 2004.[202]

Abizaid's immediate guidance to Barno when he arrived in October was to focus his efforts toward "big POL and little MIL," meaning that he should prioritize the strategic aspects of the campaign related to the Afghan government and regional politics. Barno soon realized, however, that the mission would have to involve a much larger emphasis on a broad range of nonmilitary efforts, from economic development to a battle for ideas. Success would require an integrated effort across all the elements of power.[203] He decided that his top priority needed to be countering the growing insurgency that was threatening the Afghan people and the nascent government.[204] Barno's decision to pursue a counterinsurgency strategy meant that PRTs were no longer a relatively low priority engaging only the civil affairs community. Instead, he saw them as essential to the success of the military campaign and to achieving US strategic objectives in Afghanistan.[205]

Given the lack of counterinsurgency doctrine discussed earlier in this chapter, Barno relied heavily on his own knowledge. He brought a textbook from a class on revolutionary warfare that he had taken during his second year at West Point, as well as some books on counterinsurgency from his personal library. He also acquired others while he was in theater.[206] Barno and his staff developed a counterinsurgency campaign plan that was conceptualized as a roof supported by five key pillars.[207] The roof showed the Afghan people as the center of gravity in the campaign—meaning that anything that endangered popular support for the Afghan government or the US military put the entire mission at risk. The five pillars were the operational concepts that would support the roof, which explicitly involved the entire US interagency effort and not just military operations.[208]

PRTs were critical to the success of Pillar 3 and were the central element of Pillar 4.[209] Pillar 3 was called "sustain area ownership," which Barno later described as "the most important, though the least visible, change on the ground."[210] Previously, CJTF-180 had kept responsibility for the entire battlespace, rather than assigning subordinate units distinct areas of operations. Depending on available intelligence, these units might operate in one part of the country on one operation, and a completely different area the next time they sortied from their fortified bases. This raiding approach meant that tactical units had no responsibility for achieving any long-term results.[211] They also had no ability to build lasting relationships with the Afghan population or their local leaders. Sustaining area ownership meant that maneuver units at brigade, and often battalion, level were permanently assigned to a specific area of operations, usually for their entire yearlong deployment. Where possible, they followed classic counterinsurgency principles by operating from small outposts in Afghan villages instead of from large bases. The new strategy enabled them to build trust with local leaders, gain better intelligence, and see the long-term results of their efforts.[212] The new counterinsurgency approach made the PRTs more important and more effective, since they could coordinate more easily and focus on longer-term projects that the local populations wanted most.[213]

PRTs were also the central element of Pillar 4, "enable reconstruction and good governance." Barno saw them as a "powerful offensive weapon in our strategic arsenal," since so much of his mission focused on connecting the people to the new Afghan government.[214] By this time, many PRTs included representatives from the Ministry of the Interior, which helped extend the reach of the central government.[215] When Barno arrived in Afghanistan in late 2003, only four PRTs were operating across Afghanistan. He soon decided to establish eight more before the spring 2004 fighting season. He deployed these new PRTs into some of the most dangerous parts of the country, in order to accelerate security and reconstruction efforts in the areas at most risk.[216] Since the number of civil affairs personnel in theater remained limited, he reorganized the CJCMOTF headquarters and reassigned many of its personnel to the new PRTs.[217] By the summer of 2004, 12 PRTs operated across Afghanistan. The United States led nine of them, including all of the new PRTs in the Taliban-contested southern and eastern areas of the country.[218]

When a second brigade headquarters arrived in spring 2004, Barno assigned each of them major portions of contested territory in the south and east to firmly establish area ownership. He designated them as regional commands and gave the commanders operational control over virtually all of the military forces operating in their battlespace—including the PRTs.[219] The regional commanders welcomed this decision since it became far easier for them to integrate civil and military efforts in their areas of responsibility—which made both much more effective.[220] Though many of the early PRTs had been staffed on an ad hoc basis, a basic model emerged in 2004 that helped standardize their approaches while still allowing variation based on local needs.[221] The PRTs also gained more funding and flexibility in early 2004, as the new Commander's Emergency Relief Program expanded from Iraq to Afghanistan, and Congress soon appropriated additional funds for their efforts as well.[222]

The number of PRTs continued to steadily expand. Nineteen PRTs were operating by the end of 2004, and four additional ones were established in 2005.[223] Most of the 23 PRTs were initially operated by the United States, though several were transferred to coalition partners as NATO expanded the ISAF mission throughout the country.[224] Zalmay Khalilzad, who served as the US ambassador

FIGURE 5.2 U.S. Army Soldiers from a Provincial Reconstruction Team Meet with Afghan Villagers in Zabul Province, 2010

Source: https://www.dvidshub.net/image/309304/prt-zabul-strives-improve-quality-life-arghandab-district.

to Afghanistan throughout the expansion period, thought so highly of the PRT concept that when he became the US ambassador to Iraq in June 2005, he helped establish PRTs throughout that country as well.[225]

Despite their popularity and ever-growing numbers, PRTs continued to suffer from a wide range of unresolved problems. Many different studies pointed out their limitations, which included insufficient civilian capacity; confused streams of funding; concerns about the militarization of humanitarian assistance; and especially the lack of a clear mission and definition of success.[226] PRTs did improve US and coalition military relations with many communities, extend the reach of the central government, and complete many reconstruction projects. However, it was not at all clear that these efforts translated into stronger popular support for the Afghan government or greater local opposition to the Taliban. And even if PRTs achieved this effect, they still remained too limited in capability and too small in number to meet the virtually unquenchable needs of the Afghan population.[227]

For the purposes of our study, however, PRTs in Afghanistan nevertheless stand as the result of a successful process of doctrinal adaptation that meets all the criteria identified in Chapter 2. As military planners scrambled to put together a campaign plan in the aftermath of the September 11 attacks, they planned to establish CMOCs in accordance with existing civil-military doctrine. But they started to adapt that doctrine even before the first civil affairs officers arrived in Afghanistan—first by establishing the first CJCMOTF and then by establishing new CHLCs to support the work of humanitarian relief organizations. After the Taliban fell in mid-November 2001, US military civil affairs planners realized that few of the NGOs and IOs would soon return. As a result, the CHLCs adapted and started to provide direct aid and assistance to the Afghan population. As the military mission shifted toward stability operations in mid-2002, the CHLCs were too limited to conduct more extensive stability and reconstruction operations, so yet another adaptation was needed. The PRTs emerged as the result, drawing on new ideas not just from throughout the military chain of command but also from the State Department, USAID, and other US government agencies. When the mission changed in late 2003 to a counterinsurgency approach, PRTs could be rapidly expanded throughout the country because they were flexible enough to adapt to local conditions. The idea proved so successful that they became a key element of the US strategy in Iraq, and were soon incorporated into both army and joint civil affairs doctrine.[228] For all their limitations, the PRTs resulted from a repeated process of impressive doctrinal adaptation.

Conclusion: Doctrinal Adaptation in Iraq and Afghanistan

In both Iraq and Afghanistan, mistaken prewar doctrine led to a process of doctrinal adaptation during wartime, but the parallels end there. PRTs in Afghanistan resulted from a continuous process of adaptation that began in the very first weeks of the war, with different organizational structures emerging to provide services and outreach to local populations. These were regularly revised and quickly replaced as the war changed and local circumstances shifted. The PRT structure that emerged worked well. It combined security and reconstruction into one joint civilian and military entity on the battlefield, while remaining flexible enough to adjust to local conditions. PRTs were embraced and expanded by subsequent US commanders and were later expanded to Iraq and eventually incorporated into army doctrine.

By contrast, the US Army's post-Vietnam rejection of counterinsurgency continued to cast a long shadow over the early years of the war in Iraq. Part of that lapse was due to the leadership of General George Casey, as discussed in Chapter 8. But it was also due to the fact that during the first years of the war, the army's usual processes for developing doctrine proved far too slow and inflexible to be able to offer anything more than an interim counterinsurgency manual, which was hastily drafted and poorly disseminated. The December 2006 counterinsurgency manual was a remarkable achievement that helped change the trajectory of the entire war, but it emerged only because of a remarkably lucky convergence of events and people that unfolded outside the army's usual doctrine process. Almost four years elapsed between the army's doctrinal failure to address a burgeoning counterinsurgency in Iraq and this key doctrinal change. That vacuum cost untold lives and nearly led the United States to lose the war—and even then, doctrinal change almost did not happen. Though the creation of the new manual was a testament to the remarkable adaptability of those who drafted it, it simultaneously demonstrated the dysfunctional doctrinal adaptability of the US Army in the midst of a war.

The examples from Iraq and Afghanistan highlight the challenges of adapting doctrine after a war begins. Wars never quite match the predictions that militaries make about the next conflict, so they always must be able to adapt their doctrine in the face of the unexpected. In its future wars, the entire US military will need to be able to adapt as quickly, effectively, and continuously as its civil affairs community did to the unexpected challenges they faced in Afghanistan. But it must also do a far better job of identifying when a doctrinal change is needed than it did in Iraq. And even more importantly, it needs to dramatically improve its processes for reviewing, revising, and disseminating doctrinal changes in the middle of a war, since it proved incapable of doing so in Iraq. We return to these themes in the final chapters of the book.

6

Technological Adaptation in Iraq and Afghanistan

BEFORE THE WARS in Iraq and Afghanistan began, the US military continued to field weapons systems and other technologies that were primarily designed for the Cold War and large-scale conventional conflict. Counterintuitively, this was partly due to the fact most US military operations after the end of the Cold War involved peacekeeping and stabilization missions. These missions did not require any major new programmatic investments, so few demands for new technologies emerged. But partly driven by inertia, the services, and especially the army, continued to invest tremendous sums of money in next-generation technologies for conventional war. Many of these acquisition programs failed spectacularly during the following decade. The army canceled 22 major weapons programs between 1995 and 2010—including heavyweights like the Future Combat System family of vehicles, the Comanche helicopter, and the Crusader self-propelled howitzer—which collectively cost more than $32 billion.[1]

These repeated and costly failures raised concerns on Capitol Hill and among senior civilian leaders in the Pentagon about the health of the army's peacetime acquisition system. It also made army leaders more committed to ensuring the success of the remaining systems after these embarrassing failures. After the September 2001 terrorist attacks, this deep commitment to expensive conventional capabilities perversely affected how the army equipped its troops for the irregular wars in Iraq and Afghanistan. Grappling with these new and unforeseen conflicts, army leaders struggled to balance the need for new hardware to fight these unconventional wars with their costly investments in new conventional weapons. Every service wrestled with this balance, but it affected the army most of all since it was the most deeply engaged in both wars.

Ultimately, senior military leaders remained so wedded to their planned systems and their vision of future conventional wars that they not only ignored but *actively suppressed* demands from soldiers and marines in combat for the technologies and systems that would help defeat the enemy and save their lives. The story of technological adaptation in Iraq and Afghanistan involves noteworthy successes at the tactical level, as innovative troops found new and creative ways to use the technologies and materials at hand. But it is also a story of almost unimaginable failures at the institutional level. Military leaders and even some civilian leaders in the Pentagon ignored, undermined, and suppressed increasingly urgent requests from commanders fighting the ongoing wars in order to protect funding for the systems they wanted for the future, with untold American lives lost as a result.

As Figure 6.1 shows, this chapter begins with two related cases of failed technological adaptation at the tactical and institutional levels in Iraq. They offer striking parallels to the two failed cases of technological adaptation in World War II discussed in Chapter 3. We examine the development of what became known as "hillbilly armor" in Iraq—a makeshift way to protect vulnerable vehicles from the growing threat posed by improvised explosive devices (IEDs). Soldiers and marines demonstrated great adaptability in their efforts to create battlefield workarounds to improve their desperately needed protection. But just as US Army troops in 1944 discovered that no amount of innovation could address the fundamental design flaws in the M4 tank, hillbilly armor could provide only limited protection against IEDs. Both of these examples of tactical adaptability were ultimately failures—not because the efforts of the troops were found wanting, but because the problems with their technology were so significant that they required an institutional solution.

By the middle of World War II, the army needed to develop a new tank that could withstand steadily improving German firepower, but failed to do so in time

	Iraq	Afghanistan
Tactical Adaptation	Hillbilly armor (failure)	Apache helicopters and close air support (success)
Institutional Adaptation	Mine-Resistant, Ambush-Protected (MRAP) vehicles (failure)	DCGS-A versus Palantir (failure)

FIGURE 6.1 Technological Adaptation in Iraq and Afghanistan

to affect the war. The challenge facing army leaders in Iraq was far simpler, since vehicles that could withstand deadly roadside bombs already existed. They were first developed in the 1970s by South Africa and Rhodesia, and the US military already had limited numbers of these mine-resistant, ambush-protected vehicles (MRAPs) in its inventory by 2003. Yet even as casualties from IEDs mounted over the following years, the US Army and the Marine Corps both continued to deny battlefield requests for greater numbers of these life-saving vehicles. The number of fielded MRAPs remained very low until 2007, when the secretary of defense personally intervened—against the advice of all of his military and civilian advisers—and demanded that large numbers of MRAPs be produced and sent to Iraq as quickly as possible. Unfortunately, most of those new vehicles did not arrive in Iraq until the overall security situation had much improved, and IEDs no longer posed as much of a threat. Countless lives had been unnecessarily lost by that time.

For Afghanistan, we examine an extraordinary example of tactical adaptation under pressure that not only helped to save lives in the middle of a battle but also pioneered a major shift in the way the army used one of its major weapons systems. The army's AH-64 Apache attack helicopter had been designed during the Cold War to kill tanks during deep attack missions behind enemy lines. But in March 2002, during their first combat mission in Afghanistan, Apaches were the only American aircraft that could provide sustained close air support to embattled US troops fighting in the Shah-i-Kot Valley. High altitudes and the proximity of the enemy prevented the Apache crews from using their standard combat tactics, so they immediately adapted and improvised a creative new way to accomplish that mission. They quickly devised new ways to employ their aircraft's powerful capabilities under fire in an entirely new battlefield environment. These new tactics proved so effective that Apache crews continued to use them throughout the rest of the war in Afghanistan—and the army now considers close air support to be one of the Apache's most important missions,[2] a function far removed from the helicopters' primary mission before the Afghan conflict.

The story of institutional technological adaptation in Afghanistan, however, is just as damning as it was in Iraq. The army insisted that its forces would employ the service's cumbersome new intelligence system, known as the Distributed Common Ground System–Army (DCGS-A), to track insurgents and predict enemy attacks even though it had been designed for conventional warfare. The system worked poorly; it was hard to use, crashed often, and was too inflexible to keep up with a rapidly changing insurgent environment. Army intelligence analysts discovered that their marine counterparts were using software from a company called Palantir to accomplish the same key battlefield tasks far more easily and effectively. As with the MRAPs, repeated requests from field commanders for Palantir's software were rejected, while the army insisted that soldiers simply needed more training on DCGS-A. The army's continuous efforts to block requests for the Palantir software in order to protect the ineffective DCGS-A program was a sustained and unnecessary failure that almost certainly cost the lives of soldiers.

Technological Adaptation in Iraq

Countering the insurgency that emerged in Iraq after the 2003 invasion did not require a tremendous amount of sophisticated technology. US forces easily overpowered Saddam Hussein's military during major combat operations, and the insurgents that emerged afterward lacked advanced military capabilities. Yet the insurgents were soon well armed with conventional explosives that they had raided from unguarded Iraqi military facilities in the chaos that followed the end of full-scale combat. As the insurgency grew, it also received increasing support from neighboring Iran. Because the insurgents could not successfully confront US forces directly, they turned to IEDs. These deadly unconventional bombs quickly became the signature weapon of the war. Between March 2003 and the final withdrawal of US troops in December 2011, 1,756 US troops were killed by IEDs, which accounts for almost 40 percent of the total American combat fatalities during the war.[3]

Adapting to the unexpected IED threat became the major technological challenge of the Iraq War. At the tactical level, US forces adapted as well as they could, experimenting with all sorts of improvised armor to add protection to their light vehicles. At the institutional level, the army and the marines initially adapted fairly well by developing and fielding armored kits that troops could bolt on to their vehicles, though the sheer numbers of vehicles involved meant that many units continued to operate without them into the third year of the war. That early institutional success, however, was vastly outweighed by the striking institutional failure that immediately followed. As the IED threat became ever more deadly, most senior Pentagon leaders, both military and civilian, spent the following years fiercely opposed to buying adequate numbers of the safer but expensive MRAPs.

Tactical Adaptation: Hillbilly Armor

As the war transitioned into an insurgency during the summer of 2003, US troops started to conduct foot patrols among the population. This required them to leave behind their heavily armored tanks and fighting vehicles and operate out of lighter, and thus far more vulnerable, wheeled vehicles. Humvees and other light trucks not only transported troops, but also were used extensively on the long supply lines that reached as far back as Kuwait.[4] The insurgents quickly began targeting these unarmored, "thin-skinned"[5] vehicles, by ambushing them directly or by placing IEDs along the roads on which they traveled.[6] The Humvees' thin metal frames and canvas doors offered virtually no protection for the army soldiers and marines riding inside.[7]

The IED threat grew quickly. In June 2003, General John Abizaid, who oversaw the war in his role as commander of US Central Command, declared that IEDs were his "No. 1 threat."[8] In October 2003, Abizaid made an urgent request to Deputy Secretary of Defense Paul Wolfowitz for a "mini-Manhattan project" to address the IED problem.[9] That same month, the army assigned 12 personnel to form a new Army IED Defeat Task Force, to develop ways to counter the threat.[10] By the end of the year, IEDs became the single largest cause of US deaths from hostile action in Iraq.[11]

US forces in Iraq tried to adapt to this emerging threat. The troops were increasingly galvanized as, in the words of one reporter, they "watched one blood-stained vehicle after another come back from patrol."[12] At the time, the army's official guidance on how to harden Humvees and other vehicles remained much the same as it had been during World War II: adding sandbags to provide an additional layer of protection.[13] Soldiers and marines put sandbags in vehicles everywhere they could—on the floorboards, under seats, above dashboards, and in cargo areas.[14] But the sandbags added so much additional weight that the vehicles moved more slowly—which could actually *increase* their vulnerability to roadside IEDs.[15] Many soldiers were skeptical that the sandbags added much protection at all. As one young soldier said, "Do you think sandbags are going to save you? No. Do they give everyone warm fuzzies? Yes."[16] Many years later, former secretary of defense Robert Gates acknowledged that adding sandbags "didn't help much."[17]

Since the limits of sandbags were immediately apparent, soldiers in the field started to improvise their own ad hoc armor. Maintenance crews scrounged up scrap metal wherever they could—borrowing or trading with other units, dismantling destroyed civilian vehicles, buying old armor from Soviet vehicles from the local population, even stealing manhole covers—and bolting or welding the metal onto their vehicles.[18] This approach was incredibly adaptive, though far from aesthetically pleasing. These modifications soon became known as "hillbilly armor," because some soldiers thought the resulting vehicles resembled the jalopy featured in the television show *The Beverly Hillbillies*.[19]

Hillbilly armor posed some safety challenges, just as the sandbags did. When hit by the explosive force of a bomb, low-quality steel could turn into shrapnel directed inward toward the soldiers inside the vehicle.[20] Yet the benefits of having at least some form of protection vastly outweighed this risk. Maintenance shops throughout Iraq responded to the particular concerns of the troops in their units and improvised the best responses that they could assemble. For example, a company from the First Marine Division asked its maintenance personnel to provide additional protection above shoulder level, a common area of vulnerability. Using steel plates that were originally intended for road repairs, a navy Seabee and another welder added new doors, higher sides, and roofs to the company's vehicles.[21] These modifications saved the lives of one marine crew after its vehicle was hit by a rocket-propelled grenade.[22]

In another example, two soldiers serving in the army's 699th Maintenance Company were assigned to force protection duties at Camp Buehring in Kuwait. They set up a small shop and managed to add improvised armor to the vehicles of an entire battalion in only three days. They dubbed their new venture the Mad Max shop, after the Mel Gibson film trilogy that included fights between up-armored vehicles in the Australian outback. Seeing its success, the company assigned more soldiers to the section so it could work on more vehicles. Soon the shop was consuming about 20 sheets of armor each day—which triggered an army investigation, since each sheet cost $1,200. As one of the shop's founders recalled,

FIGURE 6.2 Hillbilly Armor in Iraq, 2004
Source: https://www.flickr.com/photos/expertinfantry/5419932957/in/album-72157625860392287/.

the army "couldn't believe we were going through that much. They thought we were wasting metal, but then they came and saw our operations."[23] The shop operated for 12 to 15 hours every night, since the armored sheets were too hot to pick up during the day. In five months, the Mad Max shop added armor to more than 1,900 vehicles.[24]

One of the most interesting tactical adaptations was a machine shop set up by the army's 181st Transportation Battalion, which became known as the Skunk Werks. It was located at Camp Anaconda (also known as Balad Air Base), a large base about 50 miles north of Baghdad that hosted a wide range of combat and support troops.[25] Like the other metal shops springing up around Iraq, Skunk Werks developed some remarkable adaptations. Two navy explosives experts who had previously worked on armoring ship hulls were attached to the shop. They suggested adding a layer of soft steel surrounded by rubber to the armored metal sheets, to prevent the armor from cracking under the pressure of explosive blasts. This approach worked so well that they called it "dreadnought," giving credit to its naval origins.[26] What made Skunk Werks different, though, was the fact that it did not work for a single unit. It would work on any vehicle that had enough time to be worked on as it passed through the base. As the shop's crews talked with the drivers about the modifications they wanted, they also learned how individual units were adapting and improvising solutions out in the field. The crews could then incorporate these ideas into their own work, but even more importantly, share them with the other units that passed through the shop. As scholar Nina Kollars writes, "The technological adaptations *both coalesced at and diffused from this central point.*"[27]

Yet as we saw in Chapter 3, sometimes the scope of the threat is simply too big to be fixed by soldiers improvising in the combat zone. The spontaneous development of hillbilly armor throughout Iraq certainly saved some lives and demonstrated impressive tactical adaptation to the new and unanticipated technical challenges of IEDs. But just as M4 tank crews fighting in Europe after D-Day could not improvise enough additional armor to protect themselves from improved German firepower, operational units in Iraq could not effectively defeat the IED threat on their own. It required institutional adaptation from the army and the marines.

Institutional Adaptation, Part 1: The Early Success of Up-Armoring Kits

As early as September 2003, US soldiers submitted an urgent operational needs statement to the army requesting additional vehicle protection. Such statements are essentially the way to trigger an institutional adaptation; they formally notify the relevant service headquarters that its fielded forces are facing a critical capabilities gap. In this case, the army moved quickly to try to meet the need. Within a remarkable 60 days, the Army Research Laboratory designed and fielded a prototype kit that would add armor to existing vehicles, in a process that became known as up-armoring. The scientists at the lab were concerned that this initial prototype, called the Armor Survivability Kit (ASK), would provide more psychological reassurance than actual protection, but they continued to test and refine the kits even as they started shipping them to Iraq.[28] The army sent researchers to Iraq to learn more about the exact nature of the threat and the various tactical adaptations that had been developed in response. At least one researcher spent time with the Skunk Werks, and shared its lessons learned with the teams working to improve the ASK.[29] Several different types of kits were eventually developed, since not all Humvees were exactly alike.[30]

Army Materiel Command received its first order for 1,000 up-armoring kits in November 2003, and the number of orders skyrocketed over the following months. Total orders increased to 3,870 by the end of December; to 6,760 by February 2004; to 8,272 by April; and to 13,872 by the summer.[31] The army struggled to keep up with demand, and as early as January 2004, it needed more steel for armor than civilian industry was producing. At that point, Representative Duncan L. Hunter (R-CA), the chairman of the House Armed Services Committee, sent his staff members to the steel mills that produced the armor. They spoke with both managers and unions, and convinced the mills to stop fulfilling civilian orders and to deliver their full production capacity to the military. They also explored ways to increase that capacity.[32]

Yet these up-armoring kits were far from perfect. The typical kit added more than 1,000 pounds to the vehicle's weight, which placed so much stress on the suspensions and powertrains that they broke down more frequently.[33] The added weight also made the vehicles move more slowly, increasing their vulnerability in dangerous situations, and made them more likely to roll over in rugged terrain.[34] They also focused almost exclusively on protecting the sides of the vehicles, leaving the tops and the bottoms of the vehicles unprotected. This was not a major concern at the time, since most IEDs were located on the sides of the roads.[35] But it proved to be an acute vulnerability when the insurgents changed tactics, as discussed later.

Although the army had worked to speed up production of the kits, the up-armoring process remained slow because there were over 30,000 Humvees operating in Iraq.[36] Supply simply could not keep up with demand, and many Humvees and other light vehicles continued to operate without any additional armor throughout 2004. In October, soldiers in the army's 343rd Quartermaster Company refused an order to conduct a convoy mission, citing the lack of armor on their vehicles as well as the lack of an armed escort. The army's immediate response was to declare that the soldiers would be court-martialed. But one week later, 13 members of Congress asked the House Armed Services Committee to hold a hearing on the shortage of armored vehicles in Iraq. Shortly thereafter, the army relieved the insubordinate company's commander and announced that the soldiers would receive administrative and nonjudicial punishments instead of being court-martialed.[37] On November 17, 2004, the army chief of staff, General Peter Schoomaker, testified that the army was in the process of retrofitting 12,800 vehicles with up-armoring kits. He also said that the army had manufactured 5,000 new armored Humvees and would meet the full production requirement of 8,100 by April 2005.[38] While these numbers were impressive, they still only accounted a fraction of the Humvees then operating in Iraq.

On December 8, 2004, Secretary of Defense Donald Rumsfeld held a town hall meeting with US troops in Kuwait. One soldier asked the secretary, "Why do we soldiers have to dig through local landfills for pieces of scrap metal and compromised ballistic glass to up-armor our vehicles?" The other 2,300 soldiers attending the meeting erupted in spontaneous applause.[39] Rumsfeld's response became one of the most famous lines of his tenure: "You go to war with the army you have, not the army you might want or wish to have at a later time."[40] This seemingly flip response provoked outrage. The next day, President George W. Bush reassured the American people that "the concerns are being addressed, and that is, we expect our troops to have the best possible equipment."[41] Rumsfeld also added that the Pentagon had adapted "pretty rapidly" to the need for more armor, and that it "had varying degrees of success—in some cases considerable success" in addressing the problem.[42] Soon thereafter, senior army officials held a special briefing for reporters on up-armoring Humvees,[43] and continuing media attention kept a public spotlight on the issue for months to come.[44]

Despite the controversy over Rumsfeld's remarks, institutional adaptation to the IED problem was fairly successful during the first two years of the war. Up-armoring was not a perfect solution, and it progressed more slowly than forces in the field (and their families) may have wished, but adding additional armor to 30,000 Humvees and other light vehicles was an enormous challenge. Yet despite—or, more likely, because of—this success, the insurgents soon switched tactics to take advantage of the vulnerabilities that up-armoring had not addressed. This deadly shift would far exceed the adaptive capabilities of troops in Iraq and require a different institutional solution. Tragically, institutional adaptation to this new challenge was a complete and utter failure.

Institutional Adaptation, Part 2: Strong Opposition to MRAPs

The threat from IEDs continued to grow throughout 2004 and 2005. As Figure 6.3 shows, the number of US troops killed by IEDs continued to climb during the first years of the war despite the increasing number of up-armored vehicles. Some of that increase can be explained by an increase in the denominator; as the insurgency grew, US troops were exposed to more danger as they conducted more patrols and traveled in more convoys along the US military's extensive supply lines. But IEDs also became more lethal as they grew technologically more sophisticated, and as the insurgents learned how to counter US tactics, techniques, and capabilities. As the number of US vehicles with up-armored sides grew, roadside IEDs proved less effective. Insurgents began burying IEDs in the center of the roads, so that they would explode directly under US vehicles. This dramatically increased their lethality, since the flat, unarmored bottoms of Humvees and other trucks conveyed the full force of the explosion to the center of the vehicle and its occupants.[45] This effective enemy tactic spread quickly; the number of buried IEDs encountered tripled between mid-2004 and mid-2006.[46]

Fortunately, a vehicle already existed that could address this threat. Early variants of what later became called MRAPs were developed by Rhodesia and South Africa in the 1970s to protect their troops from land mines. These vehicles had two key design features. First, they sat high off the ground, which let some energy from the explosion dissipate before reaching the chassis. The second and more important feature was a V-shaped hull forming the bottom of the vehicle that

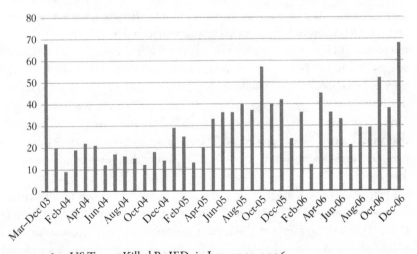

FIGURE 6.3 US Troops Killed By IEDs in Iraq, 2003–2006

Source: Michael E. O'Hanlon and Ian Livingston, "Iraq Index: Tracking Variables of Reconstruction & Security in Post-Saddam Iraq," Brookings Institution, January 31, 2011, 12.

deflected most of the remaining energy outward and away from the passengers and crew. These vehicles proved so effective that they essentially eliminated fatalities in the Rhodesian army during mounted patrols and reduced injuries by more than 70 percent.[47] The US military knew about these vehicles long before the war in Iraq began, and even owned some. In 1996, a marine captain published an article in *Marine Corps Gazette* that described MRAPs in detail. In a passage that proved sadly prescient, he wrote, "Marines have only sandbags and improvisation to rely on in an attempt to provide passive defense against landmines," and argued that MRAPs could "significantly reduce the carnage wrought on Americans by landmines."[48] The Pentagon tested MRAPs in 2000 and bought fewer than two dozen for explosive ordnance disposal (EOD) teams in the army and the marines. Some of these vehicles were even being used in Iraq, though solely for EOD missions and not for moving troops or conducting patrols.[49]

In December 2003, Wolfowitz directed the Joint Chiefs of Staff to explore options for improved armor. Analysts in the Pentagon passed along information about a particular model of MRAP made in Namibia, though little seems to have happened as a result.[50] The army also became increasingly concerned about the IED threat in early 2004. General Larry Ellis, the head of US Army Forces Command, sent a letter to the army chief of staff at the end of March cautioning that commanders in Iraq were reporting the armored Humvee "is not providing the solution the Army hoped to achieve." He also noted the lack of urgent attention to the problem, complaining that "some Army members and agencies are still in a peacetime posture."[51] He recommended accelerating production of armored Stryker vehicles instead of procuring MRAPs, but he was told that new armor kits for the Humvees would be adequate.[52] Between January and June 2004, however, an additional 103 US troops were killed by IEDs.[53] Abizaid reiterated that IEDs remained his highest priority, while the threat only continued to grow.[54] During the summer of 2004, more than 1,000 IEDs exploded in the Baghdad neighborhood of Sadr City alone, and another 1,200 buried IEDs were dug up.[55] One Pentagon analyst lamented that it was "frustrating to see the pictures of burning Humvees while knowing that there are other vehicles out there that would provide more protection."[56] Ironically, the US military did seriously consider purchasing a large number of MRAPs later that year—but for the Iraqi army, not for US forces.[57]

In the fall of 2004, marine field commanders drafted their first formal request from the field for MRAPs. The staff of the First Marine Expeditionary Brigade (I MEB), which was operating in Anbar province, drafted a request for 1,169 MRAPs.[58] The draft circulated through the Pentagon in November 2004,[59] and in December, Marine Corps System Command released a request for information on commercially available vehicles with excellent protection against mines and ballistic threats. Nine potential vendors were identified.[60] In February 2005, the commander of I MEB officially submitted an urgent universal needs statement—the marine equivalent of the army's earlier urgent request for additional armor. It stated, "There is an immediate need for an MRAP vehicle capability to increase survivability and mobility of Marines operating in a hazardous fire area against known threats," and that the increasing use of IEDs "requires a more robust family of vehicle capable of surviving the IED . . . threat."[61] It warned that without the MRAP, "personnel losses are likely to continue at the current rate,"[62] and that the marines "cannot continue to lose . . . serious and grave casualties to IED[s] . . . when a commercial off the shelf capability exists to mitigate these threats" without the "potential to jeopardize mission success."[63] This request stayed within the Marine Corps and did not become part of the joint rapid acquisition process.[64]

At the end of March, the Marine Corps held one of its regular Executive Safety Board meetings, which was attended by most of the service's senior leaders.[65] The officer who drafted the request spoke at the meeting and recommended that the service purchase as many MRAPs as possible, as quickly as possible. After some discussion, Lieutenant General James Mattis, the head of Marine Corps Combat Development Command, said, "That's exactly what we are going to do."[66] The assistant commandant of the Marine Corps, General William Nyland, then directed a study to review the feasibility of doing so and asked that the results be briefed at the next safety board meeting.[67] The briefing never occurred. By June, a status report on the field request showed that the service would instead wait for what it called a "future vehicle," which would be more mobile than an MRAP but offer more protection than Humvees.[68] This presumably referred to the Joint Light Tactical Vehicle (JLTV), which was working its way through the acquisitions process and was approved as the Humvee replacement in November 2006. (The first JLTVs were not delivered until 2016 and did not reach initial operating capability in the Marine Corps until August 2019; as of this writing, the army projected that its JLTV program would reach that milestone by the end of 2019.)[69] In August 2005, the Marine Corps staff stopped processing the February urgent request. It was neither approved nor denied; instead, some part of the staff seems to have assumed that the decision made two months earlier by the commandant of the Marine Corps, General Michael Hagee, to procure up-armored Humvees had rendered the request obsolete.[70] As a result, MRAP procurement for the marines stalled.

In February 2006, the Pentagon centralized all of the efforts to address the IED threat into a new body, the Joint IED Defeat Organization (JIEDDO).[71] Its budget ranged between $3 billion and $4 billion each year for its first four years,[72] a substantial investment. But JIEDDO did not focus on MRAPs, for two reasons. First, the new organization focused almost exclusively on preventing IED attacks—what it called "left of the boom"—rather than protecting troops against the effects of those attacks. It worked on developing jammers and other technological countermeasures, and on efforts to predict where and when IEDs would be emplaced.[73] One former defense official was so contemptuous of protection efforts that he criticized the leaders of the ground services for "focus[ing] on placing a cocoon around the soldier driving down the street in his vehicle" instead of "taking out the IEDs first."[74] Second, and more importantly, JIEDDO's mandate did not allow it to procure new vehicles. It did invest in developing better armor, since that fell within its research portfolio. But the services maintained their statutory authority under Title 10 to organize, train, and equip the force based on their own independent assessments of requirements.

In May 2006, the commanding general of the marines operating in western Iraq submitted a new urgent request for MRAPs—15 months after the original submission. This request, however, only asked for 185 vehicles, in the hopes that it would be seen more favorably.[75] What ultimately gave this new request powerful traction, however, was the fact that it was submitted through a relatively new joint process, which enabled fielded forces to bypass the acquisition processes of their individual services. Deployed units could instead submit joint urgent operational needs statements (JUONs) to the relevant combatant command (in this case, US Central Command, or CENTCOM), which were then sent directly to the Joint Staff.[76] As a result, this new marine request completely bypassed the parts of the Marine Corps bureaucracy that had stopped processing the original request. In fact, the Joint Staff responded by saying that the request was too small and encouraged the commander to submit a larger request. He did so in July, asking for an additional 1,000 MRAPs.[77] CENTCOM approved the request for all 1,185 vehicles on October 26, and the initial request for proposals for the MRAP program was issued on November 9.[78] More than three and a half years after the Iraq War began, the US military finally started the process of buying a large number of MRAPs.

Why did the joint request succeed where the service request had failed? Simply put, the joint acquisition process prioritized the immediate needs of fielded forces while the acquisition processes of both the US Army and the Marine Corps focused on the long-term requirements of their service. From a bureaucratic perspective, it made sense that the two services would prioritize the JLTV as the eventual replacement for the Humvee. They saw the MRAP as a serious threat to the JLTV program, primarily because they feared it would divert funding from their long-term wheeled vehicle priority. According to data from 2007, adding armor to an existing Humvee cost $14,000, and a Humvee built with additional armor cost $191,000—whereas an MRAP cost between $600,000 and

$1 million.[79] That meant that both services could easily spend more than $1 billion responding just to this one marine commander's request.

Moreover, in 2006 service leaders believed that MRAPs would soon be unnecessary. This was partly due to the fact that officials in the Bush administration and the Pentagon believed that the Iraq War would soon be over. But it also stemmed from the widespread sentiment among senior marine and army leaders that the expensive and cumbersome MRAPs had little postwar utility—they were just too large and too slow to employ for the types of future conventional warfare they envisioned. The bureaucracies of both services saw MRAPs as a waste of money that threatened an important long-term acquisition priority—even though Congress had approved vast amounts of additional funding for urgent wartime needs exactly like this, on top of the already-substantial Pentagon base budget.[80]

If that were the end of the MRAP story, it would already stand as an example of failed institutional adaptation in technology. But the Pentagon's bureaucracy continued to fight the large-scale purchase and deployment of MRAPs even after the joint approval of the marine commander's request. As more and more requests for the highly effective MRAPs came in, the marines and the army continued to stonewall. Both services sought to block the procurement of large numbers of MRAPs until the secretary of defense personally intervened to ensure that fielded forces received these life-saving vehicles.

Once the Joint Staff validated the marine requirement for 1,185 MRAPs, demand for MRAPs exploded. By December, the Joint Staff had approved requests for a total of 4,060 MRAPs;[81] that number rose to 6,738 vehicles by February 2007,[82] and to more than 8,000 vehicles by March.[83] The casualty data clearly explain why the combat forces wanted MRAPs so badly. In 2005, IEDs killed 408 US troops in Iraq, or 48.2 percent of the total US troop deaths in the country. In 2006, those numbers rose to 423 and 51.5 percent, respectively.[84]

In March 2007, the House Armed Services Committee held a hearing that revealed the size of the gap between the number of MRAPs requested and the number of vehicles that had been funded. The vice chief of staff of the army, General Richard Cody, stated that army forces in Iraq had asked for only 2,500 MRAPs, not enough to replace each one of their 18,000 up-armored Humvees. The service had just added a request to the fiscal year 2008 budget for the funds to purchase about 700 MRAPs, and that the remaining ones were listed as an unfunded requirement. At that point, Representative Gene Taylor (D-MS), interrupted Cody, saying in part:

> If this vehicle is going to save lives, if Humvees, as we now know, are vulnerable to mines and a hugely disproportionate number of casualties are

occurring in Humvees because of mines and we have a way to address that, why don't we address it now?[85]

Cody responded that he wanted to get the vehicles to the troops as rapidly as possible, and that the army was buying them as quickly as it could. He stated, "I think if you gave us the money today, I still wouldn't get them any faster than what we have."[86] General Robert Magnus, the assistant commandant of the Marine Corps, then added that his service did not have enough funds to pay for the 3,700 MRAPs it had approved, and that the total request by the two services for 6,700 MRAPs was underfunded by $3.8 billion. Magnus also described the total request as a "rapidly evolving requirement over the past three months."[87] While technically true, this ignored the fact that the initial MRAP request had been submitted to the Marine Corps more than two years earlier. Moreover, the flood of recent requests to which he referred only began after the Joint Staff approved the second marine request, and deployed army and marine units realized that it was now possible for them to get the vehicles as well.

On April 19, Secretary of Defense Gates read an article in *USA Today* that stated, "In more than 300 attacks since last year, no Marines have died while riding in new fortified armored vehicles the Pentagon hopes to rush to Iraq in greater numbers this year."[88] It added that the attacks had caused an average of less than one injury per attack, whereas attacks on other vehicles had caused an average of more than two injuries.[89] Gates later recalled, "I continued to think about this new kind of vehicle and asked for a briefing on it." Magnus briefed Gates on April 27, telling him that 6,000 MRAPs had been ordered, but that since the services did not have enough money to pay for them, only 1,300 would be built in 2007.[90]

Gates was stunned that the services were not moving more quickly. On May 2, he convened a meeting of Pentagon senior leaders—including the chairman of the Joint Chiefs of Staff, the deputy secretary of defense, and the secretaries of the army and the navy—to discuss the need to dramatically increase the funding, size, and speed of MRAP procurement.[91] Gates describes what happened next:

> I didn't often get passionate in meetings, but in this one I laid down a marker I would use again and again concerning MRAPs: "Every delay of a single day costs one or more of our kids his limbs or his life." To my chagrin, *not a single senior official, civilian or military*, supported my proposal for a crash program to buy thousands of these vehicles. Despite the lack of support, the same day I issued a directive that made the MRAP program the highest-priority Department of Defense acquisition program and ordered that "any and all options to accelerate the production and fielding

of this capability to the field should be identified, assessed, and applied where feasible."[92]

He followed up by giving the MRAP program priority access to the industrial components it needed in order to accelerate production,[93] and established a Pentagon task force on MRAPs that was required to brief him every two weeks.[94] The Joint Staff soon approved purchasing a total of 7,774 MRAPs,[95] but it quickly became clear that even this was not enough. During the task force's first briefing on June 8, Gates recalled, "The magnitude of the challenge became clear": the total proposed request had exploded to more than 23,000 MRAPs, costing more than $25 billion. Why did it grow so dramatically and so quickly? Field commanders had long wanted the life-saving vehicles, but saw no point in requesting MRAPs when they knew they would not get them. Now that had changed. Since the vehicles were finally going to get built in quantity, they wanted as many of them as possible.[96]

Over the summer, the embarrassing tale of the services' resistance to the MRAPs became widely known, and new efforts were made to speed production. In July, *USA Today* published the results of a lengthy investigation about these delays,[97] which Gates later said was the first time he knew the whole story.[98] A few days later, Senator Joe Biden (D-DE) entered the article into the *Congressional Record* and chastised the Pentagon for its lengthy stalling. He estimated that as many as 1,120 US troops could have been saved if MRAPs had been in the field from the start of the war.[99] Biden also resoundingly berated Pentagon leaders for having prioritized the needs of the future force over those of soldiers fighting in the field:

> Those who worry about what the military will be driving in 5 years are missing the boat here. I understand that there are great advancements being developed for our future force. But we have a sacred trust to those on the front lines today, right now. Right now, we are saying to them: If you survive this war, we will get you really good protection for the next one. Give me a break . . . Can anyone imagine Roosevelt saying, "Listen, we may not need some of those boats after Normandy, so maybe we should not build so many?"[100]

The MRAP program accelerated rapidly after this harsh criticism. In September, the Joint Staff increased the total purchase to 15,374 vehicles,[101] and the MRAPs that were already in production started being shipped to Iraq. By January 2008, there were about 2,000 MRAPs in the country; that number grew to almost 6,000 by June and over 10,000 by November.[102] The Pentagon ended up buying a total of 27,000 MRAPs—including an all-terrain version specifically designed for Afghanistan—at a total cost of almost \$40 billion.[103] Troops displayed an impressive spirit of tactical adaptation in figuring out new uses for them; some turned MRAPs into ambulances, and one inspired brigade commander put a desk into one of the vehicles to use as a mobile command post.[104] But overall, the MRAPs did exactly what they were designed to do: protect the lives of troops in the vehicles. Casualty rates of soldiers and marines in MRAPs ended up being about 75 percent lower than those in Humvees, and about 50 percent lower than those in other types of vehicles.[105] As Gates later recalled, "Time and again, commanders would walk me over to a damaged MRAP, and there would be two or three soldiers standing by it who would tell me about surviving an attack on that vehicle."[106]

In a cruelly ironic twist, MRAPs started arriving in Iraq just as the need for them dropped precipitously. As we discuss in Chapter 7, the local tribes in Anbar province turned on al-Qaeda, their previous ally, in the fall of 2006 and now started fighting it instead of US forces.[107] In January 2007, President Bush announced what became known as "the surge"—the deployment of 30,000 additional troops to Iraq under a new commander, General David Petraeus, who implemented the new counterinsurgency strategy described in Chapter 5.[108] Together, these two trends slowly changed the trajectory of the war. Figure 6.5

FIGURE 6.4 The First Shipment of US Army MRAPs Arrive in Baghdad, November 2007
Source: https://www.army.mil/article/5968/army_fields_mrap_in_iraq.

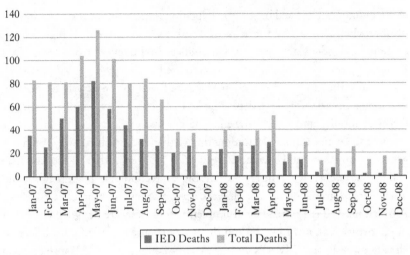

FIGURE 6.5 IED Deaths and Total Deaths of U.S. Troops in Iraq, 2007–2008
Source: O'Hanlon and Livingston, "Iraq Index," 12.

shows that combat deaths increased in the first part of 2007, reaching the highest point of the war in May.[109] But those numbers started to shrink during the summer, and the downward trend was clearly evident by the fall—just as the number of MRAP purchase approvals and deliveries began to grow. There is no definitive way to determine exactly how many lives they saved. But they clearly would have saved many more lives if the vehicles had been widely available years earlier. The leaders of the US Army and the Marine Corps are responsible for many of those losses, since they repeatedly chose to prioritize their services' future needs over the urgent needs of their troops in battle. Only the direct, personal intervention by the secretary of defense to speed the delivery of MRAPs late in the conflict kept the number of combat fatalities from being worse.

Conclusion: Tactical Success and Institutional Failure in Iraq

Hillbilly armor resulted from an impressive effort to try to adapt at the tactical level. The unexpected emergence of IEDs in 2003 posed an unanticipated problem for soldiers and marines who had just finished fighting a major conventional campaign in Iraq. They displayed tremendous creativity and ingenuity in experimenting with many different ways to better protect their soft-skinned vehicles. Units in the field set up metal workshops that focused solely on adding newly improvised armor, and their ad hoc techniques and solutions were informally shared across units. But as discussed in Chapter 3, not all technological shortcomings can be fixed by soldier adaptation in the combat zone. No amount of improvised armor could save an M4 tank crew from German high-velocity cannons, nor adequately protect US soldiers in Humvees from Iraqi IEDs. Ultimately, institutional adaptation was required.

Initially, the army and the marines adapted well to the new IED threat by developing up-armoring kits, deploying them quickly, and working with fielded forces to modify and improve them. The services worked hard to procure as many of these kits as possible, finding that civilian industrial capacity was a bigger constraint than the Pentagon bureaucracy. Yet even as they did so, the army and marines were sowing the seeds of the biggest technological failure of the war. They ignored growing evidence that the up-armoring kits were at best a Band-Aid that masked the fundamental problem: Humvees were designed to operate in safer rear areas, not to withstand the blast pressure of an IED. Both services already owned some MRAPs and knew they could survive IEDs, since many explosive disposal units were using them in Iraq. Yet even after their combat forces submitted formal requests for MRAPs, both services continued to ignore the urgent requests of their field commanders. Senior marine and army leaders believed that MRAPs threatened their long-term plans to replace Humvees with the new JLTV. Formal requests made by army and marine units in Iraq through the joint acquisition process were more successful, but even there, the pace of MRAP procurement was painfully slow given the urgency of the threat. It took the personal intervention of the secretary of defense—four years into the war, and *against* the advice of the most senior military and civilian Pentagon leaders— to significantly speed MRAP procurement.

Why did the services invest so much and so quickly in the up-armoring kits early in the war while later resisting the MRAPs in almost every way possible? Part of the reason is that senior leaders continued to expect that the war would be short, and so were convinced that MRAPs would not be needed much longer. But the much deeper reason was that the services prioritized their demands for the unknown wars of the future over the very real, but different, demands of the wars of today. The military services' principal role is to organize, train, and equip their forces for war, but the actual execution of wars falls under the purview of joint commanders, not the service chiefs. As a result, joint warfighting commands such as CENTCOM and theater commands in the field inevitably focus on immediate needs, while the services focus principally on the future. MRAPs simply did not fit into the army and marines' long-term plans. They did not want to divert any attention, and especially any resources, away from their prized JLTV program—even though Congress provided more than enough supplemental wartime funds for them to be able to procure both MRAPs and JLTVs.[110] As Kollars astutely observes, the services "were reluctant to risk future dollars on long-standing programs to answer to the war at hand. *The commanders were at war, but the acquisition programs were not.*[111] Tragically, hundreds of US troops undoubtedly lost their lives as a result.

Technological Adaptation in Afghanistan
Tactical Adaptation: Apache Helicopters and Close Air Support

When the war in Afghanistan began, the AH-64 Apache was the US Army's premier attack helicopter.[112] The Apache was originally developed in the 1980s as one of the army's Big Five weapons systems[113] produced for its new AirLand Battle doctrine.[114] It was the only one of the Big Five that was specifically designed for deep battle, penetrating far behind enemy lines to attack Warsaw Pact mechanized formations that were preparing to surge forward into western Europe. It operated effectively at night and in bad weather, critical capabilities for successfully defeating a much larger adversary with extensive air defense capabilities. The Apache was the most advanced attack helicopter in the world when introduced into service in 1986. Largely because of its capabilities, the US Army began to envision its aviation branch as a maneuver arm of mechanized warfare rather than simply a supporting branch.[115]

The Apache was optimized for the deep attack mission. It was flown by two pilots seated in tandem. The pilot in front acted as the gunner and operated the highly sophisticated night vision and infrared targeting suite in the sleek helicopter's bulbous nose.[116] This suite allowed the Apache's crew to see many kilometers even in bad weather and darkness, and to employ its lethal antiarmor weapons under nearly all conditions. The helicopter's weapons were specifically designed to kill tanks and other heavily armored vehicles. Under the nose, the aircraft mounted a unique 30mm chain gun that could spew armored piercing and high-explosive cannon rounds out to 1,500 meters at a rate of over 600 rounds per minute. Stubby pylons flanked the cockpit, enabling the crew to carry a variety of lethal long-range munitions. These included the new Hellfire missile, which had an eight-kilometer range and was the first "fire and forget" long-range antitank missile mounted on a US helicopter.[117] The Apache could carry as many as 16 Hellfire missiles on its pylons, or the weapons mix could be altered to include unguided folding-fin 2.75-inch rockets carried in pods of seven or 19 rounds.

The Apache was also equipped with sophisticated electronic warfare capabilities, including missile early warning and jamming devices. The two pilots wore innovative (and expensive) new helmets that displayed targeting, navigation, and aircraft status information through a heads-up display on their visors. The gunner's helmet could direct the Apache's 30mm cannon, so he or she could aim at targets

FIGURE 6.6 The AH-64 Apache Helicopter
Source: Matthew Fogg.

simply by gazing in that direction. Taken together, these advanced systems enabled Apache pilots to fly close to the ground, following the contours of the terrain to help mask the aircraft as it flew deep into enemy rear areas. The machine itself was also heavily armored against enemy fire. The Apache's systems were hardened to withstand multiple hits by small arms, machine guns, and in critical places, anti-aircraft cannon fire up to 23mm in caliber. Layers of reinforced armor protected the two pilots from small arms and most heavy-caliber machine-gun fire. Even the aircraft's rotor blades were built to withstand multiple hits from 23mm cannons and keep flying. Observers soon dubbed the newly deploying Apache "the flying tank."

Even though the Apache was designed for battle on the plains of northern Europe, its first major combat operation was the 1991 Gulf War.[118] It performed superbly across the open, wind-swept desert terrain. During the 100-hour ground war, Apaches ranged the breadth and depth of the sand-filled battlefield in front of advancing army forces. According to a postwar report by the US General Accounting Office, Apaches destroyed 278 Iraqi tanks, over 600 light and armored vehicles, over 100 pieces of artillery, and a variety of other targets.[119] Only one Apache was shot down in the conflict, falling victim to a short-range Iraqi rocket-propelled grenade (RPG); the crew was quickly rescued.[120] After the war, army leaders concluded that the Apache had delivered on its technological promise. They also believed that the Gulf War had fully validated the Big Five programs,[121] and that Apaches were among the key capabilities that would help the army dominate its future adversaries.

Apaches did not see combat in any US military operations during the remainder of the 1990s, which were mostly peacekeeping or stabilization missions that did not require its capabilities. They were deployed to support NATO's 1999 air war over Kosovo, but were not committed into battle for a variety of reasons.[122] The army did upgrade the Apache's already impressive technology during that decade, however, by introducing the AH-64D Longbow in 1997. This new model mounted a prominent dome above the main rotor assembly that housed a new Longbow millimeter-wave fire-control radar target acquisition system, among other advances.[123] These upgrades made the Apache an even more efficient long-range tank-killing platform that could conduct independent operations ranging far beyond friendly front lines. Yet with the demise of the Soviet Union, the massed armored threats for which the Apache was originally designed were now largely gone. It was not clear what new war the modernized Apaches were being built to fight.

The September 2001 terrorist attacks thrust the US military into a new and wholly unanticipated war. As discussed in Chapter 7, US special operations forces (SOF) were among the first to respond. Arriving in Afghanistan in October, they quickly linked up with local militias and provided them with the US air support needed to help drive the Taliban from power. By December, the Taliban and al-Qaeda fighters across the country had been routed. The army soon deployed a brigade of about 4,000 conventional troops to help SOF hunt down remaining pockets of al-Qaeda and Taliban fighters around the country. These troops, from the 10th Mountain and 101st Airborne Divisions, brought the first Apaches into Afghanistan. They began a series of offensive operations in early 2002 designed to cut off al-Qaeda fighters attempting to flee into neighboring Pakistan.[124] These offensive operations continued episodically for several years as both enemy groups reconstituted themselves in sanctuary areas across the border.

The war in Afghanistan looked nothing like the deep attack, tank-killing battlefield that the Apache was designed (and later upgraded) to dominate. Afghanistan's climate, mountainous terrain and altitude, and unconventional enemy challenged the Apache's capabilities in ways that could not have been envisaged. Apache aircrews had to rapidly adapt to this entirely new kind of war with its dangerous and very unforgiving operating environment. The aircraft's advanced capabilities were significantly degraded in the rugged Afghan mountains. The helicopter could normally carry a full load of fuel and ammunition for hundreds of miles, but the high altitudes of the Hindu Kush mountains sharply curtailed its power and range. Since fuel was consumed more quickly, aircrews often needed to replace some of their Hellfires or rocket pods with an external fuel tank. The performance of the aircraft at high altitudes could also be sluggish, which required pilots to fly the helicopter in different ways. Pilots were trained to hover the aircraft to find targets and to fire its weapons, but hovering a heavily armed aircraft was aerodynamically impossible at high altitudes.

The Apache's weapons also had to be used much differently against a new type of enemy. The aircraft's 30mm chain gun and Hellfire missiles were designed to kill massed formations of tanks and other armored vehicles at long range, but neither the Taliban nor al-Qaeda possessed that kind of heavy equipment after 2001. The Apache's nose-mounted chain gun fired dual-purpose, high-explosive, armor-piercing rounds, which were also effective against softer targets.[125] The aircraft could carry up to 1,200 rounds for this lethal 30mm cannon. Later in the war, crews replaced some of this ammunition with an additional internal fuel tank in order to extend the Apache's range and loiter time over the battlefield. The Apache's 2.75mm rocket pods could also be armed with antipersonnel rounds that spewed deadly flechette projectiles.[126]

As noted earlier, the Apache was designed to fight at night, with tactics and equipment that were optimized for night operations. But combat operations in Afghanistan often took place during the day, as the Taliban and al-Qaeda used the cover of night to move furtively into their fighting positions. Daylight made the helicopter more visible and negated many of the benefits of its infrared and night vision sensors. Moreover, the helicopter's advanced target acquisition radar and multiple target engagement capabilities proved largely useless against small groups of individual enemy fighters. The helicopters' thermal vision systems could sometimes detect the heat signature of enemy fighters at a distance when they contrasted with colder ground. But pilots quickly found that they had to visually identify their targets from dangerously close ranges. These close-in reconnaissances often drew substantial ground fire, which could cause serious damage. Fortunately, the Apache's extensive armored protection meant that the pilots could nearly always fly out of danger, even when badly damaged.

Since the US Army forces deployed to Afghanistan in early 2002 without their supporting artillery, all available air power had to fill the gap.[127] The US Air Force and Navy flew a handful of round-the-clock patrols over Afghanistan primarily to respond to US troops in contact with the enemy. However, once troops on the ground were in a firefight, the enemy was often so close that the blast effects from bombs dropped from these jets could often put friendly troops at great risk. But the precision provided by the Apache's weapons systems (and the ability for ground troops to speak directly to the army pilots) allowed for more controlled "danger close" fire support where Taliban units maneuvered in close proximity to US forces.[128] Thus, the Apache's biggest adaptation challenge was also its most critical: Apache aircrews had to figure out how to conduct close air support missions near friendly troops in a helicopter that had been optimized to kill tanks at long range. Close air support soon became one of the Apache's primary missions, and the aircraft began acting as flying artillery for US troops battling the Taliban at close range.

The first and best example of Apache pilots adapting to this new mission occurred in March 2002, during Operation Anaconda.[129] A US Army brigade-sized infantry force working with Northern Alliance ground forces planned to secure the Shah-i-Kot Valley in eastern Afghanistan. Intelligence officers believed that the area was a long-standing al-Qaeda and Taliban redoubt, dominated by a ring of mountains and pockmarked by numerous deep caves. The operation was designed to kill or capture fleeing al-Qaeda elements and their Taliban affiliates to prevent their escape into Pakistan. Anticipating only limited opposition, the Americans brought along only six of the 18 Apaches then deployed in Afghanistan. These aircraft would lead a larger force of CH-47 Chinook helicopters bringing the infantry brigade into the valley. The Apaches belonged to the Third Battalion of the 101st Aviation Brigade based in Kandahar, over 300 miles away from the Shah-i-Kot Valley. One of the six allocated Apaches encountered mechanical problems before the operation began, leaving only five available on the first day of the operation.

On the morning of the assault, the Apaches were used to reconnoiter the infantry landing zones before the helicopter assault force of Chinooks landed troops on the ground. US and Afghan forces quickly ran into a buzz saw of opposition. Well-concealed enemy forces attacked the Apaches near the landing zones and the US special operations forces accompanying the Northern Alliance forces advancing overland in the valley. Two of the reconnoitering Apaches diverted from the brigade's landing zones to support the embattled special forces, and began firing rockets to suppress Taliban mortar positions. Both helicopters were quickly damaged by ground fire, and the Northern Alliance forces moving up the valley floor were turned back by the Taliban.

Before the helicopters delivering the US troops arrived, US Air Force and Navy strike aircraft dropped bombs to suppress suspected Taliban positions. But they had little impact on the concealed, dug-in enemy redoubts. Once the assault began, US troops spreading out from their landing zones faced heavy fire and furiously battled the enemy at close range. Since they did not have their supporting artillery, they had to rely on the Apaches for fire support. The helicopters that had been supporting the aborted Northern Alliance advance also rejoined the fighting around the US landing zones. Air force and navy aircraft continued to drop precision-guided bombs at targets that were safely away from the exposed Americans on the ground, but the five Apaches loitering above the battlefield now were getting continuous calls for close air support from the embattled troops below. The Apache pilots now had to rapidly adapt to an entirely new situation and mission.

The Shah-i-Kot Valley lies at an altitude of 8,500 feet, and is surrounded by mountains that reach as high as 12,000 feet. Since fully loaded Apaches cannot hover at those altitudes, the pilots had to abandon their standard tactics of firing while at a stationary hover.[130] They quickly improvised a new running fire approach that involved shooting while simultaneously diving at the target, which created the lift necessary to stay aloft. Yet this approach, which resembled some tactics of the Vietnam era, was extremely hazardous. The helicopters also had to fly parallel to the front lines of US troops in order to lessen the chances of hitting them by accident, but this flight path made them extremely vulnerable to enemy fire.

As the Apaches shifted back and forth across the exposed battlefield, they were repeatedly hit by small-arms fire and even RPGs. One helicopter was badly damaged and was forced to land just off the battlefield in a dry riverbed, still in dangerous territory. Its wingman soon landed nearby, and both crews dismounted to assess the first machine's damage. An RPG had blown three Hellfires off the wing stub of the first Apache, and it was so riddled with shrapnel and bullet holes that all of its transmission oil had leaked away. Rather than abandon the stricken aircraft, the two crews devised a novel solution. They determined that if they filled the badly damaged aircraft with extra transmission oil from the spare cans always carried in each helicopter's cargo bay, it would just be able to make it to the nearest refueling point 50 miles away. After completing the unorthodox oil change, and less than 10 minutes after they had landed, both helicopters took off for the refueling point.[131] They arrived with only minutes to spare before the damaged Apache's rotors stopped turning, in an extraordinary example of nail-biting battlefield adaptation.

Two additional Apaches from Kandahar had joined the battle in the middle of the day, and all seven helicopters eventually took repeated hits from enemy fire. That evening after the battle, ground crews counted bullet holes in 27 of the 28 rotor blades of the seven aircraft.[132] Two Apaches had also been hit by RPGs, and the infrared jammers aboard a third had deflected at least one enemy shoulder-fired surface-to-air missile. Two helicopters required extensive repair before returning to combat, but six out of the seven Apaches had been able to keep flying throughout the day despite substantial battle damage. Because the Apache pilots adapted effectively in the heat of combat, the embattled infantry task force spread across the valley was able to stave off ferocious Taliban assaults until US reinforcements could arrive in the days to follow.

After the battle ended, Apache units in Afghanistan employed many of the battlefield adaptations pioneered in the Shah-i-Kot Valley for the remainder of the war. These tactics were refined and disseminated across the entire US Army in the following years, and close air support has now become one of the Apache's primary missions.[133] For a helicopter designed and built for a much different battlefield, this new mission continued to prove its value time and again in the tough ground battles fought by army troops in Afghanistan and Iraq over the next decade and beyond.

The Apache crews in Operation Anaconda met all three criteria for successful tactical adaptation during the battle. First, they approached the problems they encountered with creativity and a willingness to experiment. Recognizing that they could not hover at high altitudes and fire their weapons, they immediately improvised a new approach that enabled them to accomplish their mission even though it put them in much greater danger. These crews demonstrated notable courage as well as remarkable flexibility. Second, the Apache pilots and their supporting ground crews were able to manufacture or improvise solutions to their combat requirements in or near the combat zone. Few examples show this better than the ad hoc repair of a badly damaged Apache using spare transmission oil pulled from spare cans carried in the aircraft ammunition bays. This impromptu battlefield fix allowed a badly damaged helicopter and its crew to make it back to base for desperately needed repairs. Third, Apache units disseminated their new close air support techniques throughout the force after the battle, which eventually led the Apache to transform from a deep attack, tank-killing helicopter into a highly effective provider of close air support. Few examples of technological adaptation in combat have been more rapid, effective, or consequential.

Institutional Adaptation: DCGS-A versus Palantir

The Battlefield Failures of DCGS-A

As the internet and digitization began to sweep the telecommunications and commercial sectors in the 1990s, the US military services started to develop digitized command-and-control capabilities that could share intelligence and other battlefield information. Since the services operate jointly in combat operations, they agreed to develop a common umbrella system called the Distributed Common Ground System (DCGS). This initiative became a major acquisition program, requiring multiyear budget outlays and eventually costing billions of dollars. The army's portion of this mammoth project was called DCGS-A (pronounced D-sigs-A), which became an army program of record in 2005.[134] That designation meant that it was the army's sole official intelligence program, approved and funded across the Future Years Defense Program. Moreover, given the repeated army procurement failures discussed at the beginning of the chapter, the army needed to ensure that DCGS-A was successful in order to avoid a similar embarrassment.

DCGS-A had a breathtakingly ambitious scope. Though the program was initiated in the late 1990s and awarded its first contract in 2001,[135] it gained momentum after the September 11 terrorist attacks as a way to share intelligence more effectively. According to the Government Accountability Office (GAO), DCGS-A was "intended to be the Army's primary system for collecting, processing, integrating, and displaying intelligence, surveillance, and reconnaissance information about potential adversarial forces, the weather, and the terrain to Army Commanders at all echelons."[136] The first versions of DCGS-A started to be fielded to army units in Iraq and Afghanistan in 2003.[137]

DCGS-A evolved slowly as enemy activity in Afghanistan gradually increased. Insurgent activity remained relatively low for a number of years after the 2001 defeat of the Taliban, despite occasional pitched battles with the small number of US troops or Afghan security forces thinly spread across the country.[138] By 2005, however, the Taliban and associated remnants of al-Qaeda had reorganized from sanctuary areas in Pakistan and started to conduct an increasing number of attacks in Afghanistan.[139] The United States and its NATO allies responded by gradually deploying more troops, who brought additional intelligence systems with them to try to analyze and stop the threat. Yet the attacks continued to grow in number and lethality, as the Taliban started employing more IEDs in addition to suicide attacks.[140]

Figure 6.7 shows that the number of US military deaths caused by IEDs in Afghanistan, as well as the total number of deaths, had dramatically increased by 2009. The trickle of intelligence data being collected in the first years of the war tuned into a torrent, as analysts fought to make sense of a plethora of information from Predator drones, communications intercepts, prisoner interrogations, and infantry patrols. In order to grasp the enemy's patterns and avert deadly attacks, analysts needed to be able to integrate and synthesize these myriad reports into a comprehensive mosaic of enemy activity. Understanding the patterns of this new and unconventional threat became the highest priority of US commanders in Afghanistan.

DCGS-A was the only tool that army commanders in Afghanistan initially had available to categorize and synthesize the confusing jumble of incoming information and intelligence. Finding patterns of enemy activity was the key to preventing greater losses, particularly from the deadly IEDs. Intelligence analysts soon found that the cumbersome DCGS-A system was not up to the task. Largely built upon technology from the 1990s, the army's software was difficult to use, crashed easily, and could not quickly upload information from disparate

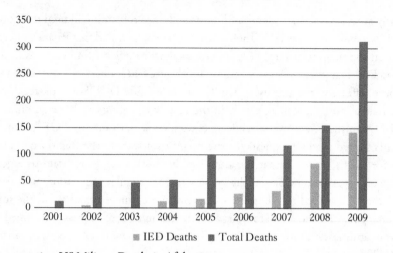

FIGURE 6.7 US Military Deaths in Afghanistan, 2001–2009

Source: Ian S. Livingston, Heather L. Messera, and Michael O'Hanlon, "Afghanistan Index: Tracking Variables of Reconstruction & Security in Post-9/11 Afghanistan," January 29, 2010, 6.

databases.[141] Despairing analysts often reverted to building intelligence products with PowerPoint and tools like Google Maps and Google Earth.[142]

While the army struggled to find effective ways to use its unwieldy intelligence program of record in Afghanistan, the marines and SOF found a different way to integrate the mountains of information. A Silicon Valley start-up named Palantir that specialized in data analytics had produced data integration software called Gotham that was quick and easy to use.[143] Palantir hired several former military personnel who immediately understood that the powerful software would prove invaluable to deployed troops who sought to predict where enemy attacks would occur.[144] Palantir scheduled several appointments with Pentagon officials to test the software, but all were canceled before the tests took place. In January 2009, Palantir representatives pitched their software to the army's intelligence directorate (G-2). But the follow-up appointment to test the software was canceled without explanation, as the others had been, and was not rescheduled.[145] At one point, a defense official reportedly suggested that it would be easier for Palantir to arrange a test if it worked as a subcontractor to Northrop Grumman—one of the companies involved in making DCGS-A.[146]

Frustrated by the Pentagon bureaucracy, Palantir dispatched representatives to Iraq and Afghanistan, hoping to demonstrate the utility of the software to those who needed it most. A special forces unit in Iraq was the first to buy the software, spending less than $100,000 from its discretionary budget to do so. By early 2010, several marine and SOF units and one army unit had used their own funds to buy the software.[147] Their experiences demonstrated that Palantir offered an off-the-shelf, relatively cheap way for analysts to quickly find patterns in the flood of data pouring into battalion and brigade headquarters. The software's speed, effectiveness, and user-friendly interface made DCGS-A look fecklessly archaic by comparison, especially in the rapidly changing combat environment of Afghanistan. As word of its effectiveness spread, many marine, SOF, and army units started sending requests for the Palantir software to the Rapid Equipping Force (REF), an office that the army established in 2002 to respond to urgent capability gaps on the battlefield by bypassing the regular acquisition process.[148] As the REF started fulfilling those requests, more units learned about the software's potential and asked for it as well. As the number of requests continued to grow, army leaders soon started seeing Palantir as a threat to DCGS-A, its official intelligence system.

Over the next three years, the army turned down requests for Palantir software from 17 army brigades deploying to Afghanistan. Some requests were denied numerous times before the army relented to its commanders' insistent demands.[149] The army refused to recognize that its official intelligence system was failing to meet the needs of commanders and their intelligence analysts battling the growing Afghan IED threat. Instead, the service leadership blamed the soldiers complaining about DCGS-A for being poorly trained and not using it properly. The army responded to the flood of complaints coming from deployed units by adding more training, while insisting that soldiers make the existing system work—regardless of the human cost.[150] The army leadership was irrevocably committed to DCGS-A, apparently whether it worked in Afghanistan or not.

By July 2010, even the top US military intelligence officer in Afghanistan recognized the seriousness of the problem. Earlier that year, Army Major General Michael Flynn, the intelligence chief of US and NATO forces in Afghanistan, published a report that utterly lambasted the intelligence community for not providing adequate intelligence for US forces in the field.[151] Palantir representatives soon reached out to Flynn, who was extremely impressed with the software after he saw a demonstration.[152] On July 2, Flynn sent the Pentagon a joint urgent operational need (JUON) statement—the same type of request that the marines used to request MRAPs in May 2006—asking for a better analytic tool. The first paragraph of the cover letter bluntly stated, "Intelligence analysts in theater do not have the tools required to fully analyze the tremendous amounts of information currently in theater."[153] Flynn then wrote:

> The impact of this shortfall is felt in almost every activity that intelligence
> supports. Analysts cannot provide their commanders a full understanding

of the operational environment. Without the full understanding of the enemy and human terrain, our operations are not as successful as they could be. This shortfall translates into operational opportunities missed and lives lost.[154]

Flynn did not mention Palantir by name, because JUONs are only supposed to describe a needed capability rather than proposing a specific solution. But the description essentially matched what Palantir's software could already do, making his intentions quite clear.

When informed of Flynn's request, the army responded that DCGS-A already had the ability to do much of what Flynn needed. It also said that the system would soon be upgraded to fulfill the rest of his requirements, but those upgrades did not occur.[155] Yet Flynn's JUON was also sent to the Pentagon's Combating Terrorism Technical Support Office (CTTSO), which resided within the Office of the Secretary of Defense rather than in any of the individual services.[156] Unbeholden to any specific service program, CTTSO concluded that Palantir's software best met Flynn's requirements. The office soon sent a dozen Palantir servers to troops in Afghanistan who were not required to use the army's DCGS-A, mostly marines and special operators. These analysts found that the Palantir software was far more intuitive and could quickly create useful products that were much easier to share. Major General John Toolan, who commanded the Second Marine Expeditionary Force and Regional Command-Southwest in Afghanistan, later wrote that "Palantir reduced the time required for countless analytical functions and streamlined other, once cumbersome, processes . . . The innovative and collaborative capabilities of Palantir have proven their mettle and effectiveness for conventional and special operations forces in combat."[157]

By the beginning of 2011, more than three dozen SOF and marine units were using Palantir's software, as were a few army units that had quietly paid for it with their own funds. Later that year, a marine colonel who was requesting funds for more Palantir servers directly stated, "Marines are alive today because of the capability of this system."[158] Despite the proven success of Palantir, the army continued to resist providing it to its formations. The army grew increasingly frustrated that Palantir was marketing its software directly to deployed units,[159] and internal army documents accused Palantir of ghostwriting Flynn's JUON and the other requests for the software that kept pouring in.[160]

While this bureaucratic battle raged, deployed army units were regularly losing soldiers to IEDs in an increasingly deadly fight with the Taliban. In one stark example from 2011, the army's 82nd Airborne Division was dutifully using DCGS-A to support its operations in Kandahar but continued to suffer steady casualties from roadside bombs. In frustration, the division's intelligence analysts requested more information on Palantir's software through the firm's website. Palantir quickly dispatched a representative to Kandahar to provide a demonstration. As one of the division's soldiers later recalled: "We had made an inquiry about Palantir [to the website] at the time, and their [field representative] showed up in the middle of our crisis to do a demo. We gave him a piece of the problem to solve, and what he was able to turn around in a couple of hours compared to what we had spent days on made our jaws drop."[161]

Major General James Huggins, the division commander, directed his chief intelligence officer to formally request the Palantir software from the army. Comparing DCGS-A to Palantir, the request stated, "Solving very hard analytical problems takes several days when using existing tools against these data sources. In our experience in using the Palantir platform against these same problems, we were able to reduce this time to a few hours. This shortfall translates into opportunities missed and unnecessary risk to the force."[162] Despite this urgent request from one of its division commanders in combat, the army denied the request.

Congressional Oversight, a Suppressed Assessment, and a Lawsuit

Immediately after Flynn submitted the JUON statement, members of Congress began investigating why soldiers in Afghanistan were unable to get effective analytical tools to uncover the roadside bombs that were grievously wounding and killing dozens of their comrades. This marked the beginning of several years of increasing congressional criticism of the army's stubborn support for the unpopular and operationally troubled DCGS-A system. On July 19, 2010, Representatives Gabrielle Giffords (D-AZ), Mike McIntyre (D-NC), and Adam Smith (D-WA) wrote to the Republican leaders of the House Defense Appropriations Subcommittee, asking them to "competitively source and secure the necessary platform sets for all intelligence and special operations units that have urgently requested the system [Palantir] but have not yet received it."[163] The chairman of the subcommittee reached out to the army, which responded a few days later, saying in part:

> The DCGS-A Cloud, together with a software upgrade planned for systems already in theater, provides the capabilities required. We will continue to look for opportunities to employ Palantir where the Cloud is not immediately available or where the employed Cloud will not provide adequate coverage.[164]

The army response also stated that DCGS-A "is effectively free for CENTCOM as it already resourced and on track for delivery to theater." That statement, while technically true, ignored the fact that the army had spent $1.6 billion on DCGS-A during the previous four years.[165]

Giffords and Smith (who was by then the ranking member of the House Armed Services Committee) were so dissatisfied that they sent the army a written request for additional information on how DCGS-A would address the gaps that Flynn had identified, to which the army did not respond. In May 2011, Smith sent the army chief of staff, General Martin Dempsey, a letter suggesting that funds for DCGS-A could be cut if the army did not address these congressional concerns.[166] Later that year, Senators Kay Bailey Hutchison (R-TX), Dianne Feinstein (D-CA), and Tom Carper (D-DE) told the CENTCOM commander that they were concerned that the army was denying requests for commercial software even though DCGS-A efforts had "been funded in the billions of dollars over more than a decade and have yet to meet the requirements of the users."[167]

In the face of this growing uproar, the army directed its own assessment of Palantir's performance in Afghanistan, "in order to assist with a procurement decision and to underpin an information paper to the Army Chief of Staff."[168] The army's Test and Evaluation Command (ATEC) deployed a team that conducted a detailed and comprehensive study of how army, marine, and SOF units were using Palantir throughout Afghanistan. It conducted extensive interviews with members of 15 headquarters and field units. The team found unequivocal support for the Palantir software among the users it surveyed. According to the report:

- Ninety-six percent "agreed that Palantir was effective in supporting their mission."
- Ninety percent agreed that "Palantir saves time when performing analysts' tasks."
- Eighty-nine percent "agreed that using Palantir allowed them to visualize trends based on time, location, or individuals."
- Ninety-three percent of intermediate supervisors "agreed that Palantir increased their ability to support the intelligence requirement of the maneuver commander."[169]

The report also noted some limitations of the system, including that some users had trouble accessing servers on other bases because of connectivity issues and bandwidth limitations in Afghanistan. The report then stated: "Recommendation: Install more Palantir servers in Afghanistan at multiple locations to mitigate issues."[170]

The ATEC report was released within the army in April 2012.[171] But in a stunning and inappropriate intervention, the army's G-2, Lieutenant General Mary

Legere, ordered it to be rescinded and for all existing copies to be destroyed.[172] As the army's senior intelligence officer, Legere was the leader most responsible for DCGS-A, and was deeply invested in the army's multiyear, multibillion-dollar flagship program. She further ordered that the report be rewritten. The new version, which was released in May, did not include the recommendation just quoted or a later recommendation to add additional training on Palantir. It included a new statement that the units using Palantir "do so to augment their intelligence estimations, but do not necessarily employ Palantir as their primary tool for analysis." And it omitted the following quotation from a survey respondent: "DCGS is overcomplicated, requires lengthy classroom instruction, and is an easily perishable skill set if not used constantly."[173] In a subsequent sworn statement, the leader of the ATEC assessment team, Colonel Mark Valeri, wrote that he was pressured to change the report.[174] In yet another slap to its soldiers in combat, the army misrepresented its own independent assessment of Palantir to help protect its far more expensive and poorly performing program of record.[175]

Congress soon found out about the army's blatant attempt to suppress and rewrite its own independent and positive assessment of Palantir. Public pressure from Capitol Hill and the media forced the army to investigate the circumstances surrounding the altered report, but the service ultimately held no one accountable.[176] In the meantime, ATEC conducted a similar assessment of DCGS-A, which found many critical deficiencies. In a limited test, DCGS-A servers failed so often that they needed to be rebooted every 5.5 hours. Its reliability proved to be so poor in the relatively easy test conditions that it prompted serious concern about the software's ability to perform on a high-tempo battlefield. Users were frustrated that even simple tasks required opening multiple screens and demanded repeated data conversions to complete. The software coding was so poor that it placed excessive strain on the system's processing power, which caused workstations to frequently freeze up.[177] ATEC's Threat Computer Network Operations Team also found and exploited several software vulnerabilities in DCGS-A during the evaluation—one of which was so serious that the team immediately issued a warning alert to all units in the field using the system.[178] The final ATEC report concluded that the army's program was "effective with significant limitations, not suitable, and not survivable.[179]

ATEC's critical evaluation of DCGS-A landed with the force of a "bombshell," as one reporter wrote, and was "shattering [to] the DCGS-A monopoly."[180] Yet even though the report seriously undermined many army arguments in support of DCGS-A, army leaders continued to deny most requests for Palantir. In May 2012, for example, the colonel who led the army's Rapid Equipping Force responded to a unit in Afghanistan that had requested Palantir as follows: "While I don't disagree with your need, I cannot buy

Palantir anymore without involving the senior leadership of the Army, and they are very resistant."[181]

In February 2012, Representative Duncan D. Hunter (R-CA) learned that the army had denied a request for Palantir from soldiers from the 82nd Airborne Division who were serving in Afghanistan.[182] Hunter was a former marine who had served in Iraq and Afghanistan, and who started supporting Palantir after participating in a congressional trip to Afghanistan. He later said that he "saw DCGS-A turned off under people's desks. I asked about Palantir, and they all said they heard about it from units that had it. They all wanted it."[183] In June 2012, after six troops from the 82nd Airborne were killed by IEDs within a two-week span, Hunter called upon the House Committee on Oversight and Government Reform to investigate the reasons for those deaths.[184] As Hunter emerged as one of the most outspoken congressional critics of DCGS-A, he began to receive a torrent of feedback from soldiers in Afghanistan complaining about the system.[185] Yet the army persisted in forcing DCGS-A upon its reluctant field commanders to keep its expensive program of record alive.

In August 2012, Representatives Darrell Issa (R-CA) and Jason Chaffetz (R-UT) of the Committee on Oversight and Government Reform sent a letter to Secretary of Defense Leon Panetta requesting any documents and information pertaining to DCGS-A, noting that they were exercising oversight of the army's use of the intelligence system. The letter cites numerous reports of user frustration with DCGS-A, including that it could not perform basic tasks. It also warned that the destruction of the initial army report "could be construed as limiting positive feedback on Palantir's system, in an effort to justify the continued use of a more expensive and less effective program."[186] The army continued to respond that the benefits of DCGS-A still outweighed the costs, and that many complaints were due to inadequate training by the users.[187]

In March 2013, Hunter sent a letter to Secretary of the Army John McHugh, which complained that the army was still refusing to fill requests for Palantir software more than three years after the first requests. An army spokesman responded that nine of 13 previous requests for Palantir had been approved. Hunter countered that army units were still forced to constantly fight the bureaucracy to get Palantir systems, and that as a result, they often arrived too late during a deployment to be of value.[188] These disagreements broke into open conflict at a House Armed Services Committee hearing in April. The hearing descended into a heated shouting match between Hunter and Army Chief of Staff General Raymond Odierno over the service's unwavering support for DCGS-A despite recurrent field complaints about its performance.[189]

By June 2013, only nine army brigades were authorized to use Palantir, while the other 129 still relied on DCGS-A.[190] That same month, the GAO published an independent report on Palantir that noted that marine and SOF units all found the Palantir software "easy to use" and "effective."[191] Army intelligence chief Legere soon countered that DCGS-A was not more expensive than Palantir. This was patently untrue, since the army was spending approximately $550 million a year on maintaining and upgrading DCGS-A.[192] Her own office estimated that the ineffective and unpopular program of record had cost more than $2.3 billion by that time, whereas it would have cost only $2.5 million to provide Palantir software to the entire 82nd Airborne Division.[193] More leaked army documents during this period confirmed that virtually none of DCGS-A's myriad problems had been fixed over the previous two years. Its habitual failings included an inability to locate files, to print, to maintain a working server, and to reliably browse data.[194] Congress responded to these continued shortcomings by slashing DCGS-A funding by 60 percent in fiscal year 2014.[195]

By the end of 2014, the number of US personnel killed in Afghanistan had dropped to a fraction of its previous levels.[196] Yet the saga of DCGS-A and Palantir was far from over. Congressional pressure on the army escalated after a December 2014 report from the GAO concluded that DCGS-A was "not operationally effective or survivable" even after the standard 80 hours of training.[197] In 2015, the army announced that it would no longer seek to improve the current version of DCGS-A, known as Increment 1, and would instead seek contracts for the next iteration, Increment 2. The army planned to award an open-ended, cost-plus contract to the selected bidder, which would then bill the army for the time its employees spent working on the project. Palantir urged the army several times to instead consider whether it and other commercial firms had existing capabilities that could be adapted through a fixed-cost contract and delivered to US troops far more quickly. The army did not respond to any of Palantir's outreach efforts.[198]

On December 28, 2015, the army officially released the request for proposals for Increment 2.[199] It offered a standard open-ended, cost-plus contract, which required bidders to demonstrate that their accounting systems could accurately record the hours its employees worked on the project. This requirement rendered Palantir ineligible for the competition, since it charged a fixed price for its software. Palantir lodged a protest with the GAO a few weeks later, claiming that the army had written the requirements to exclude Palantir and that it was thus failing to follow a law requiring agencies to structure proposals so that commercial firms with existing capabilities had an equal chance to compete.[200] The GAO denied the protest in May 2016.[201]

Palantir was not done fighting. On June 20, the company filed a lawsuit against the army in federal claims court, claiming that the army "issued a solicitation that makes it impossible for Palantir to compete for the new DCGS contract."[202] It included an extraordinarily detailed chronology of the company's multiyear battle to provide its software to US intelligence analysts in Afghanistan, and copies of hundreds of emails from deployed troops praising Palantir's software and criticizing DCGS-A. It also accused the army of having "an attitude that effectively tells units in the field, 'Don't let your war get in the way of our program.'"[203] The army's defense rested on its continuing assertion that Palantir did not meet the specific requirements of the request for proposal.[204] The court ruled in favor of Palantir in October 2016, and required the army to restart the bidding process so that Palantir and others could be considered.

On March 8, 2018, the army selected Palantir and long-standing DCGS-A contractor Raytheon to compete for the next phase of the DCGS-A upgrade.[205] Both companies provided systems to the army, which tested and compared them. The army left open the possibility that it might issue long-term contracts to both companies, but in March 2019, it made its final decision. Palantir was awarded the entire contract, which could be worth more than $800 million over the subsequent 10-year period.[206] After a decade of bitter bureaucratic wrangling, and staunch army opposition that almost certainly cost soldiers' lives, Palantir finally replaced Raytheon as the DCGS-A prime contractor and its software effectively became the army's intelligence program of record.

Conclusion: Failed Institutional Adaptation in Afghanistan

The battle between the army and Palantir is an extraordinary example of failed institutional adaptation. Despite almost a decade of complaints from army commanders and troops in combat, the service simply refused to fix or even acknowledge the deep flaws in its cumbersome, ineffective, and exorbitantly expensive official intelligence software. Army leaders in the Pentagon stubbornly failed to take seriously the intelligence challenges that their commanders were facing in Afghanistan—even after an urgent request from the dissatisfied senior intelligence officer in the country. Pleas for Palantir software to address critical intelligence requirements in combat were ignored, regardless of whether they came from two-star generals commanding army divisions or junior intelligence analysts. Instead, the army's bureaucratic imperative to protect its costly intelligence acquisition program repeatedly trumped urgent battlefield requirements, even when a proven, effective, and far less expensive solution was readily available. Army leaders simply could not tolerate another humiliating major acquisition failure with its resultant multibillion-dollar losses. It had to make its $2.3 billion DCGS-A system a success. As a result, it forced its troops to use a balky, ineffective system in the midst of combat instead of an easier, cheaper, and far more effective off-the-shelf solution.

The DCGS-A example is a sobering reminder of the immense power that bureaucratic imperatives can exert to override critical combat requirements and foil perceived threats to large defense programs, even at the height of a war. Senior civilian and uniformed leaders have an obligation to their troops to seek out and identify dangerous combat shortcomings in technology that cannot be fixed in the field. In this case, over many years, the US Army actively worked *against* its deployed troops and commanders, preventing them from using a readily available technology that would have almost unquestionably saved lives. It stands as a stark and unconscionable wartime failure of the army's institutional leadership.

Conclusion: Successful Tactical Adaptation and Failed Institutional Adaptation in Technology

Our examples of the US military's technological adaptation in Iraq and Afghanistan share striking similarities. At the tactical level, soldiers and junior leaders quickly adapted to the demands of unexpected battlefield challenges. In Iraq, they fashioned add-on armor from spare materials to protect their soft-skinned vehicles from the growing threat of lethal roadside IEDs. Their creativity was impressive, but since the problem stemmed from problems with the fundamental designs of their vehicles, the scope of the challenge ultimately required an institutional solution. In Afghanistan, AH-64 Apache pilots immediately adapted their flying and their tactics during their first major US battle in the Hindu Kush mountains. Their adaptations in the heat of combat transformed an aircraft originally designed to kill tanks at long range into a weapon that could effectively provide immediate close air support to ground troops in mountainous close combat. This remarkable tactical agility and battlefield improvisation changed how Apaches were used throughout the rest of the war, and helped cement close air support as one of the Apache's new and enduring primary missions.

Institutional adaptations in both wars, by contrast, were costly and disastrous failures. In Iraq, senior army and marine leaders in the Pentagon spent years stalling urgent requests from their field commanders for life-saving MRAPs, lest they divert too much money from their priorities for a future light vehicle. These leaders believed that the lumbering MRAPs would not be needed after the war, so sought to buy as few of the expensive single-purpose vehicles as possible. Very few MRAPs would ever have been manufactured if Secretary Gates had not overruled all of his advisers and demanded that large numbers be produced and sent to Iraq as quickly as possible. In Afghanistan, army leaders in the Pentagon blocked repeated requests for more effective and readily available commercial intelligence software to track insurgents and predict attacks. Instead, they forced army units in combat to rely upon the chronically dysfunctional and inadequate army intelligence program of record, DCGS-A. Years of urgent requests for Palantir's software and untold complaints about DCGS-A went unheeded, with the army's chief of intelligence even rescinding an independent internal assessment because it was too positive about Palantir. Together, the MRAP and DCGS-A examples clearly show that service leaders were more concerned about protecting their high-priority conventional acquisition programs from budget cuts than investing the resources needed to meet the critical battlefield needs of their combat commanders.

Effective adaptation of technology at the tactical level has long relied upon the ingenuity and improvisational spirit of troops in the field. These characteristics have always been enduring traits of American troops in war and were amply demonstrated in Iraq and Afghanistan. The US military must continue to nurture these characteristics, and given the challenges of future warfare discussed in Chapter 9, it must find ways to make them even stronger. At the institutional level, however, the US Army (and, in the case of the MRAPs, the Marine Corps as well) proved to be utterly abysmal at rapid wartime adaptation. In both Iraq and Afghanistan, they repeatedly prioritized their existing acquisition plans and programs over urgent requests from combat troops for the equipment that would save their lives and improve their prospects for victory. These appalling institutional failings must be carefully studied and understood by future generations of military leaders, in order to ensure that they never happen again.

7

Tactical Leadership Adaptation
in Iraq and Afghanistan

LEADERSHIP MAY BE the most important factor affecting wartime adaptability. As discussed in Chapter 4, battles and wars are often won or lost by the people who make crucial decisions amid the immense pressures of combat. Good leaders can offset some of the shortcomings of mistaken doctrine and poor technological adaptability. But even the most adaptable doctrines and technologies may not be able to compensate for bad decisions made by combat leaders who cannot adapt effectively to their rapidly changing environment.

In Chapter 4, we distinguished between leadership at the tactical level, which involves the officers and NCOs who are responsible for leading troops on the battlefield, and at the theater level, which involves the general officers who plan and execute the campaigns that are designed to win wars. Our key leadership examples from Iraq and Afghanistan are longer and more complex than the examples of doctrine and technology we explored in the previous two chapters, and they offer some critically important insights about the conduct and trajectories of these recent wars. The rest of this chapter focuses on tactical leadership in Iraq and Afghanistan, and saves the discussion of theater leadership for Chapter 8.

Here we examine two very different but equally critical examples of tactical leadership. In June 2006, Army Colonel Sean MacFarland and his brigade took control of operations in and around the Iraqi city of Ramadi. This key city was the capital of Anbar province and the center of Sunni resistance to US forces and Iraq's Shia-led government in Baghdad. Ramadi was the most dangerous city in Iraq at the time, and MacFarland developed an entirely new and risky approach to try to turn around the situation. He distributed his troops among the population rather than isolating them on large bases, while also seeking to improve relationships with the previously hostile Sunni sheikhs who had been supporting al-Qaeda in Iraq. His bold and innovative efforts gradually swung the sheikhs over to the US side, which shifted the tide of the war in Anbar province and ultimately throughout the rest of Iraq as well. MacFarland succeeded because he understood the urgent need for change, was willing to challenge his own basic assumptions about the enemy, and developed a radical new approach to a seemingly impossible mission.

In Afghanistan, Army Captain Mark Nutsch led one of the first two US special forces detachments into the country after the September 2001 attacks. With virtually no time to prepare, Nutsch was given one of the most strategically important missions of Operation Enduring Freedom: convincing Northern Alliance general Abdul Rashid Dostum to join with rival warlords in the region to attack the Taliban all across northern Afghanistan. Nutsch was also tasked to support Dostum's ground forces by calling in US air strikes against any Taliban opposition they encountered. He succeeded because he remained steadfast in the face of nearly insurmountable challenges, and showed striking flexibility and adaptability. Nutsch's resourceful adaptations included mounting his men on horseback to accompany Dostum's troops, and later leading a cavalry charge to rally his dispirited allies when they faltered in the face of stubborn Taliban defenders. Nutsch remained dauntless in the face of perpetual and truly unorthodox challenges, and his bold efforts were largely responsible for the rapid defeat of the Taliban in northern Afghanistan.

Tactical Adaptation in Iraq: Colonel Sean MacFarland in Ramadi, 2006–2007
Anbar Province and Ramadi, March 2003–June 2006

The insurgency that arose in Iraq after major combat operations ended in April 2003 originated in Anbar, Iraq's westernmost province. Its majority Sunni population had been steeply favored during Saddam's reign, especially the Ba'athist tribes that had constituted his power base. The US occupation meant that regime officials from the former government and the military suddenly lost the power and influence that they had enjoyed for decades.[1] The Sunnis in Anbar also knew that the free elections promised by the Americans would empower the long-repressed Shia majority, which might seek retribution against them. Yet they also faced real threats from an Islamist extremist group that became known as al-Qaeda in Iraq (AQI), which moved into the province during the chaotic summer of 2003 and began violently seizing control of territory.

Although al-Qaeda and AQI were both Sunni organizations, most Anbaris had no religious affinity for either of them.[2] But AQI nevertheless began to build support through the province, arguing that the only way to avoid what David Kilcullen called "a Shi'a-led genocide" was to band together to fight the

Americans, whom they saw as supporting the Shia-dominated government.[3] At the time, most Sunni tribal leaders saw AQI as the lesser of the two evils they faced. They joined the nascent tactical alliance emerging among AQI, unemployed soldiers, and former regime leaders whose only shared interest was fighting the US military occupation.[4]

By late 2003, an organized insurgency was well underway in Anbar. US military operations in the province during 2004 were dominated by two major battles in Fallujah—an abortive first US effort in April and then a major US and Iraqi combined offensive in November and December.[5] Yet the city of Ramadi was even more strategically important than Fallujah. Ramadi is the provincial capital, and it lies at the intersection of several key transit routes that link Baghdad with the western parts of Iraq, Syria, and Jordan.[6] As Marine Major General Richard Zilmer, who commanded US forces in western Iraq in 2006 and 2007, noted, "at the end of the day, Ramadi is key to Anbar Province."[7] Insurgents who had fled the fighting in Fallujah ended up in other parts of Anbar, including in Ramadi. US forces had been trying to clear insurgents from Ramadi since early 2004, but lacked enough troops to close key access points to the city, much less to secure the population.[8] By the summer of 2005, Ramadi had become the center of the insurgency.[9]

By that point, however, the Sunnis' tactical alliance with AQI no longer seemed nearly as attractive. Most senior Sunni tribal leaders had fled to Jordan or Syria after the fall of Saddam, but the more junior sheikhs who had remained had found that AQI's violent fundamentalist form of rule threatened their bases of power as well as their lives.[10] Furthermore, many Sunnis were reconsidering their opposition to the national political process after their boycott of the January 2005 parliamentary elections froze them out of political power. The junior tribal leaders started to calculate that supporting the national government could provide them with patronage networks that would strengthen their personal power against the exiled sheikhs if they ever returned.[11] Finally, they wanted to ensure that they, not AQI, controlled the province once US military forces departed. As Middle East scholar Marc Lynch notes, "The belief that the United States would leave may have been what made it an attractive partner, by reducing the fear that it would be a permanent occupation force."[12]

Between 2004 and 2006, Sunni tribal leaders in Anbar tried at least four times to align themselves with US forces.[13] They asked for assistance in securing the weapons, ammunition, and vehicles necessary to defend their local areas from AQI.[14] These met with some limited success, but most US civilian and military leaders were wary. The Americans instead prioritized the US strategic goal of empowering and legitimizing the provincial and national governments.

Arming and supporting the tribes, even against AQI, could undermine that crucial political objective.[15] Even more importantly, however, AQI fought back hard against anyone who dared to challenge its authority. AQI reliably targeted anyone who tried to work with the Americans and their supporters for violent reprisal.[16] The last of the tribal attempts to align with US forces began in Ramadi in November 2005, but within two months, AQI had killed more than half of the those involved in the effort.[17]

Starting in September 2005, US Marines operating in western Anbar started adopting elements of a classical counterinsurgency strategy. Lieutenant Colonel Dale Alford and his battalion pioneered this new approach in the city of Al Qaim and elsewhere along the Syrian border, which proved fairly successful.[18] Yet these accomplishments in the sparsely populated west did not stop the violence from continuing to escalate throughout the rest of the province. By the summer of 2006, the situation in Ramadi was dire. US military intelligence experts rated it as the most dangerous city in Iraq at the time, far more so than Baghdad.[19] AQI controlled most of the city and its outskirts, and on a per capita basis, Ramadi averaged over three times more attacks than any other part of the country. The Iraqi police in Ramadi only had 420 officers (despite an authorized strength of almost 3,400), but only around 100 showed up for duty on any given day—in a city of 400,000 people.[20] This skeletal force largely remained in its police stations because it was too dangerous to patrol.[21] The situation was so bad that the senior marine intelligence officer in western Iraq wrote a secret memo in August that starkly concluded: "*The social and political situation has deteriorated to a point that [US and Iraqi forces] are no longer capable of militarily defeating the insurgency in al Anbar.*"[22]

Learning the Lessons of Tal Afar

In 2006, Army Colonel Sean MacFarland commanded the First Brigade of the First Armored Division (often called 1/1, or the "Ready First"). His brigade took responsibility for US military operations in and around Ramadi in June.[23] But Ramadi was not the brigade's first stop during this particular deployment. It had arrived in Iraq in February 2006, replacing Army Colonel H. R. McMaster's Third Armored Cavalry Regiment (ACR) in the city of Tal Afar, the site of a remarkable turnaround in 2005.[24] The seeds of success in Ramadi were planted by McMaster in the northern Anbar town of Tal Afar around the same time that Alford's marines were independently trying a similar approach in the west along the Syrian border. In Tal Afar, MacFarland personally observed how an unorthodox approach could yield unexpected progress, and that approach became the blueprint for his efforts in Ramadi.

McMaster spent the first year of the Iraq War as an adviser at US Central Command (CENTCOM), where he watched the insurgency develop. He believed—correctly, as it turned out—that the insurgency consisted of many different decentralized groups, and argued for a more creative and nuanced approach than the kill-or-capture strategy that dominated US operations at the time (as discussed in the next chapter). In November 2004, he took command of the Third ACR in the United States and completely overhauled the training for its upcoming deployment to Iraq in order to focus on providing security for the population. (This was two years before the army published the new counterinsurgency doctrine discussed in Chapter 5.) By the time that McMaster and his unit arrived in Tal Afar several months later, the city was largely controlled by AQI and its allies and had become a key transit point for foreign fighters entering Iraq from Syria.[25]

McMaster quickly implemented his new approach, which author Thomas Ricks described as "a model of a counterinsurgency campaign, the first large-scale one conducted in the war."[26] After clearing out safe havens outside the city and blocking lines of retreat, McMaster began the attack on the city. But instead of running patrols from his large fortified base camp located beyond the city limits, McMaster's forces established 29 smaller combat outposts, or COPs, inside Tal Afar itself. These small units fought hard as they gradually cleared the city block by block, remaining at their COPs among the population despite the continual danger (and, ultimately, the loss of 21 soldiers).[27] At the same time, McMaster put one of his squadron commanders, Lieutenant Colonel Chris Hickey, in charge of reaching out to the local tribes and convincing them that US forces would stay and secure their neighborhoods. As Hickey recalled, "We tried to switch the conversation from Sunni versus Shia, which was what the terrorists were trying to make the argument."[28] By October 2005, this new experimental strategy had paid off, and the level of violence had dropped considerably.[29]

When MacFarland and the Ready First replaced McMaster and the Third ACR in Tal Afar in February 2006, they saw firsthand how much progress could be made through a counterinsurgency strategy focused on population security and tribal engagement. They built on that success by continuing to operate out of the COPs in Tal Afar. MacFarland's troops sustained operations against the remaining insurgents, while simultaneously maintaining and extending relations with local tribes and the community. MacFarland later recalled that the experience of implementing and sustaining a successful counterinsurgency approach in Tal Afar fundamentally shaped the way that he and his brigade approached their later operations in Ramadi.[30]

Securing Ramadi

In April 2006, Zilmer's command in western Iraq put together a plan to secure Ramadi. They asked the overall commander in Iraq, General George Casey, to create a temporary surge of troops in Ramadi by sending a replacement unit early, but that request was denied.[31] Casey's staff decided to redeploy MacFarland's brigade, even though it had fewer troops and more heavy equipment than had been requested.[32] Zilmer told MacFarland that his job was to fix Ramadi but not to destroy it, avoiding the devastation that had been wreaked on Fallujah.[33] MacFarland was somewhat surprised that he didn't receive more detailed guidance, since Ramadi was such a critical city, but Zilmer was known for giving his subordinates a great deal of flexibility.[34] Yet the challenge that he faced was tremendous. As MacFarland later recalled, their expectations were guarded: "When we arrived in Ramadi in June 2006, few of us thought our campaign would change the entire complexion of the war and push Al-Qaeda to the brink of defeat in Iraq."[35]

MacFarland's first task was to isolate the city, cutting off potential escape routes and securing principal lines of communication.[36] But on June 8, Abu Musab al-Zarqawi, the leader of AQI, was killed by a coalition air strike near Baqubah. Four days later, Zilmer told MacFarland to accelerate efforts to retake the center of the city, in order to take advantage of any confusion or infighting within AQI.[37] So, according to MacFarland, "We decided to employ a tactic we had borrowed from the 3d Armored Cavalry Regiment and used successfully in Tal Afar: the combat outpost."[38] Yet COPs remained a very unorthodox tactic in Iraq, despite their earlier successful use by McMaster. As discussed in Chapter 8, the campaign plan for Iraq set by Casey deliberately pushed commanders in the opposite direction: to withdraw US forces to large bases outside cities away from the population and put Iraqi forces in the lead. MacFarland and his unit forged ahead anyway and started to gradually build COPs throughout the city.[39] They adhered to one central rule: no COP would be abandoned once it had been established. MacFarland believed that tenet would help convince the sheikhs to trust US forces. These neighborhood outposts also made it much harder for to the insurgents to predict US troop movements, and improved intelligence, reconnaissance, and surveillance.[40] The COPs included Iraqi army units wherever possible, and occasionally Iraqi police, which helped build their professionalism and operational effectiveness.[41]

The insurgents started attacking the new COPs while they were being built, but MacFarland and his troops anticipated this from their experience in Tal Afar. US forces usually prevailed, but the increasing attacks and aggressive American patrolling led to major battles throughout the city in July and August.[42] MacFarland's brigade took significant casualties, losing nine soldiers during the first week of August alone. MacFarland and some of his officers grew concerned that either Zilmer or Casey would decide that these casualty levels were unacceptable and order them to withdraw to large protected bases once again.[43] But that order never came, and they continued to make progress. By early fall, insurgent attacks in Ramadi had declined in both numbers and intensity.[44]

Yet MacFarland and his commanders knew that lasting success required effective engagement with local tribal leaders. They had been relentlessly reaching out to the tribes from their very first days in the city. These efforts, however, encountered far more pushback from their higher headquarters than the COPs had. Zilmer's command was trying to work with the Sunni sheikhs in exile, who claimed that they still had influence in Ramadi and elsewhere in Anbar. But MacFarland argued that it would be far better to work with the more junior sheikhs still in Ramadi by offering their followers jobs with Iraqi police units in and around the city. This would provide them income and allow them to protect their own neighborhoods as part of an official government force.[45] Zilmer's headquarters objected, repeating the concern that this would empower them (and potentially encourage their often-illicit revenue streams) at the expense of the national and provincial governments. But there was another, more visceral, reason to oppose MacFarland's approach: it meant working with some of the same Iraqis who had been killing US soldiers and marines for almost three years.[46] MacFarland nevertheless continued pressing for his approach, pledging that none of his officers would engage the insurgents directly but would encourage the tribes to serve as interlocutors instead.[47] Perhaps because so many other approaches had failed in Ramadi, MacFarland's arguments eventually prevailed.[48]

By this point, the tribes in Ramadi and elsewhere in Anbar were strongly opposed to AQI. But the violent AQI reprisals in the wake of their earlier failed attempts to work with US forces made them extraordinarily cautious about doing so again.[49] MacFarland's brigade worked on building relations with tribal leaders both from the bottom up and from the top down. Battalion and company commanders built relationships and forged partnerships with lower-level sheikhs, even paying them through contracts to work on local projects of mutual interest.[50] MacFarland put Captain Travis Patriquin, an Arabic-speaking officer in his brigade, in charge of tribal outreach. Patriquin coordinated extensive meetings for MacFarland and other brigade leaders with the more senior local sheikhs.[51]

MacFarland wanted the tribes to provide recruits for the police, which he saw as the "center of gravity" in the city.[52] But that was a hard sell, since the insurgents

FIGURE 7.1 Colonel Sean MacFarland, Sheikh Sattar al-Rishawi, and Captain Travis Patriquin in Ramadi, Iraq, October 2006
Source: Sean MacFarland.

deliberately targeted the police. MacFarland recalled that his deputy commander came up with the solution, which was to tell the tribal leaders:

> We're looking at building a new police station. You help us put it where you think it'll do best, then would you be willing to man it? And they said well, of course, because then you and the government of Iraq would be training and equipping the people that are securing our families.[53]

The critical element, however, was ensuring the safety of the first tribes that agreed to provide the new recruits. The brigade improved security by moving the police recruiting center closer to those areas, and kept recruiting information secret from everyone other than the sheikhs who had promised to provide the recruits.[54] They also did everything they could to protect those sheikhs, including stationing checkpoints with tanks and other armored vehicles along the roads near their houses, and increasing drone flights overhead to provide as much intelligence as possible.[55] This arrangement started to build trust and confidence during the summer.

The Anbar Awakening

Two developments on August 21 caused the tribes to fully align with US forces for the long term. First, AQI attacked the first new police station built for tribal recruits, ferociously assaulting the compound with as many as 200 fighters. Instead of fleeing, as they might have in the past, the surviving police called MacFarland's headquarters. The Americans immediately responded with US reinforcements, air cover, and medical support. The survivors declined an offer to take refuge at a nearby US base, and instead resumed their patrolling later the same day.[56] Hours later, AQI assassinated a local tribal leader who was providing police recruits to the US forces. They also mutilated his body and hid it in a field so his family could not properly bury him according to Islamic tradition.[57] As MacFarland's operations officer later recalled, "As soon as they did that, the local tribesmen absolutely turned on Al Qaeda."[58]

Sheikh Sattar al-Rishawi soon emerged as the leader of the local Sunni opposition to AQI. Sattar was a charismatic yet fairly minor sheikh,[59] but organizing the local opposition also gave him the opportunity to raise his stature and potentially become the most influential sheikh in Ramadi. Given the high risks involved, most of the other sheikhs were happy to let him take the lead.[60]

On September 9, Sattar held a meeting of more than 50 tribal leaders. MacFarland expected to attend the meeting as an observer, but Sattar made him sit by his side as he officially announced the formation of the Anbar Salvation Council, also known as the Awakening Council.[61] The leaders in the room pledged to join the fight against AQI in exchange for US support, promising that they would consider an attack on US forces to be an attack against their own tribes.[62] Several days later, Sattar announced that 25 of the 31 tribes in Anbar had banded together to fight AQI with a total of 30,000 armed men.[63] MacFarland's brigade immediately provided support for Sattar and the council, and in return, the tribes started providing recruits for the Iraqi police. MacFarland was initially skeptical about the depth of the council's commitment, but the surging flow of police recruits provided the proof. In May 2006, there had been only 40 recruits; between September and November, there were 1,500.[64] The vast majority of those were men who were trusted by their tribal leaders, though US forces also screened recruits with biometric tools to try to prevent insurgent infiltration.[65]

The tribes in the council also launched their own efforts to root out the AQI members who lived among them. An informal division of labor emerged: MacFarland's forces and their Iraqi army partners went after AQI leaders and their fighters, while the growing Iraqi police and tribal forces raided their safe houses and weapons caches. AQI fought back hard as its losses began to mount, continuing to attack targets in the city as well as the new police stations and COPs. Yet even though the tribes were taking casualties in these attacks, support for the Awakening Council continued to grow. By late October, almost all of the tribes to the north and west of Ramadi had joined the movement, and the tribes in the dangerous area to the east of the city were quietly inquiring about doing so too.[66]

The turning point in the battle against AQI occurred on November 25 in what became known as the Battle of Sufia. AQI discovered that the one of the influential sheikhs in this rural area to the east of Ramadi had built checkpoints to keep insurgents out of his area.[67] They responded by sending 30 to 50 gunmen into the area to attack the defenses, loot homes, and murder members of the tribe.[68] MacFarland was on leave when the call for help came in, so it went to Lieutenant Colonel Chuck Ferry, one of the brigade's battalion commanders. Ferry had been planning a major operation scheduled for the following day, so he had to decide whether to delay that operation in order to help this sheikh whom the brigade did not know. He chose to postpone the planned operation and send his unit into the area.[69] Marine jets soon roared overhead and explosions began to pummel the AQI fighters.[70] As they started to flee, they were slaughtered by US tanks and armed Predator drones flying overhead.[71] AQI ended up taking more casualties than the tribe, and the lopsided battle reversed any remaining momentum that the insurgent group had in the province.[72] When MacFarland returned a few days later, he told Ferry, "You did exactly what I would have told you to do."[73]

Enemy attacks dropped dramatically in the following weeks. Tribes throughout the broader area of operations joined the Awakening movement, and the brigade continued to build new COPs and police stations while hunting down the shrinking number of AQI members.[74] Between July 2006 and January 2007, the number of monthly direct fire attacks in Ramadi shrank by two-thirds, and the number of attacks from improvised explosive devices (IEDs) was cut in half.[75] Almost 4,000 recruits joined the police in the same time frame, and by early 2007, 1,500 police were continuously on duty.[76] The increasing police capacity kept the virtuous cycle going: communities could provide for their own security after US and Iraqi forces cleared out insurgents, which enabled those forces to move to new areas and repeat the model.[77] Around the turn of the year, the Iraqi Ministry of the Interior authorized the creation of emergency response units, which essentially enabled tribal militias to police their own areas without formally joining the Iraqi police, further adding to the virtuous cycle.[78] In January, the brigade's deputy commander observed, "If you compare this place to what it was seven or eight months ago, there is not a place in this city that al-Qaeda controls."[79]

MacFarland and his brigade left Iraq in February 2007,[80] having lost more than 80 soldiers, sailors, and marines during their deployment.[81] But MacFarland's efforts in Ramadi provided one of the key turning points in the entire Iraq War. The Awakening movement continued to expand throughout Anbar, as more tribes saw the benefits that the movement was providing and wanted to join as well.[82] That fall, the US ambassador to the United Nations, Zalmay Khalilzad, reported that the situation in Anbar was "stable and quiet, permitting reconstruction to take place."[83] AQI did claim one more high-profile victim, however. Sheikh Sattar, who was largely responsible for the Awakening, was killed by an IED near his farm on September 13, only a few days after publicly appearing with President George W. Bush.[84] But in stark contrast to the aftermath of similar high-profile assassinations between 2004 and 2006, the Awakening proved to be both resilient and self-sustaining. Sattar's brother quickly took his place, and the movement continued to spread throughout Anbar and then on to Sunni tribes across Iraq. Outside of Anbar, the program to partner with local tribes became known as the Concerned Local Citizens and then the Sons of Iraq, and their numbers exploded—from a few thousand in early 2007 to 72,000 by December, and then to 103,000 by July 2008 (all in addition to the estimated 25,000 volunteers in the Anbar Awakening movement by that time).[85] Attacks on US forces dropped precipitously as a result, which enabled them to focus on the Shiite militias that were conducting attacks elsewhere in the country. The Awakening was so central to the reduction of violence throughout the country that one rigorous study concluded that the stabilization of Iraq by the summer of 2008 would not have happened without it.[86]

MacFarland and his brigade did not cause the Awakening; the tribes made their own calculations about what best served their interests.[87] But by pursuing the bold counterinsurgency approach that he had witnessed in Tal Afar—which directly opposed the overall US military approach at the time—MacFarland set the conditions that enabled his command to support and promote the nascent Awakening movement. His actions helped transform the war by starting a nationwide movement that played an essential role in reducing violence throughout the country. When the revised counterinsurgency manual was released in December 2006, MacFarland was asked whether he had read it. He later recalled that he said no, but that "they told me I didn't really need to read [it] since I had already done much of what the document said I was supposed to do."[88]

Tactical Adaptation in Afghanistan: Captain Mark Nutsch and the Horse Soldiers, 2001

The operations of US Army Special Forces Operational Detachment Alpha (ODA) 595 in late 2001 provides another exceptional example of adaptive tactical leadership.[89] Led by Captain Mark Nutsch, this 12-man unit was one of the first two US military elements to land in northern Afghanistan following the devastating terrorist attacks on the United States in September 2001. Nutsch and his senior warrant officer, Bob Pennington, led the team on a remarkable two-month mission that stands out as one of the most creative examples of adaptive tactical leadership during the recent wars.

Responding to the 9/11 Attacks

ODA 595 was part of the US Army's Fifth Special Forces Group, which focuses on operations in the greater Middle East region.[90] At the time, each group consisted of approximately 54 special forces ODAs (or A-teams) along with supporting headquarters and staff, totaling almost 1,400 troops in all.[91] Special forces (SF) soldiers, also known as Green Berets, are typically older and more experienced than other army soldiers. They must successfully complete a very intense selection and training program that lasts more than a year. SF soldiers are trained in at least two different warfare specialties and often are taught to speak one or more of the languages in their assigned region. A-teams are tight-knit, 12-soldier units designed to operate independently in foreign countries to advise and assist the militaries of other nations, especially those in the developing world. They are capable of conducting direct action strikes and special reconnaissance missions, but their forte is strengthening the military capabilities of US partners. The official special forces motto is *De oppresso liber*, "to free the oppressed." But their unofficial motto is "By, with, and through," reflecting their commitment to working with indigenous forces to accomplish missions that help advance US interests around the world.

On September 11, 2001, Nutsch was serving in a battalion staff job in the Fifth Special Forces Group at Fort Campbell, Kentucky. He had recently completed a two-year tour as the commander of ODA 595. On the morning of the attack, he was home with his two young sons and pregnant wife.[92] Pennington was still assigned to ODA 595 as the second in command, and had been with Nutsch and the team on its recent training deployment to Uzbekistan.[93] Soon after the attacks, Nutsch confronted his battalion commander, Lieutenant Colonel Max Bowers, and asked to be put back in command of ODA 595. Bowers initially refused, but recognizing that he would need to send his best teams into the war that was about to unfold, he eventually relented. Nutsch was placed back in command of ODA 595, and the team was soon readying for deployment.[94]

ODA 595 was an experienced team whose 12 members averaged 32 years old, with about eight years of experience. They had worked together for two years, and several had previous combat experience from operations in the first Gulf War, Somalia, or Kosovo. All but one member was married, and 10 of the 12 men had at least two children.[95] In the words of Pennington, "They were very mature, very family-oriented. And to me, that was a plus when you got saddled with a mission that was going to be as complex as this one."[96]

US intelligence agencies quickly attributed the devastating 9/11 attacks to al-Qaeda and confirmed that the terrorist group was based in Afghanistan under the protection of the Taliban-controlled government. But since the US military had no contingency plans for a conflict in Afghanistan, planning for the operation had to begin from scratch.[97] As the rubble of the Twin Towers and the Pentagon continued to smolder, the US military and the Central Intelligence Agency (CIA) started developing plans to deploy forces as quickly as possible.[98] The amount of accurate information about the situation in Afghanistan was remarkably limited. CIA officials and military planners estimated that it would take between six months and a year to remove the Taliban from power.[99] The war in Afghanistan therefore posed a daunting challenge from the very beginning and required extraordinary levels of military adaptation.

The commanders of Fifth Group selected ODA 595 and a handful of other teams as the first tranche to send into this unknown.[100] Members of Fifth Group had already started preparing for what might be coming. They raided local outdoor equipment stores for cold-weather clothing and camping equipment, and ordered more gear online for immediate delivery. They bought vast quantities of small-size commercial batteries to help power their gear, including flashlights, cameras, and night vision goggles. Since they lacked any real intelligence, they managed to get their hands on some Soviet-era maps of Afghanistan and started reading as much about Afghanistan as they could.[101] They packed enough materials and supplies to support a yearlong deployment, though most support would have to be airdropped in as the conflict unfolded.[102] On October 5, ODA 595 departed for a bare-bones staging base in Karshi-Khanabad (K2), Uzbekistan.[103]

During their few days at K2, Nutsch and his team began to assess the already-changing scope of their forthcoming mission. The team had been initially assigned to support rescue operations for downed US airmen,[104] but was soon given a much more formidable mission. Nutsch and ODA 595 would fly alone into northern Afghanistan to link up with Northern Alliance forces under the infamous ethnic Uzbek warlord General Abdul Rashid Dostum. Dostum was the most powerful leader of the Uzbek tribe in northern Afghanistan, and most of his men were also hardy Uzbek tribal fighters.[105] Nutsch's daunting task was to convince Dostum to join with rival warlords in the region to attack the Taliban all across northern Afghanistan, and to support Dostum's ground forces by calling in air strikes against any Taliban opposition they encountered.[106]

Even for an experienced special forces team, the challenges of this mission were staggering. ODA 595 would be the only US military unit on the ground in the area; there would be no backup, reinforcements, or timely rescue in the event of trouble. No one knew how Dostum and his fighters would receive the arriving Americans. If things went badly, they could quickly become hostages, valuable for trades to the powerful Taliban forces nearby.[107] They also knew that the Afghans had turned on and slaughtered their long-standing Soviet advisers without compunction in a bloody apocalypse during the 1980s. The winter weather and mountainous terrain were formidable, and the team would initially have only their legs for transportation across the frozen mountains, carrying only a handful of supplies. Finally, they knew the local Taliban had a sizable supply of tanks, armored personnel carriers, rocket launchers, and heavy machine guns, along with thousands of fighters with ample ammunition. With little guidance and the barest of preparation at K2, ODA 595 got ready to go, knowing they would have to improvise mightily upon arrival.

The First Days of Battle

On the night of October 19, 2001—slightly more than five weeks after the 9/11 attacks and less than two weeks after the opening US air strikes of Operation Enduring Freedom[108]—ODA 595 flew into northern Afghanistan aboard a single special operations MH-47 Chinook helicopter in the middle of a winter storm with extremely limited visibility.[109] It was one of the first two US military units to deploy into the country.[110] Nutsch and his team flew without their usual heavy body armor to reduce the weight in the already overburdened aircraft, and also to send a message that the team would take the same risks as their unprotected Afghan partners.[111] The team landed in a blinding snowstorm in the mountains at 2:00 in the morning, and linked up with three CIA officers and a few soldiers from the Northern Alliance to await Dostum's arrival after dawn.[112] The weather was bitter cold, and the team stayed on edge all night. Nutsch did not know if he could trust Dostum and was concerned about the vast unknowns surrounding their first meeting. But in the morning, Nutsch found Dostum welcoming and enthusiastic about working with the newly arrived Americans to battle the Taliban.[113]

Dostum and his approximately 2,000 troops were largely mounted on horseback, and far more lightly armed and equipped than their well-supplied Taliban opponents. While the Taliban were warmly ensconced in the main cities and villages dotting the north, Dostum's troops lived outdoors in the unprotected wilds of the mountainous countryside. He and his men were fully exposed to the harsh effects of the icy Afghan winter.[114] Dostum's troops were a motley collection of young and old men, in widely varying states of health and fitness. Some rode and fought on despite missing limbs. Their only common traits seemed to be an unwavering commitment to Dostum and their hatred of the Taliban. The Americans were taken aback to find that their allies' primary military capabilities were assault rifles and their stolid Afghan horses—an enormous disadvantage when fighting against Taliban tanks and heavy artillery.[115]

Nutsch began to think through how he could utilize the advanced US air power at his disposal to support the quasi-medieval tactics of his newfound partners. But Dostum had provided the Americans with only six horses, forcing Nutsch to divide his already small team into two parts.[116] Alpha cell, led by Nutsch, would ride north into battle with Dostum, while Bravo cell managed the logistics pipeline for Alpha cell and then moved south to fight the Taliban in a nearby valley.[117] At Dostum's request, Nutsch's team asked for an airdrop of blankets to help shield Dostum's freezing troops from the bitter cold.[118]

Nutsch and Alpha cell soon faced an unanticipated challenge. Rather than ride along with their Afghan allies in pickup trucks as they had imagined back at K2, they found that the only way to accompany Dostum's forces in the mountains was to ride on horseback.[119] Nutsch was the only American who had ever ridden a horse before, so he conducted an impromptu horseback-riding class for his surprised men.[120] As Stephen Biddle vividly describes:

> Their knees in their chests, balancing heavy rucksacks on their backs, they were instructed by their commander to keep their downhill foot out of the stirrups and to lean uphill so if the pony lost its balance they would fall onto the trail as the pony went into the gorge. On particularly rocky stretches the team commander ordered his men to travel with weapons out and a round chambered to shoot immediately any pony that bolted before it could drag its rider to his death over the rocks.[121]

Dostum's troops also gave the Americans their smaller horses, ostensibly as a rite of passage, and the beasts nearly buckled under the weight of the heavier US soldiers and their equipment.[122] Even worse, the Afghans rode on wooden saddles that were agonizing for the Americans to straddle for hours at a time.[123] Nutsch sent an urgent request back to the United States requesting an immediate airdrop of two specific types of saddles (though they did not arrive until after his team reached Mazar-e Sharif).[124] Pentagon officials had no idea that US special forces were riding on horseback until they received Nutsch's first field report.[125]

Nutsch had to determine the best way to support Dostum and his forces with his tiny 12-man force. He knew that he and his men had the communications gear that would enable them to talk to US bombers to coordinate air strikes on any Taliban locations they identified. With their GPS-guided navigation devices,

Nutsch and his team could also provide relatively precise and current enemy information to the pilots. That meant that US bombers and their dizzying array of munitions could now directly target the heavily armed Taliban forces as they battled the outgunned Northern Alliance militias.[126]

Nutsch and Alpha cell started riding north toward the village of Chapchal, alongside Dostum and his men. During their first battle together, Dostum pointed out a Taliban bunker to be destroyed, and even unexpectedly contacted the Taliban by radio to confirm their location for the wary Americans.[127] But Dostum initially kept Alpha cell far away from enemy locations, because, as he told Nutsch, "Five hundred of my men can die, but not one American can even be injured or you will leave."[128] The Americans called in the first air strikes against the Taliban while they were approximately eight kilometers away from the targets, which, given the very limited visibility, was too far away to correctly estimate the grid coordinates of the Taliban location.[129] After the GPS-guided bombs dropped by a US B-52 bomber missed their targets, a frustrated Nutsch convinced Dostum to allow his team to approach within two miles of the Taliban positions so they could provide better target location estimates.[130] The air strikes the next day yielded far better results, destroying several tanks and armored carriers, as well as a cluster of Taliban pickup trucks.[131]

Soon after the bombing, Dostum organized his ragged horsemen into formation for a cavalry charge against the remaining Taliban. The mounted fighters leapfrogged in waves from ridgeline to ridgeline, with one group of horsemen pinning down the Taliban while another group got closer. Dostum's men seemed poised for victory when they drew close enough for a final charge on the Taliban lines, but a pair of Taliban armored vehicles suddenly emerged and began to mow down the exposed troops. Nutsch tried to call strikes from a nearby B-52 bomber, but it did not have enough fuel to come to the rescue. The attack faltered, and Dostum was forced to retreat.[132]

That evening, Nutsch realized that he had to find a better way for US air power to effectively support the unconventional Northern Alliance cavalry tactics on the ground. That meant convincing Dostum to synchronize his often-impulsive battlefield moves with available US air power.[133] Dostum always put himself at grave risk during his battles by riding well forward with his men. Nutsch told Dostum that he wanted to coordinate their operations so that the bombs would hit just before the cavalry charged, to try to disrupt the Taliban's defenses at the final moment before the attack. Despite Dostum's fears about American casualties, Nutsch managed to convince him that they needed to ride together into the next battle so they could coordinate the tricky timing.[134] They did so the next morning, and this time their better-coordinated attack overran the Taliban line.[135]

FIGURE 7.2 US Army Special Forces Soldiers on Horseback in Afghanistan, 2001
Source: https://www.army.mil/article/181582/first_to_go_green_berets_remember_earliest_mission_in_afghanistan.

Nutsch also decided to break with established doctrine in order to ensure that surprise Taliban counterattacks would not happen again. Although special forces teams normally avoid sending team members out alone on the battlefield, Nutsch sent Sergeant First Class Andy Marchal to work with an Uzbek contingent well to the north of the Taliban line and to direct air strikes against the Taliban's concealed tanks.[136] On the morning of October 24, Marchal set out with 30 of Dostum's men toward a village named Oimitan, to provide targeting information against Taliban tanks that the rest of Nutsch's team could not see. After those targets were bombed, Dostum's men would rush into the Taliban bunkers to kill any survivors.[137]

Nutsch and his men were pioneering an entirely new way of fighting, as they continually adapted to the ever-changing situation on the ground. They learned as they gained experience, and constantly updated and revised their tactics. As one military analyst later noted, "It was an odd combination of medieval-style cavalry charges delivered by the Uzbek descendants of Genghis Khan supported by state-of-the-art twenty-first-century aerial killing technology."[138] In a few short days, Nutsch's team had called in air strikes that destroyed 12 Taliban command posts and over 100 armored and support vehicles, which sent the Taliban reeling. Nutsch observed that the novel combination of horse cavalry and US bombs had "scar[ed] the hell out of the Taliban."[139]

The Fight for Mazar-e Sharif

After Nutsch's first air strikes failed to hit the correct targets, commanders at CENTCOM decided to send two US Air Force special operations combat air controllers to join ODA 595.[140] They arrived in Afghanistan on October 28 and left to join Nutsch's team on October 30.[141] These specialists were trained to designate targets using a special operations laser marker (SOFLAM), so that precision bombs could follow the laser beam to the desired point of impact. Nutsch's team had possessed a SOFLAM, but did not take it to Afghanistan since it was too bulky for a team that needed to travel light. Laser-guided bombs were far more accurate than the GPS-guided bombs that had been used until that point. The laser designators used by the airmen did not require determining the exact grid coordinates of a target. Instead, the airmen could mark where the bombs should fall simply by pointing the laser at the exact spot on their intended target.[142] This feature made them especially deadly, and even more helpful in providing close air support for Dostum's forces.

More ODAs had also arrived to join the ongoing fighting across the country. ODA 534 joined up with Dostum's rival ethnic Tajik warlord, Mohammed Atta, south of the provincial capital of Mazar-e Sharif. Dostum and Atta were bitter rivals for power in northern Afghanistan, but shared a mutual hatred of the mostly ethnic Pashtun Taliban. Despite their mutual animosity, the two warlords and their accompanying ODAs soon began to coordinate their efforts to defeat the remaining Taliban forces around the provincial capital.[143]

The two ODAs and their combat controllers began to coordinate their air strikes. On November 5, the team accompanying Atta called in an air strike near Taliban lines in the desert that involved two of the largest conventional bombs in the US inventory. Both of the 15,000-pound BLU-82 bombs (nicknamed "Daisy Cutters") set off giant mushroom clouds in the air. The bombs were deliberately dropped in the middle of a deserted area so that no one would be killed, but the strike successfully achieved its objective of terrifying the nearby Taliban forces and sending them fleeing in their pickup trucks.[144]

The air controllers with Dostum's forces now regularly employed US air power to smash enemy positions whenever the Northern Alliance troops ran up upon Taliban defenses. B-52 bombers from the Cold War era, now armed with precision weapons, conducted close air support missions in combat for the first time ever.[145] After less than three weeks of fighting together, Dostum's troops and ODA 595 were now so well synchronized that, according to historian Brian Glyn

Williams, the combat air controllers were within danger close range, "almost calling bombs down on their own positions" because "Dostum's fighters knew how to calibrate their ground movements with the incoming bombs."[146] In the following days, the combination of Northern Alliance ground attacks and coordinated US air strikes forced the Taliban to steadily retreat. They dug in along their final defense line in the Tiangi Gap, located on the southern approaches to Mazar-e Sharif.

The Tiangi Gap was a key mountain pass that led out of the Balkh valley onto the open plains surrounding Mazar. The Taliban had heavily defended the pass with thousands of troops and cannons dug into the hills surrounding the valley.[147] Before dawn on November 9, Dostum asked three members of ODA 595 to climb a mountain ridge that was 4,000 feet above the gap in order to direct fires onto the Taliban artillery concealed on the northern side. After the team knocked out about a half dozen targets, the Taliban defenders retaliated with heavy rocket fire. Dostum's troops and continuing US air strikes managed to halt the rocket fire after a two-hour battle, but then the troops started to falter as the Taliban began to regroup. Nutsch, seeing that the attack had stopped, charged ahead alone on horseback into the fight, hoping to rally Dostum's men to follow. The Northern Alliance troops responded, overrunning the Taliban positions and clearing the gap.[148] In a remarkable act of courage, Nutsch had once again adapted to the challenge at hand—this time by leading a cavalry charge on horseback. Secretary of Defense Donald Rumsfeld later shared with the press that one of the participating soldiers described this remarkable accomplishment as "the first cavalry charge of the 21st century."[149]

The following day, Dostum's and Atta's forces took control of the holy city of Mazar-e Sharif and the nearby Qala-i-Janghi fort. While Atta and his men occupied the provincial capital, Dostum reopened the city's famed Blue Mosque, personally stopping to pray in thanksgiving at the shrine.[150] After just three weeks of hard fighting, the Northern Alliance and two supporting US special forces teams of only 12 soldiers each had managed to seize control of the largest city in northern Afghanistan, Mazar-e Sharif, completely defeating the powerful Taliban forces in the important northern part of the country. That defeat, according to historian Williams, "brought down the Taliban house of cards," and the Taliban soon started withdrawing from Afghanistan's capital, Kabul.[151] At the beginning of the war, the commander of Fifth Group had hoped to take over one or two northern towns before winter conditions halted the fighting.[152] Yet, largely due to the remarkable success of Nutsch and his team, Northern Alliance forces took control of Kabul on November 13 and effectively ended the Taliban regime.[153]

Conclusion: Nutsch's Tactical Adaptability

Captain Mark Nutsch proved to be an exceptionally adaptive tactical leader facing a truly extraordinary set of battlefield challenges. He handily met all of the successful adaptation criteria for tactical leaders identified in Chapter 4. Nutsch had precious little time to train or prepare for his first wartime mission, which changed on very short notice. He also had to reassess and adapt to the changing battlefield virtually every day he was in combat. Nutsch's tactical mission also had enormous strategic importance. If he had failed to convince Dostum to work with his special forces team, or if he had not figured out how to provide effective US close air support to Dostum's ground forces, the United States might have had to send tens of thousands (if not hundreds of thousands) of US ground troops into Afghanistan to defeat the Taliban. Yet Nutsch did convince Dostum to work with his team and later even persuaded him to coordinate his efforts with his main regional rival and his loyal militias. Early in the campaign, he quickly adapted when Dostum unexpectedly insisted that only half of his 12-man team could accompany him, and made a risky decision to break from standard doctrine and send one of his soldiers out alone on a mission with a small group of Dostum's forces. He and his team developed an entirely new way of fighting that effectively melded advanced military technology with some of the most ancient warfighting tactics. Dostum later praised Nutsch's team by saying, "Commander Mark and his men fought like lions. One day if America has need of my services again it would be an honor to fight alongside those brave men again."[154]

What ODA 595 managed to accomplish in Afghanistan is a striking example of the power of adaptation in the hands of a bold and talented leader. Nutsch's accomplishments should endure as an inspiration to leaders at the tactical level facing seemingly intractable battlefield quandaries and unexpected setbacks. Few of those leaders will ever accomplish their mission more effectively, nor with a more unique and compelling example of effective adaptation.

Conclusion: Successful Tactical Leadership Adaptation

Colonel Sean MacFarland in Iraq and Captain Mark Nutsch in Afghanistan demonstrated all of the qualities of successful tactical leaders in some of the most stressful, deadly, and uncertain situations found in the recent US wars. They quickly understood the nature of the battles they were fighting and grasped that traditional approaches to their missions would almost certainly fail. They both abandoned standard procedures and doctrine when necessary, even though decisions like building small combat outposts in population centers and fragmenting a special forces team exposed their troops to greater risks. And not only did each rapidly develop creative approaches, but they continually adapted and refined those approaches as battlefield dynamics shifted over time. The tremendous adaptability of both of these tactical leaders led them to successfully accomplish their extraordinarily difficult missions. MacFarland's efforts in 2006 helped reverse the downward trajectory of the war in Iraq, and Nutsch's efforts directly contributed to the 2001 defeat of the Taliban in Afghanistan.

These two powerful examples demonstrate the power of individual leaders, even at the tactical level, to influence the outcomes of wars. MacFarland and Nutsch were exceptional individuals, and by any measure standouts among their peers for their imagination and adaptability. Their ability to correctly assess their environment and then quickly find a winning approach reflects the characteristics of adaptation that the US military seeks to inspire in *all* of its tactical leaders. The US military needs to ensure that it continues to nurture and reward tactical leaders who are as adaptable as MacFarland and Nutsch, and that it seeks to develop those traits in all of its future forces. We recommend some ways to do so in Chapter 11.

8

Theater Leadership Adaptation in Iraq and Afghanistan

UNLIKE TACTICAL LEADERS, theater leaders typically do not fight in battles. Instead, they design and execute campaigns designed to win wars. Most often, they are the most senior military leaders physically located in the area where the war is being fought. These generals are responsible for the operational level of war, and often report to more senior officers operating at the strategic level overseeing the entire region, or located in headquarters (such as the Pentagon) in a national capital.[1] But theater commanders in combat zones also have strategic responsibilities, which may include maintaining unity within an alliance or coalition, and political-military roles such as engaging with senior diplomatic representatives of the United Nations and other states involved in the conflict.

This chapter examines two theater commanders who led the wars in Iraq and Afghanistan at critical junctures. In Iraq, General George Casey served as the commander of Multi-National Force–Iraq between June 2004 and February 2007. He held that position longer than any other American commander, so he had ample time to assess and reassess his chosen strategy. Despite clear evidence that the trajectory of the war was veering sharply downward, he failed to significantly change his initial approach to the war and instead doubled down on his original failing strategy. He was simply unable to adapt to the reality of an ever-worsening conflict. As a result, the United States nearly lost the war during his tenure.

General David McKiernan led the war in Afghanistan for only 11 months, during a period where levels of violence were steadily increasing. McKiernan similarly failed to understand the changing character of the Afghan war as well as the shifting political dynamics among his superiors in Washington. Lacking any experience in unconventional warfare, he was originally assigned to Kabul as the commander of NATO forces, but subsequently was given direct command of all US troops as well. McKiernan failed to understand how badly the war was going, and how urgently his new US chain of command wanted to turn it around. As a result, he was fired by Secretary of Defense Robert Gates, becoming the first US senior commander relieved for unsatisfactory performance since the Korean War.

Casey and McKiernan represent stark cases of failed leadership adaptation at the theater level that had severe consequences for the wars they were fighting. Yet there is at least one exceptional example of successful theater adaptability from the recent wars. General David Petraeus succeeded Casey as the commander in Iraq and rescued the war from the brink of defeat by applying the new counterinsurgency doctrine he had worked so hard to develop. Since we discuss that doctrine extensively in Chapter 5, and we highlight the dramatic turnaround in the trajectory of the Iraq War under Petraeus at the end of the section on Casey, we do not provide a separate case study of his efforts. Yet Petraeus nevertheless remains a critically important example of successful theater leadership adaptability in Iraq.

Theater Adaptation in Iraq: General George Casey

In June 2003, US Army Lieutenant General Ricardo Sanchez became the first commander of postinvasion military operations in Iraq, amid a wave of violence that had been escalating since the fall of Baghdad.[2] Though Secretary of Defense Donald Rumsfeld famously dismissed those involved in the violence as "dead-enders," the roots of the insurgency were already well underway by the summer of 2003.[3] Yet Sanchez never produced a campaign plan that gave his subordinate commanders guidance about how to address the growing violence.[4] Though the strategies of those two-star commanders varied,[5] most took a conventional approach that focused on killing or capturing insurgent leaders by relying upon large amounts of firepower and heavy-handed tactics with the Iraqi population. As Army Major General Charles Swannack, who commanded the 82nd Airborne Division, proudly declared in November 2003: "This is war . . . we're going to use a sledgehammer to crush a walnut."[6] The violence alienated local populations and fueled the growing insurgency over the subsequent months. By spring 2004, the failed US offensive to retake Fallujah from Sunni insurgents, and the Shia uprising in southern Iraq, had turned the US military from simply occupiers into a common enemy for Iraq's Sunnis and Shia alike.[7] The number of attacks throughout Iraq had averaged about 200 per week during the early months of 2004, but exploded to more than 500 per week by the summer.[8]

Casey in Command, Part 1: July 2004–February 2006

On July 1, 2004, General George Casey took command of the newly renamed Multi-National Force–Iraq. Despite the rising bloodshed, Rumsfeld continued to resist characterizing the violence in Iraq as an insurgency.[9] But by that time, both Sanchez and Casey's superior commander at US Central Command (CENTCOM), General John Abizaid,[10] had already used the term "insurgency" in public.[11] Casey also seems to have understood that he was facing an insurgency.[12] On his first day in command, he reportedly asked his staff, "Okay, who's my counterinsurgency expert?"[13] He also immediately started working on a campaign plan to address the insurgency, which he released on August 5. Its objectives included containing the spreading violence as well as a number of tasks built upon key counterinsurgency principles. These included building the Iraqi security forces, enhancing the economy, and simultaneously coercing and cooperating with Sunni communities where the violence was the strongest.[14]

Yet despite these early efforts to build a counterinsurgency strategy, Casey held some deep beliefs that led him to unintentionally undermine its key principles. By far the most important of these was his conviction that the presence of US forces in Iraq was fueling the insurgency, and therefore they needed to be withdrawn as quickly as possible. Abizaid shared this view, partly because of his lifelong study of the Middle East. But the views of both men were also shaped by their experience serving together during peacekeeping operations in Bosnia in the 1990s.[15] They also believed that there were limits to what US military force could accomplish absent political reconciliation, and that the extended presence of foreign troops made local populations too reliant upon their efforts. Casey later told Congress, "I saw this in Bosnia myself as a brigadier general. I remember watching myself going out and trying to solve the problems of Bosnia and as a result my sense was that they became dependent on us and they did less."[16] This directly shaped his approach to Iraq, for as he later wrote, "Given my personal experience in the Balkans, I was concerned that the longer we waited to begin giving Iraqis responsibility for their own security, the more dependent they would become on us—and the longer we would remain in Iraq."[17] As a result, Casey's strategy prioritized withdrawing US forces from Iraq as quickly as possible despite the escalating violence. In the meantime, he sought to rapidly shrink their presence among the population. Yet these priorities undercut all of the other counterinsurgency elements of his campaign plan.[18] Rumsfeld agreed with this withdrawal strategy and pressured Casey to avoid "Americanizing" the war.[19] But Casey's convictions were not driven primarily by Rumsfeld's guidance, and he dogmatically held on to this vision for his entire tenure in command. Not only did Casey consistently emphasize that the US presence was causing the insurgency, but as discussed later, he continued to do so even after Rumsfeld stepped down.

Casey's campaign plan contained three tenets that are particularly important for our purposes.[20] First, US forces would continue to conduct kinetic attacks against insurgents and their sources of support. Operations to kill and capture key insurgent leaders—known in military jargon as high-value targets—were not a part of Casey's campaign plan, because most of these were classified and conducted by Joint Special Operations Command, which reported directly to Abizaid at CENTCOM.[21] But the conventional forces under Casey's command worked closely with their special operations counterparts and conducted attacks against insurgent sanctuary and support areas in order to limit their movement.[22] Critics later derided this approach as a game of whack-a-mole, where US forces would quickly respond to attacks by striking insurgents in one part of the battlefield, only to have them reappear somewhere else soon thereafter.[23]

Second, US forces would avoid any extended presence among the Iraqi population. American units would conduct clearing operations in major cities to root out insurgent threats,[24] but would then withdraw from those urban areas as quickly as possible. US troops would live on large, fortified bases (usually called

FIGURE 8.1 General George Casey in Baghdad, April 2006
Source: https://archive.defense.gov/DODCMSShare/NewsStoryPhoto/2006-04/
20060426090813_8iraqic-20060426.jpg.

forward operating bases, or FOBs) well away from populated areas and leave only to conduct patrols and other types of operations. If the presence of US troops was inciting insurgent violence, as Casey believed, then removing them from the population, he reasoned, should reduce that violence and stabilize the country more quickly.[25] He also believed that doing so would compel the Iraqi army and police to become more independent and learn how to operate on their own.[26] Yet this policy violated the core counterinsurgency principle that required military forces above all else to protect the population from insurgents. The FOBs soon mushroomed in size and contained every conceivable comfort from home that could be transported to a war zone, including air conditioning, internet access, movie theaters, gyms, and, at some, weekly steak-and-lobster nights.[27] Many of the thousands of US forces performing support functions never left these bases, earning them the derogatory nickname of "Fobbit."[28] Those who did leave the base interacted with the population only for short periods of time before withdrawing back to safety, often returning every night. This made it nearly impossible to gather accurate intelligence, since the insurgents threatened (and often killed) anyone seen or even suspected of talking to patrolling US forces. The vast majority of Iraqis were not involved in the insurgency, and many of them supported efforts to reduce the violence. But since US troops could not guarantee safety in their homes and neighborhoods after returning back to the FOBs, locals rarely offered any assistance.

Third, US forces would focus on training the Iraqi security forces, which included the Iraqi army and the Iraqi police, as quickly as possible.[29] That would not only enable US forces to withdraw from Iraq more rapidly, the thinking went, but would also reduce levels of violence since local populations would not see Iraqi security forces as occupiers.[30] Despite being a central facet of Casey's plan, this part of his strategy ran into delays and practical challenges almost from the start. Estimates of how many security forces had been trained were unreliable and often inflated,[31] and only a fraction of those trained were available for duty on any given day for a variety of reasons.[32] Furthermore, the US-trained Iraqi units were sometimes unwilling, and often unable, to counter the zealous insurgents.[33]

In late 2004, Casey conducted an internal review of his campaign plan. It was led by Kalev Sepp, a noted counterinsurgency expert who was later involved in the effort to rewrite the counterinsurgency manual, as discussed in Chapter 5. The review concluded that the training of the Iraqi army needed to be accelerated, and Casey's staff then developed a new plan for doing so.[34] On June 28, 2005, President George W. Bush devoted a large part of a major speech to describing the key elements of this plan. The new approach, he said, was that "as the Iraqis stand up, we will stand down"—explicitly tying the progress of the Iraqi security forces to the prospects for a US withdrawal.[35] The plan outlined two new ways that US forces would work with the Iraqi army. First, US units would partner with Iraqi units, notably at the battalion level, and conduct combined operations so the Iraqis could learn and improve through shared experiences on the battlefield.[36] Second, training teams of 10 to 12 US military personnel would be embedded in every Iraqi army battalion, brigade, and division to provide advice, assistance, and tailored training.[37]

Once again, these efforts proved far less successful than Casey had hoped. Though they did make some progress,[38] the challenges to building capable security forces remained immense—including the low quality and inexperience of recruits, the lack of needed facilities and equipment, and the country's bitter sectarian divisions. These were worsened by the recurrent mass desertions of Iraqi troops when faced with insurgent threats.[39] Back in the United States, the army staff compounded these challenges by the way it selected the approximately 2,500 field-grade officers and senior sergeants that Casey needed to serve on the training teams.[40] These active and reserve officers and NCOs were pulled from across the army and received virtually no training, nor did the army establish a formal training program for them until the summer of 2006.[41] Some officers tasked as advisers soon began to feel pressure to provide unrealistic indicators of progress in developing the security forces in order to support Casey's objectives.[42] Many of the measures of grading Iraqi units were subjective and prone to easy inflation. These subjective judgments included such crucial factors as personnel and equipment readiness, training, leadership, and command and control.[43] There were also significant problems with more objective data, such as the exact numbers of personnel authorized, assigned, and trained. Even as late as 2008, the Special Inspector General for Iraqi Reconstruction was still warning about "a continued need for caution in relying on the accuracy and usefulness of the numbers."[44]

While Casey sought to accelerate a transition to Iraqi forces, he also sought to reduce the number of US forces in Iraq. Casey briefed the president and secretary of defense in October 2005 and recommended that they begin to "off-ramp," or cancel, the upcoming deployment of two US Army brigades. That would mean that troops rotating out of Iraq at the end of their tours would not be replaced, thus shrinking the US military footprint. His briefing slides stated that doing so would "remove [the] central motivation attracting foreign fighter[s] and drawing Iraqis to the insurgency." He also warned that the reduction needed to happen quickly, because "tolerance for coalition presence is diminishing."[45] Yet many of Casey's subordinates had opposed this recommendation, as did the commanders of the army division about to rotate into Iraq and the army division about to rotate out. Marine commanders in Anbar province also opposed it and requested that any off-ramped forces be deployed instead into their area since they believed they had too few troops to execute their mission. And perhaps most damningly,

Casey's own internal red team opposed off-ramping the two brigades, assessing that the security situation was too unstable to do so until spring 2006 at the earliest.[46] Nevertheless, after the December 2005 parliamentary elections, President Bush announced that the two brigades would be off-ramped. That reduced the number of brigades in Iraq from 17 to 15, for a loss of approximately 6,700 combat troops and an additional 3,000 support troops.[47] Shortly thereafter, Casey confidently predicted that he would be able to off-ramp another four army brigades and a marine regiment by October 2006.[48]

While Casey actively pursued this strategy of transition and withdrawal, he also took some steps to better prepare US forces for counterinsurgency operations. In August 2005, for example, Casey commissioned a survey of US forces to evaluate their proficiency at counterinsurgency. The results showed that 20 percent of units were proficient, 60 percent were struggling, and 20 percent had little to no proficiency.[49] In November, he created what became known as the Counterinsurgency (COIN) Academy, in order to help the majority of his units that were struggling to adapt to the nature of the fight.[50] Since the army back at home had not yet adjusted its predeployment training to emphasize counterinsurgency, Casey sought to compensate in the combat zone for that glaring deficiency.[51] Leaders of units arriving in Iraq were required to attend a five-day course at the academy, located in Taji, before moving on to their assignments elsewhere in the country. The course familiarized them with the basic principles of counterinsurgency, as well as sharing recently developed tactics, procedures, and lessons learned from ongoing operations.[52] Casey made a point to personally address every class at the academy.[53]

Yet at the same time, Casey's policies undercut commanders who were actually conducting counterinsurgency operations throughout the country. In Chapter 7, for example, we noted that Marine Lieutenant Colonel Dale Alford started to adopt a classical counterinsurgency approach in the western Anbar city of Al Qaim, around the same time that Army Colonel H. R. McMaster was also doing so in Tal Afar. Casey hailed the developments in Al Qaim and visited the battalion five times during its seven-month deployment. He supported the battalion's effort to organize tribal recruits into a security force called the Desert Protectors, and even asked Alford to lecture at the COIN Academy.[54] In January 2006, however, the commander of a marine regiment that had just arrived in Anbar asked Casey for more troops in order to extend the successful Desert Protectors program eastward. As Michael Gordon and Bernard Trainor write,

> General Casey had applauded the progress at Qaim, just as he did H. R.
> McMaster's operation in Tal Afar, but the main thrust of his strategy was
> to consolidate the number of American bases and hand over to the Iraqis.

Partnering with Iraqi tribes across Anbar was a troop-intensive strategy and would have required expanding the American footprint just at the moment that the general was committed to shrinking it. The request was denied.[55]

In the end, none of these efforts mattered very much, because they did not provide security for the population. As long as Casey focused on withdrawing US forces as quickly as possible and operating from isolated FOBs in the meantime, the population would never feel secure enough to share the intelligence necessary for effective counterinsurgent operations. And though the idea of standing up Iraqi security forces so American forces could stand down made sense in principle, Casey's plan required them to provide a level of security that was far beyond their capabilities. No other efforts at counterinsurgency could succeed as long as the overall campaign plan continued to undermine them.

By early 2006, data showed that this strategy of transition and withdrawal was failing badly. These data were available not only to Casey and his staff,[56] but to the general public as well.[57] From the time that Casey took command in June 2004 through early 2006, monthly official estimates of the total number of insurgents ranged between 15,000 and 20,000. Figure 8.2 shows the number of insurgents detailed or killed each month from 2003 to 2006. Those numbers add up to 63,470 detained or killed—a number that is *more than three times larger* than the

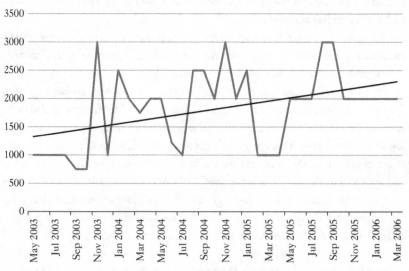

FIGURE 8.2 Total Number of Insurgents Detained or Killed in Iraq, May 2003–2006
Source: Michael E. O'Hanlon and Nina Kamp, "Iraq Index: Tracking Variables of Reconstruction & Security in Post-Saddam Iraq," Brookings Institution, April 27, 2006, 17.

estimated size of the insurgency at any point. Even after taking into account the inherent difficulties with such estimations, this wide discrepancy and the upward trendline in the figure suggest that the number of insurgents continued to grow considerably despite their apparently significant losses. Furthermore, Figure 8.3 shows that the number of daily attacks by insurgents continued to rise during this same time period, suggesting increasing capability as well as numbers.

These data clearly show that by the spring of 2006, the war in Iraq was failing badly. With such negative results after almost 18 months of effort, Casey should have stepped back and reassessed his plan. Had he done so, he could have recognized its failures and adapted accordingly. But Casey remained dogmatically committed to withdrawing US troops as quickly as possible, while isolating them from the population in the meantime. Despite the rising violence and upsurge of new insurgents, he forged ahead with plans to start withdrawing US troops from Iraq after the December 2005 parliamentary elections. He had previously decided not to replace two of the 17 brigade combat teams that were scheduled to depart in the summer of 2006.[58] Soon his plans called for slashing the number of troops by a total of more than 45 percent—from approximately 160,000 to no more than 110,000—by the fall. He also planned to close half of all US bases, to help demonstrate that the Iraqi government was in control, even though that would cause even more consolidation onto giant FOBs.[59] Casey's rigid adherence to the assumptions with which he entered command blinded him to the desperate need for adaptation to devise a new strategy.

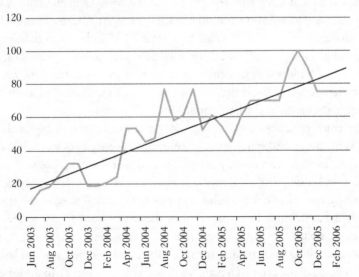

FIGURE 8.3 Number of Daily Attacks by Insurgents in Iraq, June 2003–March 2006
Source: O'Hanlon and Kamp, "Iraq Index," 22.

Casey in Command, Part 2: February–October 2006

In early 2006, two developments accelerated the downward trajectory of the war. First, the December 2005 Iraqi elections lacked a clear winner and led to months of political deadlock before the parties finally formed a national unity government in May. Casey, as well as senior White House officials, had believed that the elections would lead to some political reconciliation that would reduce the violence, but instead they increased sectarianism.[60] Second, and even more importantly, the flames of sectarianism exploded on February 22, when al-Qaeda in Iraq (AQI, a Sunni group) bombed the Golden Dome mosque in Samarra—one of the holiest Shiite sites in Iraq. Shia militias in Baghdad immediately retaliated against Sunni communities in the city through a brutal campaign of murder and intimidation that started forcing Sunnis out of previously mixed neighborhoods. More than 30,000 Sunnis fled their homes in the month after the bombing, and the violence escalated sharply as the cycle of reprisals and counterreprisals continued.[61] Between late February and May, more than 3,000 corpses were found in Baghdad alone.[62] Although the Sunni insurgency continued to rage, mostly in Anbar province, the Samarra mosque bombing transformed the conflict across the rest of the country into a nascent civil war.

Over the next two months, Casey seems to have understood that something had fundamentally changed.[63] In late February, he received an intelligence report that warned that unless the violence in Baghdad were contained, there would be "intense sectarian strife across several provinces—likely resulting in civil war." David Cloud and Greg Jaffe report that "in the margins of the report, Casey drew a star and jotted two words to himself—'must act.'"[64] And in 2008, Casey recalled that, at the time, "it was clear to me that it had shifted from an insurgency against us to a struggle for power, that it wasn't any longer totally about us, it was about them."[65] He was more equivocal in his 2012 memoirs, however, where he wrote that the increasing violence "may have been an indication of a significant change in the nature of the Iraqi conflict."[66]

Yet much of Casey's approach to the war remained remarkably unchanged even as the bloodshed spiraled out of control.[67] He erroneously concluded that this spike in violence was only temporary and that the Iraqi political process could continue despite the overwhelmingly sectarian nature of the violence.[68] He repeatedly refused to acknowledge Iraq was facing a civil war[69]—an error that an internal army study later compared to the crucial 2003 failure to recognize that Iraq was facing an insurgency.[70] In 2008, Casey told an interviewer that the Samarra bombing "didn't cause me to go back and question the fundamental underpinning of our campaign plan, which was that Iraqi success in the long term will only be achieved by Iraqis. My view then and my view continues to be now that they had to work through these sectarian problems."[71] He therefore continued to pursue his strategy of transition and withdrawal, firm in his belief that withdrawing US military forces would promote political reconciliation.

After almost two years in command, Casey still had the causality backward. He had not yet grasped that security was a necessary *precondition* for political reconciliation, because it required a level of trust among the parties that was simply impossible amid such intense sectarian violence. The passage in his memoirs where he explains this decision is worth quoting at length, because it shows how much of his approach remained unchanged despite everything that had happened during his tenure:

> My thinking was that since the fundamental problem in Iraq was over the division of political and economic power, and that this conflict was the root cause of the sectarian violence, the ultimate solution would be political and not military. Furthermore, my experience had been that the longer we remained there in force, the more the Iraqis relied on us to solve their problems, and the less they moved forward on their own. I found this to be true at both the political and military levels . . .
>
> I calculated that the specter of continued coalition reductions would reinforce the notion that the coalition was eventually leaving and create a sense of urgency in the new Iraqi government and its security forces that could spur the reconciliation that was so desperately needed for Iraq to go forward.[72]

Thus, instead of adopting a more proactive counterinsurgency strategy, where US forces would help provide the security necessary for any political reconciliation, Casey remained committed to his strategy of transition to Iraqi forces and a US withdrawal. According to an internal army study of the war, Casey "did not seem to allow for the possibility that a withdrawal of the U.S. military might actually undermine the reconciliation he believed would ultimately happen, or that the country might break into warring fragments after U.S. withdrawal."[73]

In June 2006, Casey's plan still called for withdrawing one-third of the US brigades in Iraq within six months and two-thirds within 18 months.[74] By this time, President Bush and National Security Advisor Stephen Hadley had both concluded that the transition and withdrawal strategy would not work, as we discuss subsequently.[75] Casey should have understood this as well because he had been promising to withdraw US forces for well over a year, and yet the numbers of troops in Iraq remained largely unchanged.[76] The fact that the security situation had prevented him from withdrawing any significant forces should have made clear that his strategy was indisputably failing, and prompted him to reassess this central assumption of his plan.

Two important events during the summer of 2006 should also have led Casey to revise other elements of his plan. First, the kill-or-capture approach that most of his units were employing was based on a decapitation strategy. Its core premise was that eliminating terrorist and insurgent leaders would cause their organizations and networks to grow progressively weaker and less capable. But as noted in Chapter 7, Abu Musab al-Zarqawi, the most important leader of al-Qaeda in Iraq and the primary focus of US targeting efforts, was killed by a coalition air strike on June 8.[77] Despite lopping the head from AQI and killing or capturing numerous other suspected insurgent leaders, the insurgency seemed to grow progressively stronger. The number of Iraqi civilian casualties, attacks against US and coalition forces, and US combat deaths all continued to rise throughout the rest of 2006.[78]

Second, the failure of two major military operations to retake Baghdad should have finally demonstrated that the Iraqi security forces were in no way ready for the transition element of Casey's strategy. Operation Together Forward, which was conducted in June and July, was designed as a major counteroffensive to establish security in Baghdad. It involved 42,500 Iraqi army and police forces, supported by 7,200 US forces. The operation effectively drove most of those responsible for the violence out of the city, but Iraqi forces failed to maintain security afterward—largely because many of them were direct participants in the sectarian conflict.[79] Casey himself acknowledged that the "loyalty of the Iraqi security forces, particularly the police, was the overriding issue that kept this from being a success."[80]

Nevertheless, Casey and Iraqi officials proceeded with Operation Together Forward II, which started in August and ended in October. This time, the operation was carefully planned in coordination with the Iraqi government.[81] But it followed the same basic approach, with some additional US and Iraqi reinforcements and support for Iraqi forces.[82] Predictably, the operation had the same outcome. Iraqi security forces proved even less reliable, and some were actively collaborating with those committing the violence.[83] Only 1,000 of the planned 4,000 additional Iraqi army soldiers participated in the operation, since many of them refused to fight in Baghdad.[84] As Fred Kagan later wrote, Casey's transition and withdrawal strategies were directly to blame:

> Because there were too few American troops, and *because American commanders wished to rely heavily on Iraqi forces*, US troops did not remain in cleared neighborhoods either to defend them or to support and improve

the Iraqi forces trying to maintain order there. The different Sunni and Shiite enemy groups made a point of surging into the cleared but undefended neighborhoods to demonstrate the futility of the operations, and they also attacked neighborhoods that were not being cleared by American and Iraqi troops.[85]

Violence in Baghdad skyrocketed as a result. The number of attacks grew by 22 percent during the second operation, while over 1,000 civilians were being killed each month.[86] The two offensives offered yet another bloody indictment of the failing US approach.

Over the summer, Casey decided that the escalating violence required him to "forgo the planned off-ramp of coalition forces" that he had planned. He did extend the deployment of one Stryker brigade by three to four months, which added a few thousand additional troops to the fight in Baghdad. Yet at the same time, he moved other forces out of the fight. Instead of keeping more forces available to fight the growing violence, he inexplicably "asked to keep a brigade that I had intended to send home without replacement in order to reconstitute the reserve *in Kuwait* . . . and to establish a force ready to deploy on 30-days notice *at home station* in case of larger problems."[87] In his memoirs, he states that in retrospect he took too long to decide to stop withdrawing forces from Iraq. But even then, he notes that "this caused substantial turbulence at the tactical level" without acknowledging any of the tremendous strategic implications of this decision.[88]

In mid-October, Abizaid warned Casey that "sectarian violence in Baghdad could be fatal" to the US effort in Iraq and that "the dynamic needs to change," though he did not tell Casey exactly what to do. Casey remained unyielding in his approach, saying that there were "no short-term military fixes."[89] At the time, he had an opportunity to temporarily add 20,000 US troops to the fight in Baghdad by extending the deployment of the Fourth Infantry Division so that it overlapped for several months with its arriving replacement unit. But he told Abizaid that this was "a course of action I cannot recommend until I see greater commitment from the Iraqis to solve the sectarian situation in their capital."[90] Yet his plans for 2007 assumed that political progress would be made and focused on rapidly and extensively transitioning security responsibility to the manifestly unprepared Iraqi army and police.[91]

After almost two and a half years in command, almost 2,000 US fatalities, and more than 57,000 Iraqi civilian deaths,[92] Casey still did not grasp that establishing security was a necessary precondition for any political reconciliation[93]—and that his obdurate strategy of transition and withdrawal was making it impossible for US forces to provide that security.

The Search for Alternatives: Developing the Surge

By the summer of 2006, the escalating failures of the US strategy in Iraq prompted individuals and groups, both inside and outside the US government, to develop alternative options. Though these began as separate initiatives, they all eventually coalesced around the same general idea—send more US troops and use them differently—and all reinforced each other in the waning months of the year. Casey was not involved in any of these reviews, and he consistently opposed the alternative that became known as "the surge," even as it gained momentum.

Though there had been some earlier discussions about alternative strategies for Iraq,[94] the failure of Operation Together Forward in July galvanized the first serious search for a new strategy. Its collapse led officials in the Bush administration, especially some in the White House, to finally lose confidence in Casey's leadership and his ability to deliver on his promises.[95] As Thomas Ricks later wrote, "Casey may not have known it, but the failures of the Together Forward operations were the beginning of the end for his command in Iraq."[96]

Several different organizations, including the National Security Council (NSC), the State Department, and the Joint Chiefs of Staff, initiated their own strategy reviews during the late summer. These were soon merged into a single NSC effort that Hadley, the national security advisor, chose to lead personally.[97] A surge was one of several options developed.[98] The idea intrigued Hadley enough that he asked NSC staffer William Luti to assess the capacity for a surge and determine just how many troops could be actually be deployed to Iraq if this option were chosen. Luti concluded that five additional army brigades could be sent to Iraq and recommended doing so, and also proposed some smaller tactical changes. Hadley met with Casey during a trip to Baghdad in early November, and one of several options he raised involved deploying more US troops. Although this gave Casey the opportunity to secure more US combat power with some top-level support from the White House, he remained adamantly opposed to the idea and once again argued that strengthening the Iraqi government and security forces remained the best way forward.[99]

On Tuesday, November 7, 2006, Americans went to the polls in a contentious midterm election. The Republicans were soundly defeated in their congressional races, and the Democrats regained control of both the House and the Senate. The following day, President Bush cashiered Rumsfeld as his secretary of defense and announced his intention to nominate Robert Gates as his successor.[100] Although Rumsfeld had long been a supporter of the transition and withdrawal strategy in Iraq, his sacking did nothing to alter Casey's approach. Even after Rumsfeld's dismissal, Casey continued to strongly champion his transition and withdrawal strategy.[101] This demonstrates just how deeply he believed in his approach, and makes clear that he had not been simply responding to the guidance of an over-bearing secretary of defense.

On November 13, Casey held a videoconference with members of the Iraq Study Group, a bipartisan panel chartered by Congress earlier in the year to assess developments in Iraq and provide recommendations to the president.[102] Even at this late date, he still maintained that deploying more US troops would only make the Iraqis more dependent upon them. He later wrote that he told the group that "more coalition forces at this point would give Iraqi leaders more time to avoid hard decisions on reconciliation and ultimately prolong our time there."[103] He also pointedly noted that there would "never be enough troops in Baghdad to stop them from killing each other."[104] Casey was not alone; most senior military leaders opposed a troop surge at this time as well, including the Joint Chiefs of Staff.[105] But after nearly two and a half years as the senior US commander in Iraq, Casey should have understood the war and the Iraqis better than any American. Yet he was incapable of adapting to the disastrous situation in the country, despite the clear and repeated failings of his approach.

The Iraq Study Group released its report on December 6, which started with its bleak conclusion: "The situation in Iraq is grave and deteriorating."[106] Coming so soon after the shock of the US midterm elections, the report stimulated even more discussion about a full US withdrawal from Iraq, something the Bush administration desperately wanted to avoid. It made White House officials, and ultimately the president himself, particularly receptive to other ideas. Among the most influential were those that Jack Keane, a retired army general, developed with colleagues at the American Enterprise Institute (AEI)—which provided the operational analysis needed to support the surge.[107]

Keane briefed the president on these proposals on December 11,[108] and a White House official later told Keane that the meeting decisively shifted the president's thinking about the war. Hadley and his staff had been advo-cating for a troop surge for several weeks, but this meeting immeasurably strengthened their case, largely because the AEI recommendations included a detailed and credible military plan. As Peter Feaver, an NSC staffer at the time, later wrote:

> The AEI study amounted to something of an existence proof of a coherent surge option that would pass muster with the uniformed military because individuals with bona fide military credentials, not merely civilian staffers, helped develop it. The study allowed the surge option to be debated within the internal review even though neither the military nor the key civilian departments were advocating the surge.[109]

On December 12, Casey presented his recommendations to the president and the NSC. He was not aware that Keane had briefed them on the surge option the previous day.[110] Casey told them that he "was very concerned about bringing fresh US troops into the middle of a sectarian conflict in an Arab country where there was not clear political support for their actions." He conceded that additional troops might temporarily reduce sectarian violence and could provide a bit of "breathing space" for the Iraqi government. But he also said it would take longer to transition security responsibility to the Iraqi security forces and these troops would not have "a decisive effect" without political reconciliation within the Iraqi government."[111]

By this time, Casey knew that the NSC was looking for a different approach. But he still believed in the key principles that had been shaping his strategy since 2004, despite the unremitting and horrendous intensification of violence in Iraq during the intervening 30 months. He later wrote,

> [I] believed that what I was recommending—to accelerate the transition of security to capable Iraqis in exchange for their action to solve the core problem in Iraq, that is, reconciling the interests of the different ethnic and sectarian groups—offered the most effective way to accomplish our strategic objectives in Iraq. I believed that I had asked for the troops that I needed to accomplish our operational objectives, and that, if the prime minister delivered on his pledges to the President to allow our forces and the ISF [Iraqi Security Forces] to operate freely without political interference, we would bring security to Baghdad by the summer. I felt that additional troops beyond that would risk introducing them into a very confusing and difficult operational environment without a plan for how their introduction would contribute to the accomplishment of our strategic objectives. I remained adamantly opposed to that.[112]

President Bush officially decided to proceed with the surge on December 13, during a meeting with the Joint Chiefs of Staff.[113] Ironically, the Joint Chiefs had essentially made the same argument to the president earlier that Casey had been making all along, that political reconciliation had to come before committing additional troops. They again reiterated their position with the president. But Bush now responded with what he had concluded had been the fundamental flaw in Casey's plan from the very beginning: "You can't ask them to reconcile in this security situation. Don't we have to make our bet to get the security situation in hand before, in some sense, it's fair to put them to the test?"[114] The chiefs also warned that the five additional army brigades would consume all of the military's strategic reserve, leaving the United States vulnerable to unexpected crises. Bush countered that he wasn't concerned about North Korea invading South Korea at that point or any other hypothetical scenarios: "We've got a war on our hands and we've got to win the war we've got."[115]

Despite the dramatic shift in Bush's views, Casey continued to advocate for his long-standing approach. During a meeting with the president in mid-December, Casey repeated his opposition to the surge in the absence of political reconciliation. Bush responded skeptically to Casey's claim that US troops could start withdrawing from Iraq during the summer of 2007 because the Iraqi security forces would be strong enough to operate with limited support. The debate continued the following day, when Casey told Bush that there were sufficient forces already in Iraq, and Bush replied that a temporary increase could be "a bridge to a better place."[116] Casey continued to push back in late December even after Bush had decided to conduct some form of a surge, suggesting a "mini-surge" of only two US Army brigades. And after Bush determined that the surge would include all five US brigades, Casey tried to convince him to announce that they would be deployed only as needed instead of sending all five at once.[117] In short, the commander-in-chief had now chosen to make a sweeping adaptation to the failing war when his theater commander would not. Yet even in retrospect, Casey believed that Bush decided on the surge as a way to improve domestic support for the war, rather than as a fundamentally new strategy to achieve operational and strategic success.[118]

By this point, Bush had already decided to replace Casey with General David Petraeus, who had just published the new counterinsurgency manual discussed in Chapter 5. As the new commander in Iraq, Petraeus would now get to implement the key principles of the manual while also leveraging the substantial additional forces about to be deployed. Bush made clear that he did not want to be seen as blaming Casey, telling his staff, "I am not going to make him the fall guy for my strategy."[119] A few days before Christmas, newly sworn-in Secretary of Defense Gates flew to Baghdad and told Casey that he was the leading candidate to serve as the new army chief of staff. Casey had been scheduled to rotate out of command a few months later, but when he agreed to serve in this new role, his departure date was accelerated so that Petraeus could take command sooner.[120]

On January 5, 2007, Bush announced that Petraeus would replace Casey,[121] and on January 10, he announced the new surge strategy during a nationally televised prime-time address.[122] It had taken the United States almost four years to understand and adapt to the character of the war it was fighting—longer than it took the United States to enter, fight, and win World War II.[123] In Baghdad one month later, on February 10, Casey formally transferred command authority to Petraeus, and with it the responsibility for executing the new strategy. In a private meeting that day, Casey told Petraeus about the military plans he had prepared to bring in the additional five brigades. But then, reportedly with some sadness, he added, "It is going against everything that we've been working on for the last two and a half years."[124]

Conclusion: Casey's Failure to Adapt

Chapter 4 argued that adaptive theater leaders need to understand the character of the war they are fighting, dispassionately determine whether their approach is succeeding, and to constantly challenge their assumptions about the enemy, the environment, and their own forces and plans. This requires open-mindedness and cognitive flexibility, and especially the ability to recognize and accept evidence of failure. Unfortunately, Casey displayed few if any of these key characteristics of adaptability during his nearly 32 months commanding US forces in Iraq. Most importantly, he never reexamined his two most critical assumptions: that US forces were causing or exacerbating the violence and thus needed to be separated from the population, and that political reconciliation was a precondition for improved security rather than the other way around.

The results of the surge soon demonstrated just how wrong those assumptions had been. After Petraeus took command, the levels of violence and US troop fatalities during the first few months of 2007 escalated to some of their highest points during the war. This spike had been expected, since US troops were fighting their ways into new neighborhoods and then remaining there to provide security against the fierce counterattacks that followed.[125] By the time the final surge troops arrived in June, however, the trends started to change dramatically, and violence across Iraq soon dropped to unprecedentedly low levels. Figures 8.4 and 8.5 respectively show how many Iraqi civilians and US troops were killed each month from the time Casey took command in June 2004 through the end of 2009.[126] The trend lines in both figures steadily increase throughout Casey's tenure, but then plummet sharply. Part of that decrease occurred because the Anbar Awakening was spreading to other parts of Iraq and reducing Sunni violence, as discussed in Chapter 7. But as a trio of respected scholars have conclusively demonstrated, the Awakening alone cannot account for the decrease in violence, because it would not have spread far enough or fast enough in the absence of the surge.[127] Instead, the synergies between the surge and the Awakening led to the dramatic decline of violence. Yet paradoxically, the Awakening would not have happened if Colonel Sean MacFarland had not deliberately rejected the central tenets of Casey's approach. Chapter 7 shows how he moved his troops off of large bases and established combat outposts throughout the local population, and aggressively fought al-Qaeda in Iraq while simultaneously trying to strengthen local Iraqi security forces. Thus, neither of the factors that together explain the

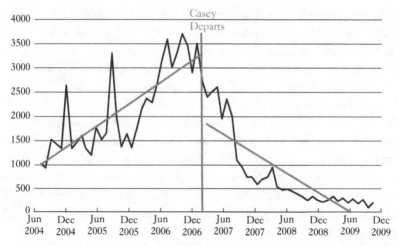

FIGURE 8.4 Estimated Civilian Casualties in Iraq, June 2004–December 2009

Source: Michael E. O'Hanlon and Ian Livingston, "Iraq Index: Tracking Variables of Reconstruction & Security in Post-Saddam Iraq," Brookings Institution, January 31, 2011, 4.

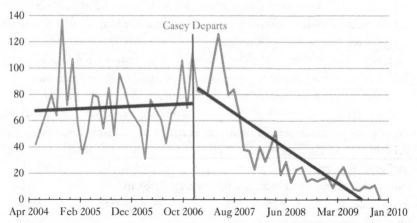

FIGURE 8.5 US Troop Fatalities in Iraq, June 2004–December 2009

Source: O'Hanlon and Livingston, "Iraq Index," 12.

reduction of violence would likely have occurred if Casey's approach had been strictly followed.

The surge was a strikingly successful adaptation that helped change the course of the Iraq War. After much hard fighting, it dramatically improved the security situation and thus enabled the warring Iraqi factions to begin talking to each other and start figuring out how to settle their differences without resorting to continuing violence. The surge did not guarantee success, but continuing Casey's approach would have almost certainly guaranteed failure.

By any measure, Casey was an accomplished and thoughtful US military leader. So why didn't he grasp the realities in Iraq as they unfolded? Why did he remain wedded to his initial assumptions despite the overwhelming, and readily available, evidence that his strategy was not working? Casey, like most of the military officers of his generation, was deeply influenced by the failures of the Vietnam War and the struggle to rebuild the army in its aftermath, as discussed in Chapter 5. According to authors David Cloud and Greg Jaffe, perhaps the most important lesson that the officers of Casey's generation drew from Vietnam was that they needed to "keep their force from becoming too deeply embroiled in messy political wars that defied standard military solutions."[128] This conclusion was strongly reinforced by the unexpectedly swift and conclusive victory in the 1991 Gulf War. US military leaders also consistently resisted involvement in the peacekeeping operations of the 1990s because of their messy political character and lack of a clear-cut military objective. In the case of Bosnia, they even managed to write their own mandate that made support of civilian tasks optional rather than mandatory.[129] Casey himself stressed the formative influence of his time in Bosnia, which seemed to reinforce his deep-seated belief that US forces in Iraq should withdraw as quickly as possible so that the Iraqi security forces did not grow too dependent upon them.

Casey was not alone in these views. Most of the senior US military leaders during the first few years of the Iraq War were shaped by these experiences, and most also failed to understand and adapt to the challenges of Iraq. Almost all of them also opposed the surge, for the same reasons that Casey did. They, too, have rightly been criticized for doing so. In May 2007, an army lieutenant colonel named Paul Yingling published a provocative and widely read article called "A Failure in Generalship," which castigated the war's leaders for having "been checked by a form of war that they did not prepare for and do not understand."[130] Retired army colonel Robert Killebrew, a well-respected military thinker, observed even more pointedly:

> Why did the American military establishment so fail to come up with a war winning strategy that it was up to a retired general and a civilian think tank, AEI, to do their job? This is a stunning indictment of the American military's top leadership.[131]

Though he was not alone, Casey nevertheless bears the primary responsibility for the failure to adapt to the changing nature of the Iraq War. As the theater commander for more than two and a half years, fighting and winning the war was his sole responsibility. As the war spiraled ever downward, his enduring inability to adapt brought the United States to the brink of losing the war and withdrawing in abject defeat.

Casey's selection to serve as the army chief of staff after his failures in Iraq upset many people, both inside and outside the military. It may have been one of the reasons that Yingling wrote what may be the most important sentence in his pointed critique: "As matters stand now, a private who loses a rifle suffers far greater consequences than a general who loses a war."[132] Casey's promotion also inevitably provoked comparisons to General William Westmoreland, who, as discussed in Chapter 4, was brought back from Vietnam to serve as army chief of staff after also rigidly adhering to a failing strategy.[133] But the two men differed in at least one critical respect. Westmoreland remained dogmatically committed to a conventional approach, while Casey seemed to support a counterinsurgency strategy in principle without understanding that his unchallenged assumptions undermined the most crucial elements of that strategy. But despite this key difference, the comparison is ultimately valid. Both Casey and Westmoreland failed as theater commanders because they never reexamined their basic assumptions and thus never adapted their original approaches to their wars, despite incontrovertible evidence that those approaches were failing.

Theater Adaptation in Afghanistan: General David McKiernan, 2008–2009

On May 11, 2009, Secretary of Defense Robert Gates held a press conference to announce that he had asked General David McKiernan to resign his position as the commander of US forces in Afghanistan.[134] It was the first time that a US wartime commander had been relieved of duty since President Harry S. Truman fired General Douglas MacArthur in 1951.[135] McKiernan had been in command for less than a year, and he not done anything that would have traditionally been considered a firing offense—he had not exceeded his orders, disobeyed instructions, suffered a major battlefield defeat, or done anything unethical or immoral. In fact, Gates acknowledged at the press conference that during McKiernan's tenure, "nothing went wrong, and there was nothing specific."[136]

So why did Gates take the extraordinary step of firing McKiernan? The answer, simply put, is that Gates and Chairman of the Joint Chiefs of Staff Admiral Michael Mullen believed that he was not sufficiently creative, innovative, and adaptable to succeed in his job. In part, they wanted a new leader to implement the chosen strategy of newly elected president Barack Obama. But in June 2008, soon after McKiernan took command and months before Obama's election, Gates had already started expressing concern that appointing McKiernan had

FIGURE 8.6 General David McKiernan in Afghanistan, March 2009
Source: https://commons.wikimedia.org/wiki/File:General_David_McKiernan,_ISAF_-e.jpg.

been a mistake.[137] Gates sensed early on what would become clear a few months later: McKiernan did not possess the traits of adaptability necessary for success-ful theater command that were identified in Chapter 4. Throughout his tenure, McKiernan failed to understand the growing urgency that Gates, Mullen, and General David Petraeus (now serving as the commander of CENTCOM) felt about reversing the downward trajectory of the war in Afghanistan. During the winter of 2008–2009, these three men all concluded that McKiernan was unre-sponsive, unable to adjust to the increasing tempo of the war, and incapable of executing the full-scale counterinsurgency strategy that they thought essential. In short, McKiernan was simply the wrong general at the wrong time in the wrong place, as his chain of command eventually and unanimously concluded.

The War That McKiernan Inherited

By the spring of 2008, virtually every indicator suggested that the war in Afghanistan was spiraling downward. During the early years of the war, the United States and its NATO allies had seemed to be making progress (albeit slowly) toward their mutual objectives of a peaceful and stable Afghanistan that no longer served as a terrorist sanctuary. US and NATO forces operated under completely separate commands in the first years of the war. The terms of the December 2001 Bonn Agreement limited the boundaries of the newly created International Security Assistance Force (ISAF) to Kabul, while US forces and coalition partners operated freely throughout the rest of Afghanistan as part of Operation Enduring Freedom (OEF). In August 2003, however, NATO agreed to take over ISAF, and its new mandate soon expanded beyond Kabul.[138] As discussed in Chapter 7, the Taliban and the remnants of al-Qaeda had been largely defeated by the end of 2001, but by mid-2005 they had started to regroup and operate again from sanctuaries in Pakistan. By October 2006, ISAF had expanded its area of responsibility throughout the rest of Afghanistan, but by then the security situation had deteriorated.[139] That meant that NATO, which had originally envisioned its mission in Afghanistan as peacekeeping and stabilization, now faced a growing insurgency. This renewed threat split the alliance, with some members pushing for a more active role against the Taliban, while others steadfastly opposed being drawn into combat. The lack of consensus was so stark that NATO could not even reach agreement on using the term "counterinsurgency," much less conducting a counterinsurgency campaign.[140]

In 2007, the number of attacks from improvised explosive devices (IEDs), US troop casualties, and Afghan casualties had reached all-time highs.[141] In February 2008, Gates decided that McKiernan would replace General Dan K. McNeill as the ISAF commander when McNeill's assignment was due to end in June.[142] McKiernan was recommended for the position by Mullen and by General George Casey (then serving as the army chief of staff, as discussed earlier).[143] In some ways, McKiernan seemed an odd choice. He was a career armor officer and had little if any experience leading light infantry or special operations, and no background with anything resembling counterinsurgency.[144] At the time, he was best known for commanding all coalition land forces during the 2003 invasion of Iraq, leading a force of primarily mechanized troops conducting large-scale maneuver warfare.[145] But he had also spent much of his career in Europe and knew NATO well. He had served under NATO command during operations in Bosnia, and in 2008 he was working closely with alliance members as the commander of US Army forces in Europe. He therefore seemed like a good choice to manage the intricacies of the alliance in its first combat operation outside Europe, and to push its members to do more in Afghanistan.[146]

Gates later recalled that he had "a very high opinion" of McKiernan before appointing him, largely because he had done such a good job of working with the NATO allies.[147] Since Mullen and Casey supported McKiernan's selection, Gates wrote that he "saw no reason to challenge his appointment." But then, with remarkable candor, Gates added: "With the benefit of hindsight, I should have questioned whether McKiernan's conventional forces background was the right fit for Afghanistan. This was a mistake on my part."[148]

McKiernan's Approach

McKiernan took command of ISAF on June 3, 2008, as the security situation across Afghanistan continued to worsen. The number of violent incidents in the first half of 2008 was 40 percent higher than in the first half of 2007, and more US troops were killed in action that June than in any previous month of the war.[149] In his first few months, McKiernan instituted elements of a nascent counterinsurgency strategy. He conducted an informal assessment as soon as he arrived, which became the basis of the two-year operational plan he issued in October. He determined that Regional Command–South was the critical center of gravity, and identified key districts where ISAF would adopt the same type of clear-hold-build approach that MacFarland had pioneered in Ramadi and that Petraeus had been implementing throughout Iraq. The plan also included a more systematic way for ISAF units to partner with the Afghan security forces. McKiernan's new operational plan was developed with more input from Afghan military leaders than previous ISAF plans, and it was rolled out at a joint meeting of ISAF and Afghan commanders.[150]

Yet the plan remained limited by the fact that the NATO mandate did not include counterinsurgency, and McKiernan was bound by its restrictions. Because ISAF was a NATO (and not an American) command, McKiernan did not have a direct US chain of command through CENTCOM to the secretary of defense and the president. Instead, he reported through NATO's Supreme Allied Commander Europe to the North Atlantic Council, the alliance's political governing body. If the United States wanted to officially influence the direction of ISAF in Afghanistan, it had to work through NATO. This initial alliance command arrangement shaped how McKiernan viewed his role and strongly influenced his priorities throughout his tenure.

Soon after McKiernan was appointed, however, the United States reorganized its command arrangements in Afghanistan so that the ISAF commander would for the first time also report through a parallel US chain of command. Until then, fewer than half of the US troops in Afghanistan were assigned to ISAF and therefore reported to the ISAF commander. The majority of US forces were deployed as part of OEF, and reported directly to CENTCOM or US Special Operations Command, effectively bypassing McKiernan even though he was the senior American officer in Afghanistan.[151] This convoluted structure made it extraordinarily hard for ISAF and OEF forces to coordinate, and it meant that the United States could not give orders directly to McKiernan before the reorganization.

In the fall of 2008, the United States created a new US headquarters to oversee all American forces in the country, and through an arrangement known as dual-hatting, named the ISAF commander as the commander of US forces as well. On October 2, 2008, McKiernan was confirmed as the commander of the newly established US Forces-Afghanistan.[152] He became the first officer to concurrently command both ISAF and OEF forces, and now reported through a US chain of command as well as the NATO chain of command. This dual status enabled McKiernan to describe the war as a counterinsurgency in his role as the US commander—even though he could not make that same claim in his role as the ISAF commander. He deliberately inserted the term into his campaign plan to push the NATO allies into seeing the conflict that way, and started emphasizing counterinsurgency principles in discussions with local and national political leaders as well as with the media.[153]

McKiernan also issued two important directives during his early months that helped promote a counterinsurgency strategy. First, he sought to minimize the number of civilian casualties caused by ISAF operations. These tragic incidents inevitably hurt popular support for the international forces, especially since Afghan president Hamid Karzai made a point of publicly denouncing each one in order to bolster his domestic standing (particularly when the casualties were fellow Pashtuns).[154] On August 22, a US air strike against a suspected Taliban compound in the western district of Shindand killed more than 30 civilians (though initial reports estimated that 90 people had been killed, including 60 children).[155] Less than two weeks later, McKiernan issued a tactical directive that reiterated procedures designed to reduce civilian casualties, and added new restrictions on the use of firepower and close air support.[156] Civilian casualties declined after that, although they started increasing again at the start of the spring 2009 fighting season.[157] But the directive also generated some frustration among US troops who felt that it unduly constrained their ability to achieve their mission.[158]

Second, McKiernan issued further guidance on how to conduct counter-insurgency operations. It was designed to be easily understood by individual soldiers as well as senior leaders from the more than 40 national contingents within ISAF. The directions were also translated into Dari and Pashto so that the Afghan army and police could understand what NATO forces were trying to accomplish.[159] The guidance clearly stated, "Our operational imperative is to protect the population while extending the legitimacy and effectiveness [of the Afghan government] and decreasing the effectiveness of insurgent elements." It then identified 14 principles of operations, including defending the population; focusing on governance, development, and security simultaneously; maintaining presence in key population areas; and protecting civilian lives.[160]

These were all important steps toward implementing a counterinsurgency strategy, but none of them proved enough to change the overall trajectory of the war. Gates and Mullen shared what one reporter called a "ferocious intensity"[161] about rapidly improving the war's direction through innovative and creative approaches. McKiernan was moving at a pace that his superiors in Washington thought was far too conservative and far too slow. They essentially wanted McKiernan to do something analogous to what Petraeus had recently accomplished with the surge in Iraq—and grew increasingly frustrated with McKiernan's ongoing resistance to doing so.

Growing Concerns about McKiernan's Adaptability

By mid-fall 2008, Gates writes, he "was openly expressing concern to my immediate staff about whether I had made a mistake" in appointing McKiernan.[162] But two key developments in October seem to have accelerated Gates's concerns that he was not the right man for the job. First, in early October, a draft National Intelligence Estimate on Afghanistan leaked to the press. It concluded that Afghanistan was in a "downward spiral" and that it would be very difficult for the Afghan government to counter the growing influence of the Taliban.[163] This conclusion, and the fact that it had been made public, fueled the growing sense of urgency in Washington about the worsening war.

Second, and more importantly, Petraeus became McKiernan's boss when he took command of CENTCOM on October 31. This was the first of several changes that greatly altered the strategic and political environment in which McKiernan operated. It should have been a strong signal to the ISAF commander that Pentagon leaders now expected a new and dynamic approach in Afghanistan based on the principles that Petraeus successfully implemented in Iraq.[164] But since Petraeus had been one of McKiernan's subordinates during the 2003 US invasion of Iraq, this new relationship also had great potential to be awkward.

Furthermore, both men now were four-star generals, so the usual deference to higher-ranked superiors was absent.

Petraeus soon initiated his own assessment of the CENTCOM region. Unsurprisingly, it called for a more proactive counterinsurgency effort in Afghanistan involving new ideas, operational concepts, and organizational structures.[165] McKiernan resisted many of these changes, partly due to his concerns that NATO remained divided over such an approach.[166] This led leaders in Washington to conclude that he was too deferential to the alliance and either unwilling or unable to exert the leadership necessary to bring the allies along.[167] But tensions existed between Petraeus and McKiernan over other issues as well. The two men reportedly disagreed about most elements of the Afghan strategy,[168] and McKiernan was fairly outspoken about his view that it was "very helpful to dissociate Iraq from Afghanistan" and that there were "huge differences" between the two wars.[169] Even in hindsight, McKiernan continued to emphasize his differences with Petraeus:

> I don't think in retrospect Dave Petraeus and I probably saw things the same way in Afghanistan. I always thought that there was "here's how we did it in Iraq and let's do it that way in Afghanistan." There were some things I think that made a lot of sense to do. There was a lot of experience that the military had gained that applies to both theaters, but in many other ways I see Iraq and Afghanistan as very, very different environments, very different conditions and we didn't always agree on that.[170]

Petraeus was well aware that Gates and Mullen were deeply concerned about what they saw as the faltering military effort in Afghanistan, and understood that a key part of his new job at CENTCOM was to turn that effort around. But he had to work through McKiernan to get that done. While McKiernan was unquestionably correct that Iraq and Afghanistan were very different, he failed to understand that Petraeus essentially had a mandate from Washington to apply counterinsurgency lessons forged in Iraq to the war in Afghanistan. As he repeatedly pushed back against attempts to interject new ideas into the Afghan war, McKiernan came to be seen by his US superiors as too resistant to change, too conventionally minded, and not sufficiently adaptable.

One early example of these dynamics involved McKiernan's ongoing request for additional troops. In early September 2008, approximately three months after taking command, McKiernan told Gates that he needed four additional combat brigades and their enabling forces (such as helicopters, logistics, and transportation assets) in order to fight the war effectively. This request of approximately 30,000 troops represented far more continuity than change, since it essentially endorsed his predecessor McNeill's unfulfilled request for three brigades and added a fourth to address heavier fighting in the east.[171] Such a large request, however, was unlikely to be approved since the US military still had 148,000 troops deployed in Iraq at the time.[172] Only one additional brigade for Afghanistan was approved in the fall.

Gates remained skeptical that simply sending more troops was the right approach, since it risked furthering a sense of occupation among the Afghan people.[173] There were already almost 90,000 US and international troops in Afghanistan at the time, more than at any previous point in the war.[174] By November, Gates and other Pentagon officials were directing McKiernan to focus on better utilizing the forces he already had in a more robust counterinsurgency campaign, and move them off their bases into local areas to provide better security for the Afghan population.[175] Newly elected president Barack Obama did approve the deployment of 17,000 additional troops in February 2009, but as discussed later, that was less an endorsement of McKiernan's approach than a practical necessity to strengthen ISAF before the start of the summer fighting season and the Afghan presidential election scheduled for June.[176]

One of the clearest examples of McKiernan's lack of adaptability was his stubborn opposition to establishing a three-star operational headquarters, which would take charge of day-to-day counterinsurgency operations against the Taliban. The tens of thousands of troops that McKiernan commanded were divided among five regional commands, each approximately the size of an army division. The regional commands possessed a great deal of autonomy over how to conduct operations in their respective areas, and their approaches often varied widely depending on which individual country happened to be in charge.[177] This essentially caused ISAF to be less than the sum of its parts; whatever advantages it obtained in local flexibility were outweighed by the lack of anything resembling a consistent approach throughout the country.

Shortly after taking command of CENTCOM, Petraeus suggested establishing a new three-star operational headquarters, which would supervise and coordinate the day-to-day combat operations by the two-star regional commands and enable McKiernan to focus more on strategic issues.[178] This proposed division of responsibilities would reproduce the successful command structure in Iraq between the four-star strategic headquarters that Petraeus led from 2007 to 2008 and its subordinate three-star operational command.[179] But it had also been the model that had been used in Afghanistan from 2003 to 2006, when the United States established Combined Forces Command–Afghanistan (CFC-A) to take strategic responsibility for the war while its subordinate combined joint task force took on day-to-day operational responsibilities for fighting.[180] Even though there were far fewer US troops in Afghanistan at that time, Pentagon leaders realized even in 2003 that the scope of political-military responsibilities for the overall US commander in Afghanistan required separating strategic and operational responsibilities into two separate headquarters.[181] Both Gates and Mullen supported Petraeus's idea, for as Gates later recalled,

> In trying to solve the command and control problem for coalition forces in Afghanistan, Mullen and I agreed that the best alternative was to replicate the structure we had in Iraq—a four-star commander of all forces, McKiernan, with a subordinate three-star commander to manage the war on a day-to-day basis. McKiernan, like McNeill before him, spent a significant amount of time with Karzai and other Afghan officials, coalition ambassadors, and visiting government officials, and on NATO-related issues—diplomatic and political duties. The role was critically important but made apparent the need for someone else who would be totally focused on the fight.[182]

Yet McKiernan adamantly opposed the idea. This was partly because he did not think that NATO would give him enough strategic tasks to make this change worthwhile, since his NATO mandate still excluded missions like ministerial development, security force training, and support for the rule of law.[183] But he also continued to believe that the decentralized approach remained better, and that creating another layer of bureaucracy over the regional commands would create too much tension between the United States and its NATO partners.[184] McKiernan continued to fight this idea throughout his tenure. Perhaps drawing from his first months in command, he seemed to continue to see his NATO role in managing the alliance effort as his primary responsibility, without fully grasping that Petraeus and Gates were pushing for dramatic changes in the way he ran the war. McKiernan's enduring opposition to forming an operational headquarters to fight the Taliban's increasing strength exacerbated growing concerns among his superiors that he was too deferential to NATO and too resistant to change.

The Decision to Fire McKiernan

By May 2009, there were more than enough data to demonstrate that McKiernan's approach to the war was not working. Compared with the same period one year earlier, the number of attacks in the five regional commands between January and May 2009 increased between 21 percent and 78 percent, depending on the region; the number of attacks on government infrastructure increased by 156 percent; and the number of complex attacks increased by 152 percent.[185]

Yet the sequence of events that led McKiernan to be fired began much earlier, shortly after Obama's inauguration on January 20, 2009. On the campaign
trail, Obama repeatedly argued that the unnecessary war in Iraq had distracted
the United States from the far more important war in Afghanistan, and pledged
to finish the fight against al-Qaeda and the Taliban.[186] Once in office, Obama
asked Bruce Reidel, a former CIA analyst, to lead a 60-day review of US policy
toward Afghanistan that would help him develop a strategy for accomplishing
that objective.[187] However, Obama faced a problem of timing: he had to decide
whether to deploy more forces to Afghanistan before the review was complete if
he wanted them to arrive before the June election and the Taliban's usual summer
offensives. After a robust debate between the Pentagon and the White House
about the number of troops that could deploy in time, Obama announced on
February 17 that he would deploy 17,000 additional troops to Afghanistan.[188]

By the middle of March, both Gates and Mullen understood that the Reidel
review would recommend a far more comprehensive counterinsurgency strategy.
They had been seriously considering removing McKiernan since the beginning of
the year, and concluded that the new strategy required a far more aggressive and
adaptable commander. Mullen reportedly decided that the ISAF commander
had to be replaced after a meeting where McKiernan fumbled the answers to his
questions about reconstruction and counternarcotics operations.[189] The Reidel
review was officially released on March 27, around the same time that Gates and
Mullen started seriously discussing replacing McKiernan.[190] Gates later wrote
that Michèle Flournoy, the undersecretary of defense for policy, returned from
a trip to Afghanistan in early April with the same concerns about McKiernan's
leadership that he had long held, which further reinforced his decision to act.[191]

They did not have to look far to find his replacement. Army Lieutenant
General Stanley McChrystal worked for Mullen as the director of the Joint Staff.
He was best known for the five years he spent as the innovative commander of the
military's secretive Joint Special Operations Command, in which he completely
transformed the way it operated.[192] He was an obvious choice to reinvigorate the
Afghan command and implement a new counterinsurgency strategy, and his long
experience with special operations forces and light infantry contrasted sharply
with McKiernan's career in mechanized and armored forces.[193] Gates and Mullen
both thought the advice about Afghanistan they were receiving from McChrystal
in the Pentagon was better and more insightful than what they were hearing from
McKiernan in Kabul. They felt the same way about Lieutenant General David
Rodriguez, who had served as a military assistant to Gates and as the commander
of ISAF's eastern regional command (and who, like McChrystal, supported the
proposal for the three-star operational command that McKiernan so strongly
opposed). Gates and Mullen started quietly discussing replacing McKiernan with

McChrystal and naming Rodriguez as a new deputy commander. They soon ran the idea by Petraeus, who supported it enthusiastically.[194]

Throughout the wars in Iraq and Afghanistan, senior commanders who were deemed a poor fit for their positions had been moved to other billets or even promoted, instead of being penalized or relieved.[195] Casey was the most obvious example, who was then serving as the army chief of staff even though his performance overseeing the war in Iraq was far worse than McKiernan's performance in Afghanistan by any objective measure. But in the spring of 2009, there were no vacant four-star positions into which McKiernan could be shifted. That meant that Gates and Mullen either had to convince McKiernan to retire or they had to fire him, even though he had not done anything definitive that would typically warrant such a drastic action. But as Rajiv Chandrasekaran later reported, "The secretary and the chairman had come to believe that the war in Afghanistan required *immediate innovation and creative risk-taking*, even if it meant drumming out one of the Army's most-senior leaders."[196] Gates, who had previously shared his concerns about McKiernan with Obama, told the president in mid-April that he, Mullen, and Petraeus all recommended replacing McKiernan with McChrystal. Obama quickly approved their recommendation.[197]

In mid-April, Mullen flew to Afghanistan to try to convince McKiernan to retire. McKiernan knew that he no longer had the full support of his leadership, but he refused to agree to leave quietly. He told Mullen that Gates would have to fire him, since he had done nothing wrong and believed he was executing the guidance he had received.[198] He later said, "I suppose that would have been an easy, painless way out—just to say, 'Well, I've been here for a year and I'm rotating out.' But I told a lot of people that I was staying for two years. I couldn't look myself in the mirror if I said that."[199] Although admirable in one sense, this outlook further demonstrates how little McKiernan grasped about his strategic and political responsibilities as commander of a wartime theater. All senior US military leaders serve at the pleasure of the president and the secretary of defense, who are their civilian superiors in the chain of command. McKiernan ignored this fundamental aspect of civilian control of the military when he refused to step down at Mullen's request, regardless of any promises he had made to his staff or to his Afghan colleagues.[200]

On May 6, Gates flew to Kabul and told McKiernan that he had decided to replace him.[201] After returning to Washington, Gates took the highly unusual step of holding a press conference with Mullen at his side to announce that he had asked McKiernan for his resignation. He emphasized that the administration's new strategy required different military leadership, and that he recommended appointing McChrystal as the new commander and Rodriguez as McChrystal's deputy. Mullen also spoke at the press conference and strongly endorsed the need for fresh leadership. When a reporter asked what McKiernan had done wrong, Gates replied, "Nothing went wrong, and there was nothing specific."[202] Yet numerous reports on the decision stressed that McKiernan was seen as too cautious and too conventional in his approach, especially in contrast to Petraeus.[203] Gates technically only had the authority to replace McKiernan as the commander of US Forces–Afghanistan, since NATO officially appointed the commander of ISAF, but his decision left NATO no choice but to follow suit.[204]

Gates later wrote that firing McKiernan was "one of the hardest decisions I ever made. He had made no egregious mistake and was deeply respected throughout the Army."[205] But he also stressed the urgency he felt, saying that he couldn't wait until McKiernan's tour ended in the spring of 2010 to make the change. He had told Mullen in March, "I've got kids out there dying, and if I don't have confidence I have the best possible commander, I couldn't live with myself."[206] Mullen used virtually the same language in explaining his decision. "There are those who would have waited six more months" to ensure a smoother transition, he said. "I couldn't. I'm losing kids and I couldn't sleep at night. I have an unbounded sense of urgency to get this right."[207]

Conclusion: McKiernan's Failure to Adapt

McKiernan had an interesting perspective on his dismissal, which differed greatly from the perspectives of those who fired him. In an interview for an army oral history several months after leaving Afghanistan, McKiernan said that he probably should have communicated more effectively with his senior leadership. When asked what he would have done differently, he said that he should have developed better strategic communications.[208] He also stated:

> One of the reasons I fell out of favor, one of the reasons why, I don't know if "fell out of favor" is the right term, one of the reasons [Petraeus] and the Chairman [of the Joint Chiefs of Staff] wanted to change commanders, looking back, probably is because I had not "fed the beast" enough. If I were to do it over again I think I would have initially come in with a pretty robust capability to work strategic communications back to Washington and back to NATO.[209]

He also explained that he had tried to focus on operations in Afghanistan and perhaps had not spent enough time on maintaining relationships with his senior leaders.[210] In a separate interview, he said that he believed that "an operational commander needs to spend the vast majority of his energy and time and efforts focused inside the theater of operations and not on his trips to Washington."[211]

These statements reveal just how fully McKiernan failed to grasp the reasons for his dismissal. The ISAF commander was *not* just an operational commander. He was the four-star theater commander in Afghanistan, responsible for tens of thousands of US and NATO troops and the overall conduct of the war. McKiernan's problems had nothing to do with strategic communications or the ability to better manage his bosses; they had everything to do with his lack of adaptability in finding a more effective approach to winning the war. He did not understand the urgency that Gates and Mullen (and, after October 2008, Petraeus) increasingly felt about changing the trajectory of the war, nor that the success of the surge in Iraq had created even greater pressures for an innovative and more aggressive counterinsurgency approach in Afghanistan.

McKiernan lacked many of the key characteristics of adaptive theater leaders identified in Chapter 4. He did not fully understand the situation he was in, on several levels. He failed to grasp the dangers that the Taliban's growing success posed to his mission, and how urgently his superiors wanted to reverse the trajectory of the war. He continued to see his role as overseeing battlefield operations, and opposed changing the command structure to enable him to focus on broader strategic issues. He did not modify or abandon elements of his approach even when the evidence available strongly suggested they were not working. He did modify some existing procedures, through his directive on civilian casualties and counterinsurgency guidance, but these mostly consolidated and reemphasized previous guidance rather than offering a new approach. And he did not systematically organize or lead battlefield change, which is what his Pentagon superiors wanted most. He continued to believe that he was constrained by the divisions within NATO about counterinsurgency, rather than using the newfound US urgency about the war to push the alliance toward a new consensus to confront the Taliban gains.

In his memoirs, Gates offered his only public explanation of why he fired McKiernan. In a remarkable passage, he writes:

> To this day, it is hard for me to put a finger on what exactly it was that concerned me [about McKiernan in the fall of 2008], but my disquiet only grew through the winter. Perhaps more than anything it was two years' experience in watching generals like Petraeus, McChrystal, [Peter] Chiarelli, [David] Rodriguez, and others *innovate* in blending both counterterrorism and counterinsurgency operations, and observing their *flexibility* in embracing new ideas, their willingness to *experiment*, and

their ability to *abandon an idea* that didn't pan out and move on to try
something else. McKiernan was a very fine soldier but seemed to *lack the
flexibility and understanding* of the battlespace required for a situation as
complex as Afghanistan.[212]

In other words, McKiernan failed to demonstrate the adaptability required to
succeed as a theater commander in a complex counterinsurgency conflict—and
he was fired as a result.

Conclusion: The Failures of Theater Leader Adaptability in Iraq and Afghanistan

Although their wars were very different, Casey and McKiernan both failed to
exhibit the most important qualities of adaptive theater commanders. First and
foremost, they did not fully understand the nature of the wars they were fighting. Both men joined the army in the last years of the Vietnam War and were
profoundly shaped by the army's post-Vietnam focus on large-scale conventional
operations, discussed in Chapter 5. In Iraq, Casey recognized that he was fighting
an insurgency, but despite an overwhelming amount of evidence that the United
States was heading toward defeat, he never reexamined the critical assumptions
that shaped his entire campaign plan. In Afghanistan, McKiernan's tenure was
short but largely reflected the same approach as his predecessor despite a rapidly worsening situation. Notwithstanding ever-increasing pressures from his US
superiors, he failed to recognize and adapt to the growing evidence that this strategy was not working. Both Casey and McKiernan were responsible for wars that
were moving along an accelerating trajectory toward failure. The US military was
on the cusp of losing the war in Iraq when Casey was replaced, and Afghanistan
was plunging toward failure when Gates fired McKiernan. Yet neither commander proved capable of adapting his strategies to reverse the downward spirals
in his war, and to provide the United States any prospect of meeting its wartime
objectives in either country. These were momentous failures of adaptation, and of
generalship, at the theater level of war.

PART III

Looking to the Future

9

The Challenges of Future War

ONE OF THE biggest mistakes militaries make is to assume that future wars will resemble wars of the past. Very often this assumption is not made purposefully, since history clearly shows that no two conflicts are the same. But militaries are large bureaucracies, and as discussed in Chapter 1, bureaucracies naturally prefer incremental changes over deeper reform. Militaries are also more inherently conservative than other types of organizations, as they seek to control the tremendous uncertainties of warfare to the greatest degree possible, because the consequences of failure are so extraordinarily high. Even if individual military leaders explicitly acknowledge that future wars will differ from those in the past, their ability to translate that vision into specific changes within their forces may remain limited.

Furthermore, a military's recent combat experience may unintentionally make it harder to prepare for the challenges of future war. Combat imprints searing lessons among those who experience it, especially about how to fight and adapt effectively. But it can also produce subconscious biases and blind spots, which may inhibit the ability of these combat veterans to think clearly and creatively about the types of wars they will fight in the future. Starting in the early 1970s, Daniel Kahneman and Amos Tversky began a set of psychological experiments that demonstrated conclusively that systemic and unconscious biases affect how people process information, especially when trying to make sense of complexity.[1] Their work became the foundation of the field of behavioral economics, and earned Kahneman the 2002 Nobel Prize in Economics.[2] One of the earliest and most important biases they identified is called the *availability heuristic*: the more easily an example comes to mind, the more likely we are to think it will represent the future.[3] Recent experiences are usually easier to remember than past ones, and intense experiences like combat impress themselves into our memories far more than ordinary experiences do. As a result, those who have served in combat may

subconsciously assume that future conflicts are likely to resemble linear exten-
sions of their past battlefields.[4] This bias holds particular dangers for the leaders
of military organizations, who are responsible for preparing for future fights.

All militaries face challenges of accurate prediction, but these debates are
likely to affect the US armed forces even more than others in the years ahead. As
a global power, the United States cannot easily surmise when, where, or against
whom it will fight. Its security threats are not confined to its national borders or
even its hemisphere. It may be able to anticipate the rough outlines of some future
conflicts—with rising powers, rogue states, or malevolent nonstate actors—but
it can never be sure of who its next opponent will be. On September 10, 2001, for
example, no one serving in the Pentagon could have possibly imagined that less
than 24 hours later, their building would be directly attacked and the US military
would start preparing for a war in Afghanistan—a scenario for which US Central
Command had not even developed contingency plans.[5]

Furthermore, the US military will likely have a harder time predicting and
adapting to the character of its future conflicts, for at least three important reasons. First, strategic uncertainty today is higher than it has been in decades, certainly since the end of World War II and arguably since the early 1930s. Second,
any future conflict with a capable adversary will surely be fought in the cyber
domain, and possibly in outer space as well. These two domains of warfare are
entirely new and in many ways unknowable, since the military has virtually no
past experience or lessons to draw upon. Third, the scale and ever-increasing
speed of technological change in the 21st century is causing unprecedented disruptions in most human endeavors, and warfare will be no exception. The rest of
this chapter examines these three reasons in greater detail.

Increasing Strategic Uncertainty

The strategic landscape of today's world is undergoing a tectonic transformation.
First and foremost, great power competition has returned. After the Cold War
ended and the Soviet Union collapsed in 1991, armed conflict among major powers seemed to be a relic of the past. The Soviet collapse was famously characterized as "the end of history," and the United States enjoyed a "unipolar moment"
as it ascended to a position of unchallenged global preeminence.[6] The US military shrank by approximately 40 percent from its peak Cold War levels,[7] and its
subsequent operations shifted toward peacekeeping, stabilization, humanitarian
assistance, and other "military operations other than war"[8] as it worked to manage the consequences of civil wars and internal conflicts around the world.

Yet that unipolar moment did not last long. Only a decade later, the
September 11 attacks abruptly ended the halcyon era of promised peace,[9] and the

United States embarked upon what President George W. Bush described that same day as a war on terror.[10] For the next two decades, the United States found itself embroiled in a seemingly unending series of unconventional US military operations against the Taliban in Afghanistan, and against al-Qaeda and its successors across the broader Middle East and around the world. The United States also elected to invade Iraq in 2003 to topple Saddam Hussein, in a war that US leaders assumed would last only weeks or months, but that lasted more than eight years. The disastrous consequences of the war in Iraq, and its second- and third-order effects, continue to roil the Middle East to this day.

As the United States became deeply preoccupied with its wars in the Middle East and central Asia, events in the rest of world continued to evolve. The axis of global geopolitics began to shift steadily eastward to the Pacific Rim as prosperity in the region grew rapidly. China emerged as the world's most prominent rising power, leveraging an increasingly interconnected global economic system to rapidly expand its middle class and grow in technological knowledge, economic power, and international influence in ways that would have taken decades in previous eras. Since 1979, China's economy has grown by an average of almost 10 percent a year—which the World Bank described as the "fastest sustained growth by a major economy in history."[11] China had the world's seventh-largest gross domestic product in 1999, but it rose to third place in 2007 and has been second only to the United States since 2010. And China's purchasing power parity surpassed that of the United States in 2014, and has remained significantly higher ever since.[12]

Today, China rivals US global influence in nearly every category. Successive Chinese leaders have invested in sustained military modernization and reform, and many of its military capabilities now match (and some may even exceed) those of the United States and its allies.[13] Its military and civil technological prowess continues to grow, through sustained domestic investments by its rulers, as well as by pilfering foreign intellectual property.[14] China's military and economic power may not continue to grow as rapidly as it has in the recent past. It may even decline, because it faces many domestic challenges and demands from its population of 1.4 billion people. But China will nevertheless remain a major global competitor for the United States, with serious implications for potential future armed conflict. In any future military clash with China, the United States would face "a deadly trifecta of cutting-edge technology, advanced military capabilities, and substantial financial resources."[15] Furthermore, the Chinese military is larger than the US military, and could be expanded far more easily given China's tremendous population and authoritarian government.[16]

The US and Chinese economies remain deeply interdependent, so the outbreak of war between the two countries would have devastating economic as well as military consequences. That may help deter both countries from choosing to resolve their conflicts through the use of force. Yet war between the two countries nevertheless remains possible, more likely as a consequence of inadvertent escalation than deliberate intent. China's international behavior has grown increasingly assertive, especially in and around the contested South China Sea. China claims the area as its own, in contravention of international law. It is building civil and military infrastructure on some of the islands—and even building brand-new islands—that China claims are for peaceful purposes but could also be used to support military operations. China has also increased its maritime patrols in the area, typically using ships from its coast guard and maritime militias in order to appear less provocative. The United States has responded by conducting what it calls freedom-of-navigation missions in the South China Sea, sailing US warships within 12 miles of these islands to demonstrate that it rejects China's interpretation of international law.[17] In such a politically charged environment, it is not difficult to imagine how a small incident or accident at sea could quickly escalate into a far broader conflict that neither country deliberately sought.

Russia has also returned to the global stage as an influential and aggressive actor. The collapse of the Soviet Union in 1991 plunged Russia into a decade of political and economic chaos. The rise of Vladimir Putin to become Russian president in 2000 (and enduring strongman thereafter) signaled a new direction for Russia in the 21st century. Although Russia continues to struggle economically and demographically, Putin has effectively consolidated his internal power and is modernizing Moscow's conventional and nuclear forces in an effort to restore Russia's past role as a great power.[18] He has also used the increasingly capable Russian military to project power in Russia's near abroad and beyond, to include invading Georgia, annexing Crimea, and deploying significant Russian forces to fight in Syria.[19] Under Putin's leadership, Russia appears intent on undermining the solidarity of NATO, disrupting democratic states, and aggressively advancing Russian interests around the world. Russia exercises its power most effectively through conflicts that fall just short of the threshold of overt aggression that could trigger a Western response.[20] Moscow's interference in the US electoral process and its reflexive alignment with US adversaries around the globe make it a particular threat to US interests.[21]

Both Russia and China have also started conducting a relatively new form of low-level warfare that is often called hybrid warfare or conflict in the gray zone. This 21st-century form of unconventional war blends subversion, propaganda, political warfare, and paramilitary forces acting in ways that are unlikely to provoke a military response.[22] Russia has conducted this type of warfare against Crimea, Ukraine, and Estonia, sometimes followed by an outright occupation or annexation.[23] Iran has employed similar tactics in Syria, Lebanon, and Yemen.[24]

And China's use of its coast guard and maritime militias to exert its claims in the South China Sea and its island-building campaign also fall within this category.[25] These conflicts short of outright war also typically feature aggressive cyber intrusions and the use of active campaigns of disinformation and media manipulation. Given the legal and political difficulty in marshalling effective responses to these actions, this form of conflict will likely occur more frequently in the future.

Renewed great power competition is not the only driver of today's strategic uncertainty. Hostile and unpredictable regional actors such as North Korea and Iran continue to present a military and geostrategic menace to their neighbors and to the United States. North Korea already possesses some nuclear weapons,[26] and the US withdrawal from the 2015 nuclear deal with Iran raises the prospect that the Islamic Republic's nuclear weapons program could resume.[27] Both states have long exported their nuclear knowledge and covertly supported international terrorist groups, which pose less overt but no less real nuclear threats to the United States and its allies. Nonstate actors such as al-Qaeda, the Islamic State, and their ever-changing constellation of global offshoots present more unconventional challenges. These violent extremist groups, often distorting religion to motivate their followers, continue to foment significant unrest and conduct lethal attacks around the world. They are also seeking to develop weapons of mass destruction, though they face significant obstacles in doing so.[28]

Strategic uncertainty is also increasing because several international trends are interacting in ways that will increase social instability and global migration. First and foremost, global climate change is rapidly changing the physical environment of the planet (which will eventually pose an existential threat to the entire human race). According to NASA, global temperatures have increased by 1.4 degrees Fahrenheit since 1880, with two-thirds of that increase occurring since 1975.[29] That rapid rise is melting the polar ice caps at an unprecedented rate. In the Arctic, for example, the amount of sea ice has decreased by more than 13 percent each decade since the 1970s,[30] and many scientists forecast that the Arctic Ocean could be seasonally ice-free by the 2030s.[31] As it becomes increasingly navigable, the Arctic will emerge as a new area of geopolitical competition for resources and influence, especially with Russia.[32] The ice melt is also raising the levels of the world's oceans, which, when combined with increasing temperatures, may be causing more highly unpredictable weather patterns—such as severe storms and devastating floods in some areas, with droughts that are destroying agriculture and promoting wildfires in others.[33]

Climate change is already causing significant resource shortages and population displacements around the world. But those trends will continue to be exacerbated by the ever-increasing process of urbanization. More than half of the world's population lives in urban areas today, with that number rising to

60 percent by 2030. The number of megacities with populations over 10 million will also increase, from 33 in 2018 to 43 by 2030.[34] Most of these cities lie at or near sea level, which makes them particularly vulnerable to coastal flooding. That liability increases the likelihood of massive humanitarian disasters, triggering more internal displacements and refugee flows. Furthermore, megacities pose a range of challenges for humanitarian relief efforts and military operations that will make it even more difficult to address these problems.[35]

These destabilizing trends will be further compounded by changing demographics and increasing global inequality. Most economically advanced countries have declining birth rates, which will become a challenge to their continuing economic growth.[36] These nations will have fewer workers in the future, and will have to dedicate more of their national resources to care for aging populations.[37] Most of the world will have the opposite problem, though, as childbirth rates continue to increase and job creation rates struggle to keep up. Exploding populations have long been a problem in the developing world, but increasing global inequality is rapidly making this situation worse. Between 1980 and 2016, the richest 1 percent of the world's population captured 27 percent of global economic growth, while the poorest 50 percent captured only 12 percent.[38] According to one recent study, the richest 1 percent owns half of all global wealth, while the world's poorest 3.5 billion people control just 2.7 percent of that wealth.[39] Another study suggests that this already enormous gap will only continue to grow, with the richest 1 percent owning as much as two-thirds of global wealth by 2030.[40] Growing income disparities and fewer available jobs in an increasingly interconnected and turbulent world may exacerbate refugee flows,[41] leading to greater global economic and political unrest.

Last, but certainly not least, the crumbling of the post–World War II open international order is causing far greater strategic uncertainty than would have been imaginable even a few years ago. Populism, nationalism, mercantilism, and nativism are rising forces affecting democracies and autocracies alike. These political trends partly reflect dissatisfaction with the uneven effects of globalization, but they are also a reaction to perceived assaults against national, social, ethnic, and racial identities. These trends have ushered autocrats into power in some previously democratic states (Hungary, Poland, Turkey, and Venezuela, to name a few) and put democratic norms under assault in many other states as well.[42] Many countries are also rejecting the long-established institutions of the postwar era. Some are turning to the Chinese-financed Asian Infrastructure Investment Bank, for example, instead of traditional funding sources such as the World Bank or International Monetary Fund.[43] The most dramatic example of institutional rejection to date is the British decision to withdraw from the European Union. With its final form still unresolved as of this writing, Brexit has already had and will continue to have significant economic and political consequences for both Britain and an EU that is struggling to manage the European debt crisis and the increased authoritarianism of some of its members.

Yet the greatest assault on the postwar order is coming from the country that established it in the first place. The United States created most of the international principles and institutions of governance that emerged after World War II, and actively led international efforts to maintain that order for more than 70 years. Democrats and Republicans fought frequently and often loudly about how best to exercise US global leadership, but there was a strong bipartisan consensus that such leadership was the best way to secure US national security interests. That consensus all but disappeared in 2016, however, when Donald Trump won the presidential election. Trump's pledge to put "America First" explicitly rejected this leadership role. Before the election, he campaigned to reduce the US presence overseas, challenge US commitments to its allies and partners, and renegotiate or withdraw from international agreements like the Iranian nuclear deal, the North American Free Trade Agreement, and the Trans-Pacific Partnership.[44] He robustly followed through on these campaign promises once in office, and has further upended other long-standing principles of US foreign policy such as actively supporting democracy and human rights abroad. It is far too early to tell whether these policies will be sustained by whoever succeeds Trump in 2021 or 2025, or whether they will simply represent a temporary political aberration before a return to a more traditional American internationalist outlook. Yet some of Trump's policies have already damaged, if not yet completely shattered, important elements of the current international order.[45] Tragically, this American policy upheaval may be the most important cause of growing strategic uncertainty: the framework that has governed international relations for decades is under assault by the very nation that created it. It is completely unclear what framework, if any, will emerge to supplement or replace this long-standing international architecture.

Two New Domains of War: Outer Space and Cyberspace

For thousands of years, land and sea were the exclusive domains of life on earth, and of war. Humans have always lived on land, and have frequently relied upon the sea for transport, trade, and sustenance. They also soon began to fight over those domains. The tactics, armaments, and technologies of land and sea warfare evolved across centuries, and generations of military leaders studied past conflicts to prepare for the future. For centuries, adaptation to the inevitable surprises of the next war meant adjusting to unanticipated military developments on land and at sea.

The invention of the airplane in 1903 marked the first time in history that humans could navigate a new domain beyond the land and sea, and it profoundly changed the ways in which wars were fought. By the time that World War I began in August 1914, all of the belligerents fielded rudimentary air services outfitted with unarmed, fabric-covered biplanes. None had any previous experience employing airplanes in war, so there were no lessons upon which to draw. Yet by the end of the bloody conflict four years later, every combatant wielded a sizable air force with diverse capabilities. By 1918, these nascent air arms comprised hundreds of armed aircraft capable of air-to-air combat, scouting and reconnaissance, attacking ground troops, and bombing population centers. In just four years, airplanes grew from a whimsical novelty into a powerful fighting capability. Even the most far-sighted advocate of the airplane in 1914 would have had difficulty imagining how thoroughly air power would transform warfare in just four short years. Air power, although still a new concept, had already begun to revolutionize the ways in which wars were fought.

World War I firmly established that all future wars would be fought in the air as well as on land and at sea. World War II demonstrated the full fury of total war fought in all three domains. It culminated in the use of a new and terrifying weapon dropped from the air—the atomic bomb. Just as the invention of the airplane brought the air domain into warfare for the first time in World War I, advances in technology after World War II began to open new domains that would soon impact warfare as well. When the Soviet Union launched the Sputnik satellite into orbit in 1959, space became the fourth domain of human activity. The ensuing space race kicked off an unprecedented period of technological development that also produced advanced computers, microchips, and the internet. In concert, these inventions ultimately led to the creation of the world's first virtual domain, the unique realm of cyberspace. These technological breakthroughs, and the cascade of other inventions that accompanied them, have offered new and strikingly different challenges and opportunities for society. They have also posed vexing new challenges for those charged with thinking about and preparing for war.

Space and cyberspace now constitute the fourth and fifth domains of war. And just as militaries had to wrestle with the mysteries of the new air domain in 1914, today they are trying to discern the ways in which these two new domains could revolutionize warfare in the 21st century. Though commercial and military activity in these two domains has existed for decades, the US and other militaries have primarily used them to support operations in the three traditional warfighting domains of air, land, and sea. The advent of GPS and other space-based technologies has vastly improved US military command-and-control capabilities, for example, and the development of precision weapons that rely on space and cyber capabilities has changed the way the US military fights its wars. But technologies that depend upon space and cyberspace have been so thoroughly integrated into modern life that these two domains will almost certainly become battlefields in any future war among modern adversaries.

The space domain more closely resembles its more traditional counterparts on land, at sea, and in the air. Space is a physical realm with tangible characteristics and defined boundaries. Attacking and defending assets in space involves physical platforms operating in the real (if celestial) geography surrounding the earth.[46] Protecting the nation's commercial and military space architecture fits fairly easily into more established notions of terrestrial offense and defense. Orbiting platforms, together with their earth-based control nodes, serve as both sensors and targets. Attacking and defending physical assets in the space domain mirrors many of the same characteristics as attacking and defending physical assets in the other three physical domains.

Space capabilities already support a vast range of civilian as well as military capabilities. GPS satellites and other similar satellite constellations provide essential positioning, navigation, and timing (PNT) signals that enable a vast range of services used by businesses, governments, and individuals. Mobile banking, high-frequency stock trading, and emergency response coordination all depend upon accurate GPS, for example. PNT synchronization undergirds untold other functions, ranging from internet centers to power grids to modern agriculture.[47] Even in developing countries, space-based assets support banking, communications, and the transport of goods to markets. For advanced militaries, satellite signals enable precision strikes by guided weapons, navigation across trackless deserts, and encrypted global communications that link the most remote teams to higher headquarters anywhere in the world. Satellites also provide vital intelligence, surveillance, and reconnaissance, which range from detecting ballistic missile launches to tracking threatening military buildups.

As discussed subsequently, the rapid pace of technological change means that modern societies will rely ever more heavily on space-based capabilities and become ever more vulnerable to their disruption. Yet no war has yet been fought in outer space. The 1967 Outer Space Treaty proscribes the militarization of space.[48] But the United States and several of its potential adversaries seem poised to develop and deploy weapons into space, if indeed they have not already covertly done so.[49]

As space-based vulnerabilities continue to grow, they may provide too tempting a target for adversaries to ignore. To date, neither military nor commercial space capabilities have been significantly interrupted, even in wartime. The 1991

Gulf War was the first major military operation that relied on space-based capabilities like GPS navigation, satellite communication, and precision weapons enabled by PNT signals. In the decades since, the United States has not directly faced a wartime adversary that could interfere with or destroy those growing and vital capabilities. And the long wars in Iraq and Afghanistan have reinforced an unwarranted complacency among both military and commercial space users that wars will not extend beyond Earth's atmosphere.

The cyber domain offers an even more complex and unpredictable set of challenges to those thinking about the next war. Like outer space, cyberspace represents an entirely new domain of modern warfare. No major war has yet been fought in cyberspace, although skirmishing and incursions that fall short of traditional definitions of war have caused substantial disruptions around the world.[50] And attacks in cyberspace have now begun to appear in some limited conflicts around the globe, most notably the Russian cyber offensive against Georgia in 2008.[51] But even though the US military relies heavily on its cyber networks today, it has fought against adversaries in Iraq, Afghanistan, and Syria for nearly two decades without facing any serious threat to these vital networks. Partly as a result, it has yet to come to terms with the massive vulnerabilities that ubiquitous cyber technology now poses to its military operations, and even more importantly, to the entire nation that it is pledged to defend. War, in the future, will no longer be confined to the physical battlefield and may not even involve militaries fighting other militaries. The unique aspects of the cyber domain take it far outside the traditional boundaries of conventional warfare and threaten to make cyberspace the most complex and unpredictable realm of future wars.

While the space domain somewhat resembles the traditional domains of land, sea, and air, the cyber domain is different in every conceivable way.[52] Physical differences clearly delineate air, land, sea, and space, but cyberspace is man-made and virtual. Distance and size are essentially irrelevant, and the degree of speed is unprecedented compared to any other realm. Information that would occupy thousands of pages of paper can be moved halfway around the world in milliseconds. Data move at the speed of light, unconstrained by the physical forces that govern ships, satellites, or even hypervelocity missiles. Moreover, the characteristics of cyberspace are constantly changing. The internet of things (IoT), for example, will connect nearly every modern machine to the internet. According to Intel, 200 billion smart machines are already connected, and by 2025 the global worth of IoT technology could reach as much as $6.2 trillion.[53] That ongoing transformation, and whatever transformations succeed it, will continually expand and weave the cyber domain into nearly every corner of modern life.[54]

Attempts to establish boundaries, such as firewalls for cyber protection, will never be fully effective. Cybersecurity is a problem that can only be managed, not solved, as hackers and cyberwarriors continually find new vulnerabilities and system hosts race to catch up. Furthermore, most cyber experts believe that unwanted intruders and adversaries may *already* be inside nearly all nations' important government and business networks.[55] These silent spies are likely reconnoitering their targets and vacuuming up sensitive information for exploitation, often with the host entirely unaware of their presence. Virtually all civilian and military networks are likely to have some "zero-day vulnerabilities," which are exploitable flaws in computer hardware and software of which users are unaware.[56] Hostile agents discovering such vulnerabilities can exploit them right away, or simply file that information away for future use. The United States and its potential adversaries are undoubtedly cataloging some of these vulnerabilities as potential targets during future wars.

The cyber domain may become the most important domain of warfare in the 21st century and beyond, because it critically enables modern warfare in the other four domains. It therefore poses a major and unprecedented vulnerability. And that is a particular problem for the United States, because no military depends more on the cyber domain than the US armed forces. Directing continual global operations requires highly advanced communications capabilities, nearly all of which rely upon digital systems. And virtually all major US weapons systems rely upon microchips to function—including wheeled vehicles and main battle tanks, and literally every aircraft that flies and every warship that sails. Moreover, these tanks, aircraft, and ships communicate with each other and conduct operations jointly enabled by networked military computer systems. Among their other important functions, these sophisticated military battle networks allow these disparate elements to see each other and the enemy more clearly, helping to differentiate friend from foe to avoid deadly fratricidal strikes. Space capabilities are important to US military operations, but microchips, computers, and networks are its lifeblood.

Cyber systems so deeply enable every aspect of US military operations that they will inevitably be among the first targets in any future war. Many of these systems may already be penetrated, with malicious software quietly waiting to be unleashed with a few taps on a keyboard. Weapons and equipment may not function as advertised, communications between units may be lost or degraded as networks fail, and doubt about the veracity of available information may become widespread. Since many US military technologies rely on components that originate from foreign (and potentially hostile) sources, any system relying upon microchips, circuit boards, or routers from overseas may be suspect. Some of these systems might even be turned against American forces, attacking their US military operators and units.[57] Moreover, the US military relies heavily upon many commercial capabilities for its logistics and a number of communications functions. These enterprises are likely to be even more vulnerable to wartime disruption and exploitation than more hardened military networks.[58] Widespread cyber surprises and disruptions could create utter mayhem during the opening days of the next war, and possibly well into the conflict.

Warfare in the cyber and space domains will require tremendously adaptable US forces. The challenges and effects of warfare in these two brand-new domains may be completely unforeseen, since there is little or no past experience upon which to draw. Yet the US military today remains largely unprepared for these massive potential disruptions. Maneuvering, fighting, and communicating without reliable cyber and space-based capabilities has become almost unimaginable for America's technologically sophisticated armed forces. Virtually all of its command, control, and targeting capabilities depend upon cyber or space to some degree, but the US military does little to train its forces to perform the most demanding missions in an environment where cyber and space capabilities are either degraded or entirely disrupted. Few if any exercises require US forces to operate effectively over extended periods of time with unavailable or inaccurate GPS, disrupted communications from downed networks and satellites, and malfunctioning weapons from hacked guidance systems. The lack of operational experience and training while under cyber and space attacks can only promote unrealistic expectations for the future. Adapting to these potentially staggering challenges on the first day of combat in the next war could prove extraordinarily difficult.

The Increasing Scale and Speed of Change: The Fourth Industrial Revolution

Military planners have long had to contend with the technological shifts occurring in the world around them.[59] They have had to adapt to new weapons, warships, and munitions, as well as new civilian technologies like the internal combustion engine, flying machines, and the computer. Yet the scale and speed of technological change today is simply unprecedented, and will only continue to increase. In the last 20 years, these shifts have caused tectonic changes in the relationship between technology and society, businesses, and governments. New technologies are being invented and integrated with existing technologies at breathtaking velocities. As but one example, Apple's iPhone was introduced in 2007, and within a few short years, it entirely transformed how people communicate, travel, live, and work. Smartphones are now a ubiquitous feature of daily life in all modern societies, and even in many developing countries. Similar examples are too numerous to count.

Klaus Schwab has described this tremendous constellation of changes as the Fourth Industrial Revolution—an ongoing global transformation in which the boundaries between the physical, digital, and biological domains overlap.[60] He argues that the first industrial revolution, which lasted from 1760 to 1840, supplied the steam engine, railroads, and machine manufacturing. The second revolution, from roughly 1870 to 1914, produced electricity and mass production. The third revolution, sometimes called the digital revolution, occurred during the latter part of the 20th century, and furnished semiconductors, computers, and the internet to society. The fourth revolution started around 2000 and continues today. It subsumed the third, he argues, and stands out because of the explosion of technological breakthroughs and continually accelerating rates of change. The technologies of the Fourth Industrial Revolution are most easily seen in fields like robotics, machine learning, quantum computing, and artificial intelligence. But their impact extends far beyond these technical fields. Their unparalleled breadth and depth are disrupting and transforming nearly all aspects of labor, industry, government, and society as they reach into virtually every corner of our world.

These tremendous changes will inevitably transform the character of warfare. Clausewitz argues that the fundamental nature of war always remains constant, encompassing fear, uncertainty, and chance. Those aspects of war are immutable, but the ways in which wars are fought, what Clausewitz called the "grammar" of war, constantly morphs and shifts as societies and militaries evolve. The shifts wrought by the Fourth Industrial Revolution are already disrupting and changing the battlefields of the 21st century and will continue to do so for decades to come. Along with the emergence of the space and cyber domains, as discussed already, these changes include:

- **Artificial intelligence, big data, machine learning, autonomy, and robotics.** These new technologies will enable combat operations to unfold at speeds never before experienced in war. Adversaries may decide to employ these capabilities in an unconstrained manner in order to garner their immense potential operational advantages of machine, rather than human, speed. Such decisions could include delegating lethal decision-making authority to autonomous weapons systems, which could be guided by AI algorithms rather than human decisions. Countering such attacks in time to avoid defeat may require taking humans partly or completely out of the loop for certain types of lethal responses. Allowing intelligent machines to make decisions about killing humans is fraught with moral peril. But countries that choose to disregard moral boundaries and ignore the laws of war while using these technologies may gain a decisive military advantage, which would put immense pressure on their adversaries to do so as well. In this environment, intelligent machines may be required to adapt, learn, and adapt again to other machines' actions, absent human involvement. Wartime adaptation in this setting may partly transcend human actors, which will pose entirely unknown challenges.[61]

- **The return of mass and the defensive advantage.** In recent decades, the US military has increasingly traded mass for precision, enabling smaller forces using guided weapons to fight successfully and obviating the need for the massive industrial armies of the last century. But military scholar T. X. Hammes convincingly argues that the Fourth Industrial Revolution means that mass is once again returning, but in a new way.[62] Modern technologies that were once only affordable by the most advanced and wealthy states are becoming ever cheaper and more widely available. This democratization of advanced weapons and technological systems may allow a wide range of states (and even individuals) to acquire highly destructive capabilities. Rapid advances in additive manufacturing (also known as 3D printing) may offer these combatants the opportunity to produce these weapons en masse.[63] As an example, even small states (or nonstate actors) with limited resources can deploy masses of autonomous drones capable of swarming and destroying a target and delivering great massed destructive power. Terrorists may not be able to afford to build a B-2 bomber, but they may be able to deliver similar effects with a large but inexpensive swarm of suicide drones. Hammes believes that such swarms "may make defense the dominant form of warfare," since they may make "domain denial much easier than domain usage."[64] Autonomous drone swarms could also upend the advantages enjoyed today by large wealthy states that can afford advanced military capabilities, and increasingly make those capabilities available to much smaller states and entities with far fewer people and resources. The impact of such a threat on the US military, with its limited number of highly capable but staggeringly expensive weapons systems, would be profound. Adaptation under these conditions could require a nearly unimaginable reversal—potentially sidelining generations of expensive US weaponry and quickly shifting to producing masses of easily deployable smaller systems.

- **A new generation of high-tech weapons.** The Fourth Industrial Revolution is also yielding a host of innovative new weapons systems that may radically reshape how the next war is fought. The United States and some of its potential adversaries are rapidly developing and adopting new capabilities like directed energy weapons, hypervelocity projectiles, and hypersonic missiles and railguns.[65] These new weapons will transform the range, speed, and destructive power of conventional arms far beyond anything previously imaginable. None of these weapons have been used in warfare yet, so their actual impact on the battlefield is so far unknown. Still, that very uncertainty suggests that the degree of adaptation required when these types of weapons are first used could be immense. And that process may be particularly painful for the United States if it requires technological adaptation. As discussed in Chapter 10, the US military acquisition system is obsolete, overregulated, and horrendously slow—usually taking more than a decade, and sometimes more than two decades, to field major new weapons systems.

- **The unknown X-factor.** The next war will almost certainly involve technologies and capabilities that were secretly developed by the United States as well as by its future adversaries. These new technologies could make current weapons inoperable or obsolete, and even offer a surprise war-winning capability. But neither side can be sure that its most secret weaponry will actually be a surprise when first employed—or whether the adversary is already in the know and waiting with an unexpected countermeasure. Even the most carefully guarded technology secrets can turn out to be unexpectedly compromised, and their value suddenly nullified.[66] Adaptation to entirely new and unanticipated enemy capabilities will be a significant problem for military leaders in the next war's opening battles. But an effective enemy counter to closely held US secret weaponry may prove an even greater shock, offering yet another unexpected challenge and sudden need to adapt.

Taken together, these different aspects of the Fourth Industrial Revolution will continually change the character of warfare in unprecedented ways and at dizzying speeds. They will make adaptation in the next war a monumental challenge for the US military—and that challenge will only increase in the coming decades as the vast disruptive potential of this era unfolds.

Conclusion: The Adaptation Gap

The 21st century offers a series of challenges for military planners that may present the most demanding set of problems in decades. Given the wide range of trends that are colliding around the globe, the challenge of predicting the character of the next war with any accuracy seems nearly impossible. Yet the US military will still have to make big decisions—about its warfighting doctrine, what weapons to buy, and how to train its leaders—in order to effectively fight in this unknown future.

These emerging dynamics suggest that wars of the future will look dramatically different than the wars of the previous decades. As a result, the gap between the predictions and the reality of future wars will likely grow as well—and that will make effective adaptation even more important. Figure 9.1 illustrates what we call the *adaptation gap*. The left side of the figure depicts the challenge of adaptation in the past, showing the difference between predictions about the next war and the actual war that resulted. As we have argued throughout the book, adaptation has always been necessary to bridge the gap between the type of war that was predicted and the inevitably different realities of an actual war. But all of the trends discussed in this chapter are making it harder than ever to make accurate predictions about future wars. That means that the gap between predicted wars and actual wars is likely to grow ever larger, as shown in the right side of the figure. The degree of adaptability that militaries will need to bridge that growing gap will become ever more important—even as it becomes ever more difficult.

The growing adaptation gap poses challenges for all military forces, including potential US adversaries. Yet that gap will remain a larger problem for the United States than its potential adversaries as long as it remains a global power. China and Russia are both major powers, as discussed above, but their political objectives are primarily regional and will likely remain so well into the future. They seek to establish spheres of influence in their neighborhoods, not to extend their military power around the world. Although they are clearly willing to take advantage of opportunities in other regions, as Russia has in Syria, they see the United States as the main obstacle to their objectives, and that the US military is by far the most likely adversary in their next major war. The United States, by contrast, does not know whether its next war will involve Russia or China— or North Korea, or Iran, or some other unknown adversary that may seem as unlikely a foe today as Afghanistan did on September 10, 2001. Even if the effects of strategic uncertainty, two new domains of war, and the increasing scale and speed of change were to somehow affect all militaries equally, the adaptation gap would still remain a greater problem for the United States than for its potential adversaries simply because of its global role and responsibilities.

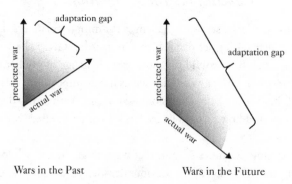

Wars in the Past   Wars in the Future

FIGURE 9.1 The Growing Adaptation Gap

Furthermore, the adaptation gap poses particular challenges for the United States because, as a status quo power, it largely seeks to defend itself and its allies. But aggressors often possess a substantial first-mover advantage. They force adaptation upon their victims, using the initial advantage of surprise to their benefit. And that initial advantage becomes even more important when aggressors choose to fight against strong powers like the United States, since it offers them the best opportunity to disrupt and perhaps even defeat superior military forces. Future US adversaries may well decide that their best strategy involves forcing the US military to adapt to a fast-moving foe attacking in unexpected ways.

The adaptation gap is therefore likely to be larger for the United States than for other militaries. In the next chapter, we examine whether the US military today is up to meeting the challenges of the growing adaptation gap.

US Military Adaptability Today

THIS CHAPTER RETURNS to the central question that motivated this book: Is the US military adaptable enough to prevail in the wars of the 21st century, especially as the adaptation gap continues to grow? Gazing into a rapidly changing and uncertain future, it is difficult, if not impossible, to know for sure. Yet substantial evidence suggests that the answer may be no. Chapters 5 through 8 show that the US military's efforts to adapt over the last two decades in Iraq and Afghanistan provide a troubling picture at best. Individual soldiers and units demonstrated tremendous creativity in adapting to the challenges they faced, by finding new ways to engage with local populations, engineering new forms of protection for their vehicles, pioneering new tactics, and finding creative ways to accomplish their missions. At higher levels, however, adaptation proved far harder and repeatedly failed, often with disastrous consequences. In doctrine, the army's standard processes for revising doctrine proved so cumbersome that the 2006 counterinsurgency manual completely bypassed them, and might never have been written if a number of stars hadn't aligned at exactly the right time. In technology, the services remained so deeply wedded to their legacy programs and procurement plans that they repeatedly refused to provide front-line troops with the weapons and other systems that would help save their lives and accomplish their missions. In leadership, the longest-serving senior commander in Iraq failed to understand the character of the war he was fighting, and never revisited the key assumptions of his strategy despite overwhelming evidence that his approach was failing. And at a critical juncture in Afghanistan, the theater commander was fired by the secretary of defense for being too inflexible, rigid, and unimaginative—an extraordinary action that last occurred in 1951, when General Douglas MacArthur was relieved by President Harry S. Truman during the Korean War.

These experiences in Iraq and Afghanistan should already raise some serious concerns about US military adaptability. Yet, as we discussed in the previous chapter, adaptation is likely to be even more challenging in the future than it has been in the past, because the adaptation gap is rapidly growing. Increasing strategic uncertainty, the completely new challenges of warfare in outer space and cyberspace, and the transformational technologies of the Fourth Industrial Revolution may all radically disrupt the future battlefield and will demand levels of adaptation that will dwarf the demands of the past. Moreover, speed may prove to be the most dominant characteristic of the next war, creating even more vexing dilemmas. Sudden and cascading events will place immense pressure on US military and civilian leaders to understand what is happening in real time, and adapt with unprecedented swiftness as battlefield dynamics continuously shift and evolve.

In such an environment, the successful tactical adaptability of the recent past cannot be taken for granted in the future, and the grave failures of adaptability at the higher levels must be addressed even more urgently. Yet almost two decades of war have caused remarkably few changes in the ways that the US military develops and updates its doctrine, how it acquires new technology, and how it educates, selects, and promotes its leaders. Changes in all three areas continue to proceed slowly if at all, with improvements often taking years. That pace is simply not sufficient to keep up with the extraordinarily rapid changes in the strategic environment that will shape the wars of the next decades and beyond.

In the rest of this chapter, we diagnose some of the key problems that currently affect US military adaptability, in order to better identify their sources and consequences. While this book focuses primarily on US land forces, as noted in the introduction, many of these problems apply to air and sea operations as well. Since the challenges of future war identified in Chapter 9 cut across all three of these domains (and extend into the brand-new domains of outer space and cyberspace), many of the limitations we discuss may affect the US Navy and Air Force as much as the US Army, Marine Corps, and special operations forces. We provide recommendations on how to start addressing these shortcomings in the next and final chapter of the book.

Adaptability in Doctrine

The US military currently faces three major challenges to its adaptability in the area of doctrine. The excessive amount of doctrine hinders flexibility and creativity; the process for revising doctrine remains agonizingly slow; and unrealistic training provides too few opportunities for commanders, particularly at the more senior levels, to test and practice new doctrine against a rapidly adapting enemy.

Excessive Doctrine

The US military today has an astonishing amount of doctrine, with a multifarious library collection that is vast in size, scope, and detail. As noted in Chapter 1, the army alone has 1,340 doctrinal publications totaling tens of thousands of pages, if not more. The other services also maintain large volumes of doctrine (though not at that scale), and joint doctrine sits as a capstone on top of all service publications. Doctrine governs every aspect of military operations, from countering weapons of mass destruction to the details of mortuary operations. Yet for the most part, doctrine is intended to be descriptive and not prescriptive.[1] It should serve as an authoritative guide, not as a dogmatic checklist to be applied mechanically. Commanders are expected to use doctrine as a starting point, but to adapt it when necessary. Yet, at the same time, deviating from doctrine is strongly discouraged. As the preface to Joint Publication 3-0, *Joint Operations*, states: "The guidance in this publication is authoritative; as such, this doctrine will be followed except when, in the judgement of the commander, exceptional circumstances dictate otherwise."[2]

But the sheer volume of published doctrine, particularly for the US Army, makes that adaptability difficult, since military leaders are expected to know and use approved doctrinal solutions in battle before trying other approaches. In some ways, the very existence of doctrine on a subject can hinder adaptability—and the more detailed the doctrine, the more problematic this becomes. Stepping outside of doctrinal lines requires the type of boldness and risk-taking that doctrine's authoritative nature suppresses. The more specific and comprehensive the doctrine, the more it may undermine the creativity and innovation of those confronting new and unfamiliar environments.

US military doctrine does emphasize the importance of adaptability. The joint publication that governs all US military operations, for example, uses some form of the word "adapt" 17 times.[3] The capstone doctrine of the Marine Corps does so 37 times, and the US Army's doctrine on leadership does so 82 times.[4] Yet despite this rhetorical emphasis, current doctrine says very little about what adaptability involves or how to develop it. This is a striking omission given the repeated emphasis on how important it is. Doctrine applauds more adaptability, but fails to provide any semblance of a road map to get there.[5] The net result is that US military personnel, especially in the army, are told to be adaptable while simultaneously being saddled with a plethora of doctrine that they are expected to follow. That is not a formula for adaptability.

Slow Processes for Revising Doctrine

Chapter 5 demonstrated that the existing processes for revising doctrine can be painfully slow. In 2004, Army Lieutenant General William Wallace directed his staff to produce an interim counterinsurgency manual in six months rather than the usual two years. Despite their best efforts, the result was far too limited and too poorly disseminated to help US forces in Iraq counter the growing insurgency. The effort to undertake a full revision proceeded at an agonizingly slow pace, especially given the escalating levels of deadly violence against US troops. When Lieutenant General David Petraeus replaced Wallace, he deliberately bypassed these existing processes and took the unprecedented step of convening a group of outside experts to do the job instead. They produced a full draft within six months and had fully revised and updated the text by the time it was published six months after that.

Little seems to have changed in the more than a decade since Petraeus decided that he needed to circumvent the entire existing process for developing and revising doctrine. That process remains almost comically slow today, especially in an age where information moves at the speed of light. Joint publications, for example, are revised every three to five years, with a standard review and approval cycle of about 18 months from start to finish. Fast-track reviews can be shortened to just over 12 months, but even changes that are deemed urgent require nearly six months to rush through the doctrine review bureaucracy.[6] Optional "test" publications covering new topics can also be requested, but lengthen the process by two additional years. Service doctrinal timelines are comparable, as most publications seem to be revised every few years.

Why is that pace so slow? In the normal review cycle, the steps are so laborious, and the coordination and staffing process so time-consuming, that only a small number of the hundreds of joint and service publications can be under revision at any given time. Changes in joint doctrine must be reviewed by each of the services, and disagreements can grind the process to a halt for months, if not years. For service doctrine, every command and internal constituency has an opportunity to provide comments and record disagreements. In the army, for example, such a review would typically include input from its 30 different branches and centers of excellence, as well as a host of major and minor commands and headquarters. Every effort will be made to reach consensus, which can involve endless cycles of revisions and additional reviews. Even if senior leaders are prepared to overrule any objections, the staffing process still absorbs vast amounts of time.

The services and the Joint Staff have tried to encourage adaptation by establishing organizations responsible for capturing lessons learned in near real time. These organizations were partly designed to shortcut the laborious process of changing doctrine by compiling the good ideas and best practices that units discover or create during training exercises and operations. They do not change doctrine, but they can capture good ideas from the field and rapidly disseminate them across the force in the form of handbooks or other informal reports.

The army created the first of these organizations, called the Center for Army Lessons Learned (CALL), in 1985. It was initially charged with collecting and synthesizing best practices from training at the army's combat training centers, and from the biennial Warfighter exercises that train its fighting division and corps headquarters. As early as 1989, CALL also began collecting lessons from army units engaged in contingency operations and peacekeeping missions. After the 9/11 attacks, CALL sent a series of assessment teams overseas to capture lessons from Afghanistan and later the Iraq conflict. The Marine Corps followed the army's lead in 2006, creating a lessons-learned program and center with essentially the same mission.[7] The joint community has also had a Joint Lessons Learned Program since the mid-1980s that was run by Joint Forces Command for many years, but is now overseen by the Joint Staff.[8]

These programs do help improve adaptability by assembling new ideas and feedback from the field, though they often focus on minute tactical details and their reports are usually posted online rather than actively disseminated by trainers to the force. But informal reports and handbooks are not a substitute for doctrine. They do not have the same authoritative force, and they do not seem to feed into the formal process of doctrinal change in any systematic way. The Joint Staff and the services' plodding and overly bureaucratic process for revising doctrine remains largely unchanged, and unequivocally remains far too slow and inflexible for the rapid adaptations that may be needed in any future war.

Flawed Training and Exercises

Military forces learn how to apply doctrine in peacetime through training and exercises. These events sharpen the combat edge of forces readying for battle and shape how they will think and adapt during actual missions. They also enable units to learn what works and what does not in complex military operations, since approaches that seem sound on paper may turn out to be impossible to execute in practice. The lessons from training and exercises also help shape and inform improvements in doctrine.

The US military invests more time and resources in its training and exercise programs than any other military in the world. It seeks to ensure that no soldier, sailor, airman, or marine will perform a task for the first time in combat. Some of these exercises are as small as teaching infantry squads the proper tactics to attack a trench line, while others can involve tens of thousands of troops, tanks, and aircraft in the field. Large war games are conducted alongside US allies and partners, such as the annual Foal Eagle exercises with South Korea or the biennial Keen Sword exercises with Japan. Many large exercises also involve computer simulations that replicate live operations with sophisticated digital representations of large-scale combat. Most large US military warfighting staffs train for their operational missions using these simulations, especially since large live events are expensive and can require immense amounts of physical space.

Rigorous training exercises are the only way to experiment with and validate military doctrine outside of war. The US military services do a very good job of providing demanding training events that teach and test their core service functions at the tactical level. The army, for example, has long provided exacting training programs for tactical formations at brigade level and below at its combat training centers, and for division and corps headquarters in Warfighter simulation exercises. The Navy's Top Gun program and the Air Force's Red Flag exercises do much the same for combat aircrews. But some of these tactical exercises, and many higher-level service and joint exercises, include artificial constraints or scripted elements that reduce the realism of the event. These parameters often shield participants from being fully tested against unconstrained and free-thinking adversaries in order to improve their skills at rapid adaptation.

Sometimes those constraints are imposed so that the services can demonstrate the value of their preferred operational concepts or weapons systems without risking a public failure. For example, the dates and times of expensive missile-firing exercises in both the army and navy, particularly those requiring the interception of an enemy missile, are well known to firing crews in advance. Missile crews also often know when and where the enemy attack will originate, and typically have ample time to "tweak" their own weapons system to peak levels of performance before the expected event.[9] After the fact, the services can then claim that their weapons systems and tactics have been "validated" and thus should be included in official doctrine. Unfortunately, these sensitive systems may not work nearly as well under the fog and friction of war. And just as importantly, carefully scripted events prevent service members from having to adapt quickly under pressure, which means that they cannot practice doing their tasks under the most demanding wartime conditions. As a result, the US military does not actually know if some of its expensive weapons systems will function effectively until the opening hours of combat. This may be far too late to fix a costly failure that could have been discovered through rigorous training beforehand.

One powerful example of this problem occurred just before the 1942 Battle of Midway in the Pacific.[10] Admiral Isoroku Yamamoto, commander of the Japanese naval forces, had devised a plan to defeat the American carrier task forces defending Midway Island. In early May, Yamamoto ordered a series of fleet tabletop war games to validate his plan of attack. His red team, playing the US Navy, devised several effective strategies to counter the Japanese plan. In two different scenarios, they simulated unexpected bombing attacks from US warplanes that sent several of the Japanese carriers to the bottom. But each time the Japanese fleet was mauled by the team playing the Americans, the exercise umpires either nullified the embarrassing results or reduced the losses. Although some of the Japanese officers participating in the exercise were quietly unhappy with this blatant intervention to uphold the fleet commander's plan, few dared to speak out. In the end, Yamamoto's blueprint for the Midway attack plan was validated with almost no alterations, despite the worrisome flaws that the exercise had revealed. Unfortunately for Tokyo, one month later the Americans struck the Japanese fleet assembled at Midway in ways that closely resembled how the Japanese red team attacked the fleet in the exercise. US Navy dive bombers and torpedo planes sank all four of Yamamoto's carriers, saving Midway from invasion and inflicting one of the greatest defeats on the Imperial Fleet of the entire war. As the Japanese found out, skewing or ignoring exercise results can lead to catastrophic outcomes in battle.

The best-known modern example of deliberately distorting exercise results occurred during the 2002 Millennium Challenge exercise, also known as MC02, which was run by US Joint Forces Command.[11] It brought together actual and simulated forces from all four military services in a complex Middle East scenario. It involved 13,500 people and cost $235 million. Paul Van Riper, a retired marine lieutenant general, commanded the opposing forces in the exercise, and his guidance gave him the freedom to play the enemy as he saw fit. During the opening move of the war game, Van Riper swarmed the US fleet with scores of small boats in a deadly asymmetrical attack. This unexpected gambit crippled the surprised US naval force in the exercise, throwing the game into turmoil. Exercise planners, worried about how the loss of the fleet would affect other exercise goals, restored the shattered armada and reset the game. They then constrained the ways in which the opposing forces could operate in the following moves so that all the game objectives could be achieved. Van Riper reported that he was ordered to pull back his opposing forces so that US units could land safely on shore, required to disclose some of his troop locations so that US forces could find them, and prevented from conducting a chemical weapons attack.[12] Van Riper quit in disgust, and the controversy became public when it was reported in the press.

Clearly the war game needed to be reconstituted in some way after the initial surprise defeat, but the way it was altered gave an overwhelming advantage to US forces. Unfortunately, distorting exercises to ensure the success of US forces happens far more often than it should. In 2017, a US Navy captain who frequently commands opposing forces in naval exercises wrote, "The reality is that we repeat this experience [of MC02] on a smaller scale multiple times each year."[13] While some exercises should and do focus on rehearsing existing plans thoroughly on a step-by-step basis, the remainder should include a realistic, free-thinking enemy that would require US forces to rapidly adapt to unforeseen conditions. That all too rarely happens.

Put simply, many US military exercises fail to prepare participants for the true realities of war. Failing to adequately validate existing US doctrine against unconstrained adversaries undermines the wartime efficacy of that doctrine. But the even larger problem is that it prevents military leaders at all levels from training as they must eventually fight, against a thinking and evolving enemy that will require US forces to adapt quickly, effectively, and constantly. As the Japanese learned at Midway, the results of this failure can be disastrous.

Adaptability in Technology

At the tactical level, US military forces in the recent wars proved quite adept at using existing technologies in new and creative ways. That success cannot be taken for granted, of course, and it is one of the many reasons why rigorous and unconstrained exercises are so important. At the institutional level, however, technological adaptability during these wars was utterly abysmal. The services often prioritized their long-term procurement plans over urgent battlefield requests and failed to energize their acquisition systems to support the needs of those doing the fighting. Few if any of the dynamics that led to those failures have changed in the intervening years, because of two main problems: the structural tension between the services and the combatant commanders, and the broken acquisition system.

Structural Tensions between the Services and the Combatant Commanders

For most of US history, the military services were responsible for all aspects of military operations. In 1944, for example, Army Chief of Staff General George C. Marshall was responsible for both raising and training the US Army and for helping devise and gain approval for the operation to invade the continent of Europe. That changed in 1986, when the Goldwater-Nichols legislation gave joint commanders primary responsibility for employing US military forces. The combatant commanders became responsible for fighting battles and winning campaigns, while the service chiefs remained responsible for organizing, training, and equipping their forces. This dramatic change created a structural conflict between the services and the combatant commands (CCMDs), since their perspectives, incentives, and especially time horizons suddenly became very different. Since the CCMDs are responsible for fighting and winning the nation's wars, they necessarily concentrate on conducting current operations and developing war plans for scenarios that could quickly emerge. The services, by contrast, focus far more on long-term priorities. At the risk of oversimplification, the combatant commands generally focus on today, while the services focus on tomorrow.

This structural tension was a primary cause of the institutional failures of technological adaptation in Iraq and Afghanistan. Chapter 6 shows that senior leaders in the US Army and the Marine Corps protected their major acquisition programs, which were primarily designed for future conflicts, at the expense of competing battlefield needs from their men and women in combat. While the commander of US Central Command focused exclusively on the two consuming wars in his region, the service chiefs were concerned that those wars might simply be temporary distractions from the greater hazards of unknown future wars. They feared that devoting too many resources to the wars of today could upend the long-term priorities of the service.[14] Senior army and marine leaders in the Pentagon refused to buy large numbers of life-saving MRAPs because they feared that they would lose congressional support for their nascent JLTV program—even though the influx of supplemental funds meant that they did not need to make explicit trade-offs between the two vehicles. And army leaders in the Pentagon repeatedly stonewalled requests from their soldiers in Afghanistan for Palantir intelligence software in order to protect the troubled DCGS-A network, denying intelligence analysts access to critical tools that could have helped soldiers achieve their mission more effectively and with fewer casualties.

This structural tension is built into the very foundation of the US military. It will continue to create perverse incentives for the services to focus more on their future needs than the needs of their troops in combat, especially when fighting wars that are not deemed existential. There are no easy fixes or specific recommendations that can help overcome this problem. It will continue to hinder wartime adaptability unless future service leaders recognize this dynamic and are willing to put some of their future programs at risk in order to ensure that combat forces get the weapons and technologies that they need in order to prevail on the battlefield today.

The Broken Acquisition Process

Institutional adaptation in technology is and will remain deeply problematic, because the Department of Defense is shackled to an archaic and glacially slow acquisition and procurement process. The pathologies of defense acquisition have been well known for decades, and its enormously detailed processes and requirements are far too complex to adequately summarize here.[15] But they have proven virtually impervious to change, because Congress, the Department of Defense

(DOD), and the defense industry all face strong incentives to keep things the way they are.

The problems with the acquisition system begin with Congress. The acquisition process is governed by the paralyzingly complex Federal Acquisition Regulation (FAR), which derive largely from congressional guidance and statute. The 2019 FAR, for example, was a staggering 2,300 pages long, and the 2019 Department of Defense supplement to the FAR added another 1,680 pages of regulations.[16] Congress has an understandable interest in effectively overseeing the defense acquisition process, given the many billions of taxpayer dollars at stake every year. But the FAR, with its drawn-out milestones and requirements for lengthy testing before a new system can move into operational use, substantially hampers the US military's ability to incorporate cutting-edge technologies. It also requires a massive acquisition and procurement workforce within the Department of Defense—more than 207,000 people in 2017, larger than the entire US Marine Corps.[17] This poses big problems for the military in general, but it also makes rapid adaptation in wartime an even greater challenge. Moreover, Congress continues to add layers of oversight and regulation nearly every year in an effort to avoid massive cost overruns and ensure successful outcomes. Since few if any of the older rules are ever reassessed or removed, the acquisition process simply continues to get slower, more bureaucratic, and less effective.

The Pentagon exacerbates these statutory problems with its own internal procedures and policies. Every step of the acquisition process requires regular coordination between the Office of the Secretary of Defense, the Joint Staff, Congress, and the services.[18] The system focuses heavily on caution and cost avoidance, following dense sets of rules, required milestones, and lengthy independent testing. Even relatively simple and inexpensive programs can take five to six years to be fully fielded.[19] Furthermore, the services often change their requirements for their new systems during the prolonged development process, which makes the process take even longer. A vicious cycle results: changing requirements cause industry delays, which in turn make more changes necessary to keep the new system current with emerging threats and technological developments.

Finally, the US defense industry also makes rapid wartime adaptation difficult. All companies naturally mold their business models to find profitability in the incentives and strictures that make up their business environment. For companies that work with the Department of Defense, that means compliance with the FAR, partnering with (sometimes inexperienced) military program managers, and responding to all sorts of change requests from their military customer, no matter how disruptive to design or production plans. There are few if any incentives in the current acquisition model for industry, rather than the military, to inject fresh ideas and innovation, especially about how to accomplish a mission in an entirely different way or at much less cost. The shrinking defense industrial base has meant less competition in recent decades, which also reduces incentives for innovation. It has also raised the stakes for each individual contract, which has led to the bizarre but now common situation where leading defense firms submit joint proposals, preferring to share profits with their competitors rather than take the risk of losing a contract altogether. The entire system is now heavily biased toward keeping risks low, for both industry and the government, which only prolongs development and delivery timelines.

There is virtually nothing in this immensely bureaucratic system that facilitates rapid acquisition in the face of sudden combat demands, especially to address complex solutions or to devise major new weaponry. During the wars in Iraq and Afghanistan, DOD and the services created rapid fielding offices to circumvent the sclerotic acquisition processes and meet urgent operational needs more quickly. In 2002, the army created a new Rapid Equipping Force (REF) that allowed soldiers of any rank to directly request new capabilities.[20] In 2003, DOD created the Joint Rapid Acquisition Cell to develop and field joint capabilities more quickly.[21] In 2006, the Marine Corps established an urgent universal needs statement process that created a streamlined structure to respond to requests from marine commanders in the field, which became formalized into a new Rapid Capabilities Office in 2016.[22] This multitude of new offices helped speed smaller items to the field, like improved body armor and some individual tactical radios. These new rapid acquisition organizations improved adaptability, but they did not and could not develop and produce larger systems like MRAPs or other major platforms. As small and relatively new organizations, these rapid fielding offices also remain bureaucratically vulnerable to pressures from senior military leaders. As shown in Chapter 6, army soldiers requested and began to receive some Palantir software from the REF. Yet senior army leaders were so committed to the approved DCGS-A system that they prevented the REF from fulfilling those requests without more senior-level approval, which was frequently withheld.[23]

In addition to rapid equipping offices, the Department of Defense and the services also created several other organizations designed to address emerging battlefield problems. The vexing challenge of improvised explosive devices (IEDs) topped this list of new challenges from the war zones of Iraq and Afghanistan, and stymied years of US efforts to find effective counters. A series of new organizations were created to take on these and other asymmetric threats arising from these unconventional wars—such as the Joint Improvised Explosive Device Defeat Organization (JIEDDO), the army's Asymmetric Warfare Group, and the army's Task Force ODIN.[24] Each new organization included some ability to develop accelerated technological solutions to the problems at hand. Yet they remained limited either by their specific charter, such as responding to IEDs, or their small size and influence that constrained their ability to impact major programs or drive substantial adaptation. As a former senior Pentagon logistician has noted, "These are point solutions versus enterprise solutions."[25]

The prognosis for technological adaptability at the institutional level therefore remains bleak. The proliferation of organizations designed to bypass the usual acquisition process in the recent wars simply reaffirms how broken that process is. These workarounds have improved adaptability to some degree, but they are not a substitute for desperately needed institutional reform. Rapid acquisition offices cannot provide large systems, and as the Palantir example demonstrates, they can face severe bureaucratic obstacles when they seek to provide capabilities that challenge a service's official program of record.

The US Army recently established Futures Command, led by a four-star general, in order to address this institutional problem. Although it is far too early to judge the effectiveness of this new organization, the command itself illustrates the immensity of the problem. It was created, according to a memo signed by the acting secretary of the army, because

> The Army's current requirements and capabilities development practices take too long. On average, the Army takes from 3 to 5 years to approve requirements and another 10 years to design, build, and test new weapons systems. The Army is losing near-peer competitive advantage in many areas: we are outranged, outgunned, and increasingly outdated.[26]

Futures Command, which reached full operational capability in July 2019, attempts to solve this problem by centralizing what had been widely dispersed army modernization and acquisition efforts into a single command.[27] Though some of its early efforts appear promising, it remains unclear whether Futures Command will succeed at taming the army's internal acquisition bureaucracy. And even if it does achieve this laudable goal, the army will still face extended acquisition timelines and stultifying bureaucratic regulations unless and until Congress thoroughly overhauls the larger defense acquisition process.

Adaptability in Leadership

During wartime, military leaders at all levels—from those who command tactical combat units to those who roam the corridors of the Pentagon—will need to quickly understand the unfolding conflict, assess what needs to be done, and rapidly adapt to unexpected events in order to prevail. Given the unprecedented speed with which future wars will unfold, many of their decisions will have to be made in days and weeks, if not hours. Yet today's military faces three big challenges in ensuring that its leaders are adaptable enough to meet these demands. They include the generational legacy of the recent wars, problems with the system of professional military education, and a growing tendency toward risk aversion that threatens the principle of mission command.

The Generational Legacy of the Recent Wars

For almost 20 years, the US military has been fighting unconventional wars in Afghanistan, Iraq, and Syria, and across the broader Middle East and Africa. In the US Army and the Marine Corps, many officers and NCOs know the sound of gunfire and the chaos of actual battle, experiences that can never be fully replicated in peacetime training. Decades of fighting in Afghanistan and Iraq mean that US military leaders, from sergeants to generals, possess a great deal of combat experience. Today's generals were young captains or majors when the attacks of September 2001 occurred, and many have spent their entire professional lives engaged in the wars that followed.

Yet this deep bench of wartime experience may also present a dangerous blind spot for the current and rising generations of US military leaders. Since 2001, the US military has been able to operate in ways that maximized its substantial advantages while obscuring its less apparent vulnerabilities. The most advanced capabilities of the US military have been largely untested and unchallenged by the limited nature of the unconventional wars in Iraq and Afghanistan. US adversaries have been bereft of virtually any high-tech weaponry. American ground forces have operated untouched by air attacks, battling adversaries who had no air forces and could offer only insignificant defenses against US air raids. Globe-spanning supply lines, logistics hubs, and communications networks have also reliably remained safe from attack.

None of those conditions are likely to occur during a future war against a more capable and reasonably competent adversary. As we noted in Chapter 9, the next conflict could feature unconstrained combat in the cyber and space domains, disrupted air and sea supply lines of communications, and contested control of the air. During his tenure as army chief of staff, General Mark Milley repeatedly warned that the future may look entirely different than the past. He observed, for example, that an entire generation of officers and senior NCOs has become accustomed to operating from large bases with air-conditioned barracks, hot showers, and even Pizza Huts—but noted that conditions in the future will likely be very different. Soldiers and units will need to move constantly in order to survive, since fixed sites will be easily targeted in an era of nearly unlimited information. In other words, Milley concluded, "If you're stationary, you'll die."[28] Furthermore, he cautioned that soldiers should expect little support on the battlefield beyond water, food, ammunition, fuel, maintenance, and medical treatment. And, with characteristic bluntness, he warned, "Learning to be comfortable with being seriously miserable every single minute of every single day will have to become a way of life for an Army on the battlefield that I see coming."[29]

That will be a profound change for a force that has spent almost two decades fighting in one particular way. The services are addressing this by overhauling their training to focus on high-end operations. The army's combat training centers, for example, are increasingly focused on restoring lost conventional skills, such as maneuvering mechanized formations against enemy tanks, and coordinating live artillery fire. But as discussed in Chapter 9, current and future military leaders will remain deeply shaped by the past two decades of irregular war. This inevitably includes strong but subconscious assumptions that will affect how they process information and make decisions. That invisible bias may limit their ability to rapidly adapt to new conditions, tending toward choosing approaches and solutions, even in very different environments, that worked in the past.

Even more importantly, the current and rising generations of military leaders are at great risk for groupthink, because almost all of them have spent their most formative years sharing largely similar experiences in Iraq and Afghanistan. Yet extensive studies have demonstrated that groups with diverse perspectives are more innovative and make better decisions than ones with homogenous outlooks.[30] In civilian settings, that often means that groups with racial, gender, and economic diversity make better decisions than groups drawn from only one segment of society. In a military setting, however, those traditional markers of diversity can pale in comparison to the shared experience of combat. Though their individual experiences certainly vary, they have all served in two prolonged irregular wars against tenacious but low-tech adversaries, with outcomes that can best be described as inconclusive. And since the military only promotes people from within, and its more senior leaders are disproportionately drawn from the combat arms, the diversity of perspectives among the military's leadership will continue to shrink as these Iraq and Afghanistan generations move up the ranks in the years to come.

In some ways, the generational effects of Iraq and Afghanistan may resemble those that occurred after the Vietnam War. That conflict was very different from the recent wars, of course, but it profoundly shaped an entire generation of military leaders. Those who chose to remain in the military did a truly remarkable job of rebuilding a broken force and successfully prepared their services for large-scale conventional operations like the 1991 Gulf War. But the lieutenants and captains, privates and specialists of Desert Storm became the colonels and generals and sergeants major who led the force into Iraq and Afghanistan. And as discussed in Chapters 5 and 8, their collective experiences made it extremely difficult for many of them to adapt to the challenges of these new and completely different types of war. As this cycle inexorably continues, the lieutenants and captains, privates and specialists of Iraq and Afghanistan will become the colonels and generals and sergeants major who will lead the force in the next war. They will need to make deliberate efforts to recognize and avoid the potential dangers of groupthink that their shared formative experiences may engender in order to adapt to a rapidly changing environment and possibly very different wars of the future.

Problems with Professional Military Education

The US military invests a great deal of time and resources in the professional military education (PME) of its officers.[31] They attend a series of lengthy courses over their careers to develop them for increasing levels of responsibility and complexity. Of these, the most important are two 10-month-long educational programs, one at the mid-career level and the other at a more senior level. At approximately 10 years of service, the best captains and majors are selected to attend a command and general staff college, and as they near 20 years of service, the best lieutenant colonels and colonels are selected to attend one of the joint or service war colleges. These programs represent a significant investment in preparing officers for future leadership roles, but they suffer from two important flaws. Many of these programs are not sufficiently rigorous, and they unintentionally exacerbate the problems of homogenous thinking and perspectives we have discussed. Both of these flaws reduce the critical thinking skills and creativity that rising US military leaders will need to address the challenges of warfare in the 21st century.

First, many PME programs, especially at the more senior levels, simply are not rigorous enough. The 2018 *National Defense Strategy*, which Secretary of Defense Jim Mattis personally shaped, blasted the PME system as having "stagnated, focused more on the accomplishment of mandatory credit at the expense of lethality and ingenuity."[32] The requirements of these programs are often far less demanding than those found in comparable civilian programs, with fewer required readings and writing assignments that do not sufficiently challenge their students. Some of the mid-level programs are so easy on their students that they have been described as "no major left behind" courses.[33] Too many of the military's mid-level staff colleges and even senior-level war colleges have long operated on the equivalent of a pass-fail basis. For many years, some of these schools administered few if any examinations, and demanded little of their graduating students save regular attendance at classes.[34] Even today, the number of students who fail any of these yearlong programs is vanishingly small, and often none at all.[35] There are some commendable exceptions, of course. The Naval War College curriculum is rigorous and demanding, for example, and the army's Basic Strategic Arts Program, a tough four-month course for new strategists, routinely drops students for poor academic performance.[36] But they remain the exception rather than the rule. Most PME courses remain essentially pass-fail programs where everyone passes.

This lack of rigor also means that students have no serious incentive to excel at their studies. Not only do they know that they are almost certain to graduate, but neither poor performance nor academic accolades significantly affect their career paths.[37] Every graduate receives an academic performance report in his or her official file, but promotion boards routinely discount them in favor of fitness reports that document an officer's performance in the field. Board members can easily identify (and therefore choose to disregard) academic performance reports, since they are recorded on a different form from the fitness reports for staff or command jobs. Promotion boards care far more that an officer has attended a program than how well he or she did in that program. Completion of schooling checks the block, which counts as much as or even more than the quality of academic performance. Peers and promotion boards alike view excelling in the classroom as much less important than performing well in a military command or staff job.[38]

Second, programs at the staff colleges and war colleges also promote homogenous thinking, and reinforce the already commonly shared perceptions among their students. In the past, the services enabled many officers to enroll in full-time civilian graduate programs, often before they attended mid-level PME courses. During the 1970s and 1980s, thousands of officers were funded to attend graduate programs at top universities in order to study with some of the world's foremost civilian scholars and students. These experiences greatly broadened their perspectives and strengthened their critical thinking and analytic skills. In the 1980s, for example, the army paid for as many as 7,000 officers to enroll in full-time civilian graduate programs annually. But by 2014, that number had been slashed to between 600 and 700 officers each year.[39] Even accounting for the fact that the army was about 40 percent smaller in 2014 than it had been during the Cold War, the number of civilian graduate opportunities still declined by a factor of six.

Starting in the 1990s, most of the services revamped their PME programs in order to offer all of their students a civilian-accredited master's degree simply by fulfilling the requirements for graduation. This change coincided with deep cuts to the amount of funding allocated for civilian graduate education. Historically, PME institutions frequently offered graduate degrees in partnership with local universities, which required often substantial additional coursework, papers, and examinations beyond the military curriculum. But today, both mid-level and senior PME programs award most of their students an accredited master's degree upon graduation, even though they may never interact with civilian academics or students.

Yet a master's degree from a PME institution is *not* the same as a master's degree from a civilian institution, for a variety of reasons. As we have noted, these programs are often far less rigorous. Furthermore, academic achievement is not a primary factor in admissions to either mid-level or senior PME programs. Instead, officers are selected to attend largely based on their military performance. Before 2019, most officers were not required to take the GREs, the standardized test required for most graduate schools in the United States.[40] It not yet clear how the GREs will factor into the PME selection process, but the academic entry bar will likely remain lower than at many civilian universities—and far lower than at the most competitive civilian universities. It will also have a far less demanding curriculum.[41]

But the most critical difference between civilian graduate schooling and PME is that the vast majority of students at PME institutions are military officers, which promotes a uniformity of thinking and a lack of openness to new ideas that discourage adaptability.[42] Unlike at civilian universities, PME classrooms typically consist of large numbers of people who look alike, shop in the same stores, and often share remarkably similar life experiences. A war college seminar that is chock-full of officers with one or more decades of common experience in uniform is unlikely to produce a healthy exchange of diverse ideas. When the instructors are retired military officers, as they frequently are, the effect is even more pronounced. These dynamics exclude divergent thinking from the classroom, depriving the military's most talented rising leaders from the give and take of the varied outlooks and sharply discordant arguments that commonly characterize civilian academia. Military students in largely homogenous PME classrooms are unlikely to readily question their embedded assumptions, or to learn how to be open to disruptive ideas. Rising military leaders need to be exposed to people who are different, think differently, and make them uncomfortable by challenging their assumptions. Today's PME institutions rarely afford them that opportunity.

A 2013 study by two professors at the US Army War College, Stephen Gerras and Leonard Wong, demonstrates why this is such a problem. They gathered data on key personality traits among the lieutenant colonels and colonels who comprised the student body at their institution. One of these traits was openness, which, in their description, seems quite similar to adaptability. According to the authors:

> Openness is manifested in a strong intellectual curiosity, creativity, and a comfortable relationship with novelty and variety. People scoring high in openness tend to be more creative and more aware of their feelings. They are more likely to hold unconventional beliefs and can work with symbols and abstractions. People with low scores on openness tend to have more conventional, traditional interests, preferring familiarity over novelty. They tend to be conservative and resistant to change . . . Leaders high in openness search for relevant and conflicting perspectives. Not only are they imaginative, but they also solicit other points of view and are comfortable debating with those whose perspectives differ from their own. *They are generally more receptive to change.*[43]

Their analysis found that the most successful army officers had lower degrees of openness than the US population in general, which already poses concerns about their degree of adaptability. But then they studied the traits of the students who were chosen to command brigades—a position of responsibility reserved for the very best colonels, and which is essentially a prerequisite for promotion to general. They found that these officers, those with the brightest futures ahead of them, scored even *lower* in openness than the rest of their war college classmates. They note this disturbing paradox: "The leaders recognized and selected by the Army to serve at strategic levels—where uncertainty and complexity are the greatest—tend to have lower levels of one of the attributes most related to success at [the] strategic level."[44]

This paradox does not bode well for the adaptability of future military leaders, especially at the theater level. And homogenous PME programs only exacerbate this problem rather than help alleviate it. The examples of Generals George Casey in Iraq and David McKiernan in Afghanistan demonstrate all too clearly the dangers of failing to dispassionately examine and, when necessary, abandon long-held assumptions and cherished strategies.[45] Yet many PME programs may be doing far too little to instill this trait in the military's most promising rising leaders. These findings do not augur well for the adaptability of those who will lead the nation in the wars of the future.

Risk Aversion and Mission Command

Mission command is a key principle of the US military's warfighting philosophy, built around decentralized command and control. It empowers junior leaders to execute actions independently in combat within the boundaries of broad guidance from higher commanders, so they can adapt in whatever ways are necessary to achieve their commander's intent. It enables them to operate effectively with only minimal guidance, so they can continue their operations or even take command if their leader becomes a casualty. It requires a bond of trust throughout the chain of command, and is rightly seen among military leaders as the most important precept of US military leadership.[46] Mission command will remain extremely important in future wars, especially as troops disperse across the battlefield in order to survive.

Yet even though mission command is the cornerstone principle of US military leadership, it often remains aspirational in practice. As the four-star commander of the army's Training and Doctrine Command recently wrote,

> The bad news is many in our Army find the idea of mission command confusing or insincere. For some, there is a difference between what mission command should be and what actually happens. Over the past decade, leaders at various levels routinely cited their personal experience in garrison, during field training, and while operationally deployed as at odds with our mission command philosophy.[47]

The army's professed commitment to mission command is often usurped by its compliance-based bureaucracy, which is especially prominent in garrison and away from the battlefield. In August 2019, for example, the army had almost 500 different publications of regulations in place, all of which provide guidance that soldiers are expected to follow.[48] Failing to do so can lead to administrative punishments and, in some cases, legal prosecution. When the army is training at home, in its constellation of garrisons across the United States, senior commanders often evaluate their subordinate commanders and junior leaders more by their compliance with these regulations than by their bold audacity and ability to improvise and adapt in training.[49] Tactical commanders are often evaluated and compared based on the extent to which their units are complying with all sorts of different requirements, from dental checkups to sexual harassment training to motorcycle safety. Those who repeatedly score poorly on these myriad compliance factors rightfully worry about their next performance evaluation. To junior leaders, this practice exemplifies an army more committed to micromanagement than mission command. A 2014 army study, for example, found that 41 percent of junior NCOs did not believe that they were empowered to make decisions, and only 59 percent were satisfied with the amount of freedom they had to perform their jobs.[50]

Furthermore, it is literally not possible for army commanders to meet all of their specified requirements. In 2015, an internal army document found that it would take an average company commander 451 days to complete all of the required annual training.[51] That represented an increase of more than 50 percent since 2002—which was already an impossible standard to meet.[52] In 2015, the same two professors at the Army War College who did the study on officer personality traits published a damning study called *Lying to Ourselves: Dishonesty in the Army Profession*. Wong and Gerras described the culture of compromised integrity created by this disconnect between mandatory requirements and finite available time. Leaders across the army felt compelled to lie and cheat in order to meet statistical scores that their command deemed satisfactory, in order to avoid career damage and censure.[53] To his great credit, former secretary of the army Mark Esper worked to address this problem during his tenure by eliminating many such rules. But it is laughable that some of the rules he eliminated ever existed in the first place—such as the ones that required soldiers to wear reflective belts during daytime physical training or that required them to fill out an onerous risk-planning worksheet before requesting leave.[54] These are laudable steps forward, but the scale of the problem remains immense and continues to undermine the trust required for mission command.

Mission command should work better in combat operations, because it is explicitly designed to improve battlefield performance. Yet again and again in Iraq and Afghanistan, junior leaders complained about micromanagement by their senior leaders during combat operations.[55] Even in missions where small units were spread across hundreds of miles of remote and hostile terrain, leaders at higher levels of the chain of command found ways to interfere with the initiative and authority of their subordinates. While commanding the army's Training and Doctrine Command, General Stephen Townsend made reinforcing the principle of mission command one of his top priorities, because he was troubled by soldiers "who don't believe that we as an Army are consistently practicing the principles of Mission Command."[56] He noted that instead of providing broad guidance to their subordinates, many senior leaders in Iraq, Afghanistan, and Syria instead insisted that extensively detailed concepts of operations be submitted for their approval before conducting any mission. As a result, he said, many junior leaders "are not buying this Mission Command thing."[57]

Furthermore, new technologies are making the tendency toward micromanagement even worse. Senior commanders far from the battlefield can often now access real-time, full-motion video footage from drones orbiting overhead of even the smallest subordinate units as they conduct operations—which means that they can issue orders directly to those units in the middle of battle. Generals and colonels now have the technological wherewithal to jump down multiple levels of the chain of command to make decisions about the conduct of tactical engagements, undermining mission command even further.[58]

Junior and mid-level commanders also face unprecedented scrutiny of their battlefield actions after the fact. In the army today, for example, every fatality—even those unequivocally the result of direct enemy action—triggers a mandatory investigation, a practice formerly reserved primarily for peacetime training accidents.[59] These intrusive investigations probe every aspect of unit leaders' actions that may have contributed to any soldier's death, even in an all-out firefight with the enemy. While this level of scrutiny may make sense to identify and correct failings that lead to preventable accidents, it sends exactly the wrong message to combat leaders. Employing such a legalistic method to dissect the cause of losses known to be caused by the enemy cannot inspire the trust and confidence in junior leaders, and thus further undermines mission command.

While investigations for significant combat actions with substantial losses may be appropriate, a policy that directs an investigation of every battlefield casualty is not. Such demoralizing investigations would have been unthinkable in any previous major conflict, partly because the large numbers of US casualties would have required thousands of such investigations every year. Few leaders will feel encouraged to be adaptable and innovative in combat knowing that any mistake that results in friendly casualties could result in a career-ending reprimand. Senior commanders on the ground can intervene to encourage initiative and bold combat leadership despite these strictures. But the message from senior military leaders, and from the US military as an institution, deeply undermines rather than encourages the adaptability and initiative they purport to champion.

Conclusion: The State of Adaptability Today

Our assessment of US military adaptability today in doctrine, technology, and leadership offers serious cause for concern. It suggests that the US military may encounter substantial difficulties adapting to the unprecedented speed and demands that will likely characterize the next war, particularly in a major conflict against a highly capable adversary. Its record of adaptation since September 2001 is decidedly mixed, as we discussed in Chapters 5 through 8. But there is little evidence to suggest that the US military today is significantly more adaptable, or will be in the future. Problems with adaptability in doctrine, technology, and leadership from the recent wars remain unaddressed, and will hinder the US military's ability to adapt to the challenges of a future environment—especially one characterized by increasing strategic uncertainty, entirely new domains of war, and the ever-increasing scale and speed of change.

As a result, the adaptation gap that we discussed in Chapter 9 is likely to continue to grow. The next and final chapter of the book provides some recommendations for ways that the US military can improve its adaptability and shrink that gap, to ensure that it is as prepared as possible for the challenges of its future wars.

Improving Adaptability in the US Military

THROUGHOUT THIS BOOK, we have argued that adaptability is one of the most, if not the most, important attributes of military forces. Our examples from the wars of the 20th century, as well as the recent wars in Iraq and Afghanistan, illustrate that militaries that can rapidly adapt to the inevitably unforeseen circumstances on the battlefield gain tremendous advantages over less adaptable adversaries. This increases their chances for victory and may reduce their casualties. We also argue that the rapidly changing 21st-century strategic environment, and its exponential rates of change, will make adaptability even more critical in the future than it has been in the past.

This poses some real challenges for the US military. Although today it is unquestionably the most advanced and most powerful fighting force in the world, it may unknowingly soon find itself in a situation similar to the one that European militaries faced at the beginning of the 20th century. In August 1914, France, Germany, and Britain believed they were well prepared for whatever the next war would hold. Their views were substantially influenced by the major wars that they fought during the previous century: the 1870 Franco-Prussian War, and even the Napoleonic wars of the early 19th century. Few had studied the bloody evolution of firepower and field fortifications in the American Civil War, which foreshadowed the steadily increasing lethality of weaponry, and the growing importance of trenches to afford some protection. Fewer still seriously considered lessons from more distant but recent conflicts, such as the Second Boer War of 1899–1902 in South Africa and the Russo-Japanese War of 1904-5. These wars introduced the machine gun, rapid-firing artillery, smokeless powder, and hand grenades. They also demonstrated the value of entrenchments for protection and barbed-wire obstacles to disrupt infantry assaults. Despite a rising tide of new military technology and important lessons from these small wars on their periphery, France, Germany, and Britain all believed that the next major war would basically be a linear extension of their past European experience. Another war on the continent would be a traditional one, though leavened with some new and heavier firepower that each army already had brought into its fighting arsenal. The warfighting doctrine of European armies had evolved little since the late 1800s; their use of military technology only in fits and starts; and their thinking about leadership virtually not at all.

World War I decisively swept away virtually all of these militaries' prewar assumptions. It has been said that a battalion commander of the German or French army serving at the end of World War I who was suddenly catapulted into the late 20th century could, with little adjustment, understand the shape of modern war. But if that same officer were transported from 1914 into 1918, a battlefield of the very same war would have been nearly unrecognizable because of the staggering changes in warfare that had emerged in just four years.[1] Wartime adaptations in doctrine, technology, and leadership demolished the vision of war that had been almost universally embraced just four years earlier. In its place was a nascent form of combined-arms warfare, built around the tanks and airplanes that had emerged by the end of the war. Combined-arms struggles remained a fundamental cornerstone of land warfare well into the 21st century.

The analogy of the officer of 1914 provides an important lesson for today's US military leaders. Arguably since 1918, and almost certainly since World War II, conventional land warfare has focused primarily on large-scale, combined-arms battles. The vast mechanized air-land campaigns fought by the US Army, the Soviet Red Army, and their allies across Europe to defeat Nazi Germany remain the archetype of massive, modern mechanized war. Even the 1991 Gulf War reinforced far more than it changed that fundamental model of armored warfare. Despite the advent of precision strike and unprecedented new capabilities in satellite imagery, navigation, and night vision, Operation Desert Storm still closely resembled the large clashes of mechanized formations in northern and eastern Europe during World War II (though against a far less capable adversary). As the US military emerges from almost two decades of unconventional warfare in Iraq and Afghanistan, where its tanks have been largely irrelevant, it is seeking to restore its capabilities to fight these large, conventional combined-arms battles. The US Army has stepped up the number of its armored brigades, is rebuilding its capabilities in long-range precision fires, and is once again training its forces in mechanized warfare for what it calls decisive combat.[2]

The next US war might involve that type of large-scale, combined-arms battles. But just as the wars since 9/11 have proven unpredictable in character, it is also possible—and, given all of the tremendous changes discussed in Chapter 9, may even be probable—that the next war will look radically different. The US military could easily be as wrong about the next war as the European militaries were about World War I. Military leaders always have to make choices, and place their bets, about the shape of the next war. They must decide what weapons to buy, what warfighting doctrines to pursue and practice, and what leadership talents to nurture and emphasize. But history teaches us that those same leaders must be prepared to get many of those choices substantially wrong.

As the US military prepares for the next war, it must balance the need for *preparation*—to ready the force for what is anticipated—with the need for better skills in *adaptation*—to ready the force for the unexpected. To that end, the US military must embrace rapid adaptation as a key element of modern war, and fully integrate it into every aspect of its planning, training, and education. Every segment of the force must be prepared for the fact that disruption, shocks, and surprises await in the opening battles of the next war. Its leaders must also embrace the precept that those who can rapidly adapt in the midst of these unexpected blows will be far more likely to survive, and ultimately turn the tide of battle to prevail. To do so, the US military must reform itself now to improve its ability to adapt under fire.

Improving Adaptability in Doctrine

Doctrine drives action in the US military. It reflects the best military thinking about war in peacetime, shapes its training and exercises before wars begin, and guides its actions under fire once conflict erupts. Agile and adaptable doctrine is flexible and not dogmatic. It must be dynamic in the face of the unexpected, able to rapidly shift when necessary to adjust to the changing demands of battle. Yet little in the US military's process of developing doctrine today encourages rapid adaptation. The exigencies of war are unlikely to alter this bureaucratic inertia, as Chapters 5 and 10 highlighted. The US military needs to develop ways to rapidly develop and disseminate new doctrine now, in order to ensure that it can rapidly adapt its doctrine during the fast-moving operations of the next war.

Additionally, service and joint doctrine must also provide practical ways for those in uniform to hone their adaptability *before* the next major conflict. As we noted in the previous chapter, current doctrine repeatedly stresses the importance of adaptability, especially in its leaders and soldiers—but it rarely explains *how* service members can become more adaptable. Wherever possible, doctrine should include practical steps to help them reach that important goal.

The US military should also

- **Adopt adaptability as a principle of war.** The US military has long recognized nine principles of war as the key tenets of successful military operations: objective, offensive, mass, economy of force, maneuver, unity of command, security, surprise, and simplicity.[3] Given the rapidly changing character of war, adaptability must be added to this list. Doing so would induce the military's training and education establishments to increase the emphasis on adaptability throughout their academic curricula and field exercises. US military education programs would be prompted to include studies and classroom exercises focused on adaptability in their courses of instruction, which would help foster much-needed discussion of this vital aspect of military operations. Adding this principle might also nudge military planners to incorporate greater adaptability into their war plans. For example, it might lead them to red-team their plans against unconstrained mock adversaries to uncover flaws and revise their plans accordingly.

- **Integrate adaptation and free play into all major exercises.** As noted in Chapter 10, the US military conducts a wide range of exercises and war games every year to prepare for future wars. Yet very few of them replicate the potentially devastating disruptions that may occur during future conflicts. Friendly Blue Force units nearly always know the kinds of missions they will encounter, when the training event is scheduled to begin and end, and what type of challenges the enemy Red Force will most likely pose. Tactical-level training events, such as those conducted at the army's National Training Center or the Marine Corps Air Ground Combat Center, offer a degree of uncertainty to participating units. But larger exercises typically conducted by the combatant commands (CCMDs) are often much more scripted and circumscribed. Yet in nearly all these events, Red Force actions that significantly deviate from the exercise design are unwelcome, since they could reveal embarrassing seams in the planners' assumptions, or even invalidate concepts of operations in which the services are deeply invested. Moreover, unconstrained opposing players might reveal that a major weapons system or other expensive capability is either ineffective or irrelevant.

US military exercises need to be largely freed of artificial constraints, and some should be specifically designed to directly challenge the ways in which the US military prefers to operate. As General David Berger, the commandant of the Marine Corps, has noted, "What is arguably our greatest deficiency in the training and education of leaders [is] practice in decision-making against a thinking enemy."[4] Robust war-gaming at all levels—from the tactical to the theater and strategic echelons—is essential to offer the kinds of challenges to US forces that the next war will immediately thrust upon military leaders.[5] Moreover, red teams playing the enemy should not exclusively consist of current and retired military personnel, as they often are today, since their tactics may unconsciously mirror the ways that the

US military fights or reflect the institutional preferences of their services. Instead, these red teams should include outside experts in digital technology and artificial intelligence, civilian strategists, and maybe even a hacker or two. The teams should also include millennials, and increasingly, members of Generation Z rising behind them. These digital natives approach problems differently than older generations—and, by definition, all but the most junior military leaders are from older generations. Those diverse perspectives would help the enemy teams think outside the US military's proverbial box and cause military leaders and their units to confront the kind of vexing and realistic adversary challenges that they may encounter during the next war. It would also help train them to become more comfortable with rapidly adapting to the enemy's unexpected thrusts—a skill that is much better to learn in practice than on a real battlefield.

- **Design, train, and test units to operate under analog and degraded conditions.**[6] As discussed in Chapter 9, the US military is deeply vulnerable to technological disruption. More than any other military in the world, it relies upon sophisticated battle networks employing advanced technology to command and control its forces. Its targeting, navigation, intelligence, and surveillance capabilities also rely on assured access to signals transmitted through outer space and cyberspace. None of these capabilities have been seriously threatened during the recent wars, which makes it easy to take them for granted. Yet any future US adversary that can degrade or destroy these capabilities will almost certainly do so. It would be far easier (and less risky) to disrupt a future attack by US armored vehicles or helicopters by shutting down many of their key digital systems or by denying their access to GPS— or, even worse, spoofing these systems with incorrect information—than to physically engage them with force.

To prepare for this eventuality, every major US military training event should include a sustained period of severe electronic and technical disruptions. Today, serious challenges like sustained broken communications, failed battle networks, untrustworthy position locations, and severely degraded space and cyber capabilities are relatively uncommon (and even then, very brief) in training. To be prepared for future battlefield threats, units should be required to operate for 48 or 72 hours under fully analog conditions, in order to practice their backup means of command and control, navigation, and targeting. This would significantly improve adaptability and resilience, and better prepare units for future battlefield conditions that challenge their reliance on digital capabilities.

The US military also needs to ensure that its organizational designs, operating procedures, and especially logistics doctrine can function effectively in combat operations characterized by limited connectivity and austere and unpredictable environments. Combat units should be designed to ensure they have the capabilities and robustness needed to operate independently in degraded, unconnected fashion on the battlefield for long periods of time. Units should be redesigned to emphasize self-sufficiency and task organization at much lower levels to permit sustained independent operations without access to robust logistics and communications.[7]

- **Emphasize resilience in doctrine and training.**[8] According to the Department of Defense, more than 5,400 US military personnel have been killed in action since September 11, 2001.[9] While every death is a tragic loss, this number is remarkably low compared to previous US conflicts. The US military lost over 416,000 US service members in World War II, and 2,500 soldiers were killed in action just on the first day of the D-Day invasion of Normandy.[10] And more than 40,000 were killed in action in Vietnam, which was considered a limited war at the time.[11]

The relatively low numbers of casualties in the recent wars means that today's US military leaders, up to and including the service chiefs, have never sustained heavy casualties in battle. Moreover, they have never dealt with the leadership stress that results when units are overrun, fragmented and disorganized, or simply unable to continue fighting due to overwhelming losses. These catastrophes often seem to be simply an artifact of past conflicts. But there is every possibility that the US military may once again be required to fight under such desperate conditions, especially in a future war against a major power like China or Russia.

Effective adaptation under this kind of extreme duress will require updated doctrine and better training on how to continue fighting despite heavy human losses and shattered units. Peacetime exercises must expose units to intense and deadly simulated attacks from aircraft and artillery, and to the effects of chemical and nuclear strikes. Doctrine from the Cold War that addressed these challenges needs to be resurrected and revised so that commanders know how to regroup and reorganize units after taking mass casualties, and how to integrate replacement units and individuals. Junior leaders must routinely train in exercises to take over more senior responsibilities, in case their superiors are wounded or killed.

Improving Adaptability in Technology

At the tactical level, the US military has a good track record of adapting to technological challenges. The creativity and innovation that have always characterized American troops seem alive and well today, though these traits must continue to be nurtured and encouraged. New technologies like 3D printing are already enabling soldiers and leaders in the field to adapt even more effectively.[12] Then, as now, sometimes senior leaders just need to get out of the way.

At the institutional level, however, rapid and effective technological adaptation has failed far too often. In Iraq and Afghanistan, service senior leaders repeatedly placed the needs of the future force over the needs of their troops actively engaging in combat. There will always be tensions between the military services, which are responsible for training, organizing, and equipping their forces for the future, and the combatant commands, which focus on fighting the wars of today. Future secretaries of defense will need to actively ensure that these tensions do not prevent essential wartime adaptation—as Secretary of Defense Robert Gates did when he almost singlehandedly forced the US Army and the Marine Corps to purchase large numbers of MRAPs and rush them to Iraq and Afghanistan. Moreover, institutional adaptation will almost certainly remain encumbered by the intense bureaucracy surrounding the acquisition process. Fundamental reform of this byzantine process may be too much to expect, but some reforms must be taken to address this enormous problem.[13] The painful failures in institutional adaptation of the last two decades resulted in lives unnecessarily lost on the battlefield.

To reinforce the successful trends in tactical adaptation and avoid repeating the recent disastrous experiences with institutional adaptation, the US military should

- **Revamp, revitalize, and fully resource rapid-adaptation organizations.** As discussed in Chapter 10, several organizations were established during the recent wars to help facilitate rapid technological adaptation. Yet the drawdowns in both Iraq and Afghanistan have enabled traditional bureaucratic politics to reemerge within the Pentagon and threaten some of these important offices. Two examples illustrate the problem. First, the army's Asymmetric Warfare Group (AWG) was originally conceived in 2006 as an innovative response to the unconventional challenges of counterinsurgency warfare and did a very good job of helping deployed commanders deal with irregular warfare threats. It combined an experienced operations cadre with a rapid prototyping capability, and has since nimbly evolved into addressing new forms of war from subterranean operations to counterproliferation missions.[14] Despite its novel success, AWG's capabilities and people were cut substantially in 2019, and it was shifted from reporting directly to a four-star army commander to reporting to a far less influential two-star general buried within the army's training and doctrine command. Its future survival is now in doubt.

Second, then–Deputy Secretary of Defense Ash Carter created the Strategic Capabilities Office (SCO) in 2012 in order to "meet urgent near-term needs for faster solutions to immediate battlefield problems," and had it report directly to him.[15] SCO achieved many successes in its first years of operation.[16] It found ways to use cheap air-deployed microdrones to jam enemy radars, for example, and redesigned an existing navy surface-to-air missile so that it could attack ships as well.[17] But in 2019, the Defense Department directed the merger of SCO into the Defense Advanced Research Projects Agency (DARPA)—an organization that generally focuses on longer-term projects rather than addressing the immediate needs of warfighting commanders. Furthermore, this move means that SCO resides deeply inside the Pentagon bureaucracy, four levels removed from the secretary of defense's personal oversight.[18]

Both of these wrongheaded decisions reflect the long-term risks that new and nontraditional organizations face inside the massive DOD bureaucracy. Both the AWG and SCO were deliberately designed for quick adaptation and rapid response. Yet these very characteristics make them threatening to existing organizations operating within the traditional slow-moving bureaucracy, which inevitably seek to take control of them. AWG and SCO both produced demonstrable capabilities and provided critical support for deployed warfighters within a short time frame, but they will not be able to continue doing so if they become absorbed into the bureaucratic processes they were designed to avoid. The army and the Department of Defense should undo these ill-founded moves, so that these innovative and adaptive organizations once again report directly to one of the Pentagon's most senior military or civilian leaders, and have both the authority and autonomy to continue their critically important work. Finding, encouraging, and protecting organizations that promote adaptability is one of the most important tasks for which every senior DOD leader is responsible.

- **Require all military technology to operate under degraded conditions.** Most advanced weapons systems, equipment, and command-and-control systems depend on software and other sophisticated technologies that often cannot continue to function when corrupted or debilitated. All military systems should be built to "degrade gracefully,"[19] so they maintain substantial functionality even if parts of the system have been attacked and destroyed. Furthermore, all systems need to be hardened against network attacks and catastrophic failures, by ensuring that they have analog backups or can work effectively in manual mode. Critical existing systems should be modified in these ways, and this type of resilience should become a key performance parameter for all new systems.

- **Charter a rapid-adaptation virtual skunkworks.** The Defense Department should charter DARPA (or a similar organization) to create a virtual skunkworks to enable rapid technological adaptation during wartime.[20] Several existing skunkworks exist today within the Pentagon, within defense firms, and even within cutting-edge commercial technology firms. This new DOD entity would virtually link the Pentagon to a network of private sector companies and could be rapidly activated when needed to improve tactical technological adaptation. This capability should span several industries to provide flexibility to make rapid wartime adjustments in current cyber, space, communications, aviation, and armament systems in time of major war. It should be directly linked to users in the field, so immediate battlefield needs can be recognized and acted upon quickly, with minimal involvement of the service acquisition bureaucracies.

  There is much the virtual skunkworks could accomplish with today's military well before the next war breaks out. Improving technological adaptation in a fast-moving tech environment requires up-to-the-minute knowledge of new and developing capabilities. Many of these developments occur out of sight of the US military's tech community, which is typically focused on the defense industry. The virtual skunkworks, with its ties to cutting-edge tech industries, could help serve as a clearinghouse and point of entry for the US military to stay abreast of these important but often obscure developments. Success in the next war may rest upon these breaking technologies and new capabilities, even some not yet invented, more than in any previous conflict. Having the ability to acquire and quickly adapt these emerging capabilities to meet the unexpected demands of the next major war is increasingly vital.

- **Sponsor an annual rapid-adaptation competition.** Rapid technological adaptation will not happen effectively in wartime unless it is practiced and tested beforehand. In order to do so, the under secretary of defense for acquisition and sustainment should hold an annual competition to help the Pentagon and industry improve their ability to collaborate under tight timelines. At the beginning of the competition, the under secretary would identify a new capability that needs to be developed or an existing system than needs to be upgraded, and participants would have 90 days to design and produce a proposed solution. For example, the competition could involve an urgent requirement to defeat off-the-shelf miniature drones being used by an adversary to gather intelligence on US forces. Extra points could be awarded for submitting solutions before the deadline. A dedicated lessons-learned team with representatives from both government and industry would monitor the competition to identify key problems and roadblocks in the process, and to help address them either during or after the competition. Winners would be recognized by the secretary of defense in a public awards ceremony, which would provide an incentive for both big defense companies and small start-ups to participate.

Improving Adaptability in Leadership

Improving leadership adaptability may be the single most important way for the US military to prepare for the future. Its tactical leaders demonstrated remarkable adaptability in the recent wars, consistently finding ways to accomplish their mission, no matter how difficult. Theater commanders, however, have often proven far less adaptable. Adaptability in doctrine and technology is important, but if the most senior leaders of the American military cannot adapt to the changing character of war, the US military will have great difficulty prevailing in future conflicts.

In order to improve leadership adaptability, the US military should

- **Evaluate adaptability in annual fitness reports.** Every officer and NCO in the US military receives an annual written evaluation, commonly known as a fitness report, that assesses their individual performance.[21] These reports evaluate a wide range of skills and attributes, rank service members relative to their peers in some areas, and discuss overall potential for promotion.[22] Those with stronger evaluations are more likely to get promoted, which creates clear incentives to perform well in the areas that are assessed on the reports.

  Adaptability should be added as a specific attribute to be assessed on all fitness reports, and evaluators should provide examples of demonstrated adaptability in the narrative portions of the reports. Instructions to promotion boards should also stress that adaptability is a critical attribute that officers and NCOs need to demonstrate in order to be promoted, and that the boards should be looking for specific examples as well.[23] Both measures would send a strong message to rising leaders that they are expected to adapt effectively under pressure, and would encourage them to continue developing their skills. Ensuring that officers and NCOs are formally and repeatedly assessed on their skills in this area will help ensure that the military's most adaptable leaders are recognized and promoted.

- **Strengthen mission command.** As discussed in Chapter 10, US military leaders collectively share a strong commitment to the principle of mission command, which relies upon decentralized operations and trust in subordinates. Yet these same leaders also tolerate a widespread compliance-based culture that undermines virtually every element of mission command. Much of this dissonance arises from deep aversion to risk at all levels of the chain of command, a trait that seems even more pronounced in the "frozen middle"—the midgrade officers who are worried that any flaws among their subordinates would jeopardize their promotion potential.[24] Many DOD and service leaders exacerbate this problem by promulgating (or at least tolerating) the deluge of rules, regulations, and policies that constrain junior leaders. These expansive regulations send a message that bureaucratic compliance is valued, and that noncompliance will be punished—which strongly deters innovation and risk-taking.

  To their credit, the air force's senior leaders have publicly discussed the need to address the frozen middle in their service, and to increase risk-taking and trust in subordinates.[25] And Mark Esper and General Mark Milley, while serving as the secretary of the army and the army chief of staff, respectively, made some overdue but impressive strides in reducing the number of service-originated rules and regulations.[26] Those efforts need to continue, and the rest of the Department of Defense needs to fol-low suit. But commanders must also change their behaviors in order to ensure that mission command remains more than just words on a page. Commanders must be willing to accept more risk, even if it has the poten-tial to damage their careers, and communicate that philosophy to their subordinates. Commanders of deployed units must also stop the intense micromanagement that is now too common on operations. As discussed in Chapter 10, replacing the detailed mission approval system with one that lets lower-level commanders approve small-unit combat operations would be an important step in the right direction.

  Deployed commanders must also exercise restraint in initiating investi-gations for deaths due to enemy action in a combat theater. Senior leaders should avoid pressuring subordinate commanders to investigate every battle resulting in fatalities, and step in to prevent multiple investigations by differ-ent commands. Automatically investigating all battlefield engagements cor-rodes the key principles of mission command; investigating only those that clearly warrant inquiry strongly reinforces it.

- **Reform professional military education.** As discussed in the previous chapter, professional military education (PME) programs provide critical opportunities for officers to discover new ideas and ways of thinking. Yet most of these programs need to do a far better job helping their students develop the skills that they will need to adapt to the fast-moving and complex environment of the future. Their curricula need to be more rigorous, with substantially higher standards for writing and critical thinking skills. Grading standards also need to be stiffened, and students who do not perform satisfactorily should receive failing grades—and even potentially fail to graduate. Academic performance in PME should also be valued more highly by promotion boards. One way to accelerate that shift would be to use the same fitness report for school performance as for duty in command or staff billets, so academic reports cannot be ignored as easily as they are today. Students would have far more incentives to do well in PME programs if promotion boards value academic performance instead of simply crediting attendance. The army recently revamped its academic evaluation reports to highlight PME performance (though it is not yet clear how this will affect promotion decisions), and the commandant of the Marine Corps has also suggested doing so.[27] All four services should institutionalize this change as rapidly as possible, to help ensure that their best and brightest are recognized and promoted.

PME curricula should also include a multitude of stressful decision-making exercises, placing students in increasingly difficult tactical, operational, and strategic scenarios. Immersing students in realistic exercises that involve intense time pressure and flawed information would promote habits of rapid adaptation. It would also strengthen resilience in officers who may face untold shocks and surprises in future high-intensity combat. Senior PME also needs to better prepare its students for their future theater and institutional responsibilities. In the words of one retired general, senior PME provides a unique opportunity to "radically promote those habits of thinking that will counter habits we consider essential to tactical success."[28] Senior PME should offer a variety of demanding decision-making exercises that require rapid adaptation in order to transform students who arrive skilled in tactics into graduates who are equally talented in strategy.

Finally, PME institutions also need to find creative ways to provide their students access to more diverse viewpoints, in order to open their minds to outside thinking and reduce the homogeneity of perspectives in the classroom. One way to do that would be to significantly expand the number of students attending civilian graduate programs, as we recommend below. But even so, most rising officers would still attend purely military programs. In order to help broaden student horizons, PME institutions should partner with local colleges and universities to connect military and civilian students of different ages and from different backgrounds.[29] And wherever possible, military students should be allowed—and perhaps even required—to fulfill some of their electives with classes offered at local universities or through carefully selected online programs. These nontraditional alternatives would provide mind-altering experiences for military officers who have spent decades surrounded by people in uniform, and improve their adaptability by exposing them to fresh ideas and disruptive outlooks.

- **Send more officers to advanced civil schooling.** In a fast-changing world, the US military needs to significantly increase the number of officers it sends to fully funded civilian graduate programs. This shift would help avoid groupthink among military officers who typically have little exposure to outside ideas, and provide more adaptable and well-informed future leaders for the armed forces. As noted in Chapter 10, many PME programs are not sufficiently rigorous, and their students have extraordinarily similar backgrounds and experiences. Military students rarely encounter classmates who think differently, or make them uncomfortable by challenging their assumptions. Partly as a result, the US Army has found that its senior PME students are far less open to new ideas than the general US population.[30] Yet that very openness is an essential attribute for effective leadership at the strategic level. Civilian graduate programs can provide immense value by exposing cloistered military students to a universe of new ideas and sharply differing worldviews that is notably absent from most military classrooms.

  In the last two decades, the number of opportunities for military officers to attend civilian graduate school has been slashed, as graduate degrees have now become embedded in many PME curricula. That trend must be reversed. In order to be adaptable enough for the demands of the 21st century, more military officers need an educational experience that transcends the comfortable cocoon of classes in camouflage.

- **Establish DOD adaptive leadership awards.** In order to promote and reward adaptability among military leaders, the secretary of defense should institute an annual award for adaptive leadership. This award could mirror the US Army's long-running General Douglas MacArthur Leadership Awards, an annual competition that recognizes over two dozen junior officers from nominations submitted by all of the army's major commands.[31] Criteria for this honor could include innovative suggestions, creative writing on adaptation challenges, courage in advocating for needed changes, and leadership of innovative projects that have implemented novel ideas. The award could be named after a prominent military leader who was famed for his or her skills at adaptation, such as

Air Force Colonel John Boyd or Navy Admiral Grace Hopper. Services could also consider adaptability awards programs tailored to their needs, and provide immediate rewards to those demonstrating noteworthy adaptability in operations and exercises.

The adaptive leader awards competition should be open to both NCOs and officers, with separate categories for junior, mid-level, and senior leaders up to and including flag rank. A separate category could also be established for DOD civilians as well. As the army does with its MacArthur awards, winners should be brought to Washington for a week of public recognition and professional development hosted by the secretary of defense. Such a highly publicized award would help incentivize leaders at all levels to push beyond the bounds of their daily duties to discover and promote fresh and innovative ideas. Its high visibility would also reinforce the importance of adaptability as a key military leadership trait, and would also help promotion boards easily identify those who have been recognized as leaders in this area.

- **Expand the technological literacy of future commanders.** As discussed in Chapter 9, the technologies of the Fourth Industrial Revolution are driving immense changes in nearly all aspects of government, business, and society. Yet many US military leaders have limited, if any, background in technology. This is particularly true in the ground forces, where technology is often seen as an adjunct to rather than a driver of combat operations. Relatively few army, marine, or special operations officers in operational fields have studied scientific or technological subjects since their undergraduate years—and most had little exposure to these subjects even then.

  Officers from these operational career fields disproportionately fill senior leadership positions across the US armed forces today, and are likely to continue doing so in the future. Yet they need a far better understanding of both the potential and the dangers of modern technology in order to serve effectively in such a fast-changing global environment. Some of these leaders should be tapped for brief career intermission assignments to high-technology firms such as Amazon, Google, or Apple. They also should receive priority for advanced schooling or immersion courses in rising technology areas such as artificial intelligence, big data, and machine learning. Officers who rise to high levels of command must understand the ways in which modern technology affects war and society—but too few of them understand technologies beyond their service's newest weapons systems. The military should ensure that every officer selected for general officer rank has either had one of these immersion experiences or attends an executive course designed to familiarize them with global technological trends. Rapidly evolving technologies are such a central facet of the 21st century that senior military leaders who cannot grasp their fundamentals will not be able to adapt effectively to the ever-changing strategic landscape.

General Recommendations for Improving Adaptability

There are also several ways that the Department of Defense can help improve adaptability that transcend the categories of doctrine, technology, and leadership. DOD should

- **Expand the focus on talent management.** Recruiting, developing, and retaining highly skilled service members may be one of the most critical means of ensuring adaptability in the future—yet the military services largely continue to rely on the centralized, industrial-age personnel systems that they have used for decades. Many of the personnel reform initiatives offered in former Secretary of Defense Ash Carter's Force of the Future initiative[32] should be implemented by the services to ensure that more diverse and talented people continue to be brought into the force, and that policies are in place to encourage the best of them to stay for extended careers.

    The army has already undertaken some important steps in this direction. Since becoming the army chief of staff in August 2019, General James McConville has stressed that talent management is his highest priority,[33] and the army's Talent Management Task Force has already substantially changed some of the processes through which soldiers are selected, assessed, and promoted. In 2019, for example, it introduced a new Army Talent Alignment Process for officer assignments, which uses a regulated marketplace to help match officers interested in particular assignments with units that are interested in having them.[34] The army is also experimenting with a new process for selecting battalion commanders, a job that McConville described as "one of the most consequential leadership positions in the Army."[35] This new process puts the top candidates through four days of intensive physical, cognitive, written, and leadership assessments, in order to provide a far more holistic evaluation of their individual strengths and weaknesses.[36]

    These initiatives seem quite promising, but it is far too soon to tell whether they will ultimately be successful.[37] Yet even if they are, much more remains to be done. Providing more opportunities for lateral entry of uniquely qualified civilians into uniform at higher grades, and taking full advantage of the changes to the up-or-out promotion system allowed by the 2019 National

Defense Authorization Act would add needed flexibility to attract and keep the best talent. Additionally, placing greater senior leader support behind programs that permit career intermissions, especially for women who often leave the military when starting families, would also help retain a greater breadth of talented leaders. The services should also make a much greater effort to recruit and retain geographic and ethnic diversity from across the nation, to ensure a diversity of viewpoints and experiences that will better prepare the US military to deal with uncertainty in war.

- **Get more young voices in front of senior leaders.** Three- and four-star military leaders often have 35 or more years of military experience, but few have ever spent time in civilian industry, or even in other departments of the US government. Most of them have spent years deployed in combat zones far distant from the United States. Their outlook is inevitably cloistered by virtue of these life experiences. Military leadership is also a rigid hierarchy delineated by age; by definition, the oldest and longest-serving members hold the highest ranks and have the most responsibility. This guarantees that senior leaders are nearly always surrounded by people that look, think, and have experiences that mirror their own. As noted in Chapter 10, this is a dangerous recipe for groupthink and detached decision-making in a world marked by disruptive changes.

  Senior military officers must include greater numbers of younger and more diverse individuals on their staffs and in their decision-making processes. Highly talented junior officers (especially those with key technological skills) should be assigned to positions that place them in close proximity to senior commanders. Talented young civilian experts from outside the Pentagon should also be able to apply for two- or three-year assignments working with a specific senior commander. They would help provide outside perspectives and recommendations to often-insular military staffs, especially in the areas of social media, AI, and other disruptive and fast-moving technologies. In each case, this rising young military and civilian talent could be assigned to senior leaders' initiatives and planning groups[38] to ensure that senior leaders from much older generations have access to up-to-date advice on problems and decisions reflecting the best and most diverse thinking. Alternatively, they could serve as direct advisers to senior leaders, much as the State Department assigns political advisers to military leaders today.

- **Increase the role of the combatant commands in ensuring adaptability.** In 1986, the Goldwater-Nichols legislation placed the services in charge of training, organizing, and equipping the military while charging the CCMDs with employing those forces in peace and war. Given this division of responsibilities, the CCMDs rely upon the services to train the forces that the commands then employ. The CCMDs routinely provide readiness and other requirements to the services for the forces deploying into their theater of operations, but they have few if any formal means of providing feedback on the adaptability of those forces. CCMDs should require the services to provide forces trained to specific adaptability requirements, to ensure that when forces arrive for employment, they are already as resilient as possible to anticipated battlefield disruptions. The CCMDs should then exercise and test these forces in unpredictable scenarios to evaluate their abilities to adapt to harsh and unforgiving potential battlefield conditions. CCMDs should report back to the services on the adaptability of these forces as measured by this exercise program. Brittle forces and unadaptable leaders will fail in the chaotic test of future combat; rigorous peacetime practice under the most difficult of projected wartime conditions will help to mitigate this risk.

- **Charter a Defense Adaptation Board.** Department of Defense federal advisory committees have long provided the secretary of defense vital outside feedback on his key areas of responsibility.[39] Formed from diverse groups of experts in the relevant fields, these advisory boards provide independent advice and recommendations to the secretary on issues ranging from science and emerging technology to manufacturing, logistics, business management, and the reserve component.[40] The secretary of defense should charter a Defense Adaptation Board to focus specifically on identifying, assessing, and solving problems that would hinder rapid adaptation during the next major war. This new board would inform and advise the secretary on key issues, catalyze adaptation efforts throughout the services and joint staff, and serve as a resource for congressional committees examining this issue. It would complement the efforts of the Defense Innovation Board by focusing exclusively on wartime adaptation,[41] and helping the US military better design its doctrine, technology, and leadership so it can rapidly adapt during the next conflict.

- **Assign all CIGs and CAGs wartime adaptation responsibilities.** The US military is replete with commander's initiative groups (CIGs) and commander's action groups (CAGs) across many of its headquarters, especially at the two-star level and above. With large, bureaucratic, and cumbersome staffs the norm across the senior levels of the US military, these small teams of hand-picked, highly talented officers were formed to directly share creative insights on particularly vexing problems with busy commanders. CAGs and CIGs provide a ready source of intellectual talent for senior leaders. In times of war, CAGs and CIGs should be assigned to provide independent judgment of ongoing events and help catalyze rapid adaptations throughout their organizations. They should serve as red teams to critically review reports and actions by subordinate units, and offer the commander a range of creative alternatives. These wartime roles should be regularly practiced in the peacetime exercises that involve their supported commander. Leveraging these rich pools of intellectual talent will help promote rapid adaptation among the most senior leaders of the military and throughout their organizations.

- **Assign combat assessment teams (CATs) to major combat formations in wartime.** During the 1991 Gulf War, the US military established a wide range of liaison teams to provide reliable command, control, and communications with its Arab coalition partners. Army liaison teams providing communications were large, averaging 35 soldiers each, but US Central Command also established more than 100 three- or four-person teams to facilitate coordination on a range of critical issues. Yet these teams ended up providing much more than just technical support. They also gave the theater commander an independent and objective assessment of local situations, which helped him manage his far-flung and widely diverse coalition units.[42]

  The US military should plan to add similar teams at the outset of any future conflict to every major combat formation slated for an operational mission, in order to facilitate rapid wartime adaptation. Augmentation personnel for CATs could be drawn from the reserve component and expanded from the core teams that exist today in the service's lessons-learned organizations.[43] These small teams could provide immediate, independent reporting back to the combatant commands and service headquarters on critical shortfalls and needed combat adaptations. Since they would not be responsible for overseeing forces in battle, they would be able to rapidly assess and report on areas needing immediate change.[44] They could also serve as a conduit of recommendations from field commanders to institutional headquarters in the United States responsible for analyzing, revising, and disseminating rapid doctrinal and technological changes. Both the army and the marines made efforts to do this in Iraq and Afghanistan, but these teams generally had a limited doctrinal mandate and lacked independent communications. CATs could help accelerate the adaptation process by providing dispassionate reporting on the performance of doctrine and technology from the battlefield directly to those responsible for making rapid institutional changes.

Conclusion

Many trends are converging to change the character of warfare. Yet the US military remains at risk of assuming that the next war will largely resemble those of the past—even if that assumption is made inadvertently rather than intentionally. European militaries had great confidence in their vision of future war in 1914, even as they stood on the precipice of a devastating conflict that confounded nearly every one of their assumptions. The US military must do everything it can to ensure that it does not find itself in a similar position in the next war.

We hope that this book will spark a long-overdue conversation about how to make the US military more adaptable. As we noted at the beginning of this chapter, as the US military prepares for the next war, it must effectively balance preparing for what is anticipated with readying the force to adapt to the unexpected. The US military has always invested heavily in preparation and continues to do so today, but it devotes little time and effort to honing skills and devising effective procedures for rapid wartime adaptation. We have argued that the next war has every likelihood of looking dramatically different than any war of the past, bringing inevitable shocks and surprises to military planners and civilian leaders alike. Ad hoc fixes cobbled together in the midst of that war will not suffice. Such an approach borders on negligence, given the potential consequences of defeat.

Historian Corelli Barnett has observed that wars are the "great auditors of institutions."[45] The next war will bring a momentous audit to the US military. Years of rigorous preparation and training may prove far less useful than imagined, especially if that next war turns out to be very different from the one that the United States has prepared to fight. Preparing to adapt in the next war is just as important as preparing to fight itself. Only by devoting serious time and energy to closing the adaptation gap can the US military truly be prepared to prevail in the wars of the 21st century. Failing to recognize and address this shortfall will almost certainly make the next war longer and more costly, and may well lead the nation into a disastrous defeat.

Acknowledgments

WE OWE A great deal of gratitude to many people for the help we received in researching and writing this book.

Several people generously reviewed the entire manuscript and gave us helpful feedback. These include Major John Trimble, USA, Jim Goldgeier, and Colonel (Ret.) George Topic, USA, who provided many valuable and thoughtful comments and insights. Colonel (Ret.) Jim Shufelt, USA also read the entire draft, and gave us some especially useful recommendations for building future adaptability. We are also grateful to the two anonymous reviewers of the manuscript, whose thorough reviews also contained many helpful suggestions.

We thank Amy Zegart for her insights on organizational theory—and her breathtaking three-hour turnaround on our draft chapter. Adam Grissom provided terrific thoughts about the literature on innovation and adaptation, and David Aronstein helped us understand non-linearity from a scientific point of view. We are also indebted to Lieutenant General (Ret.) Jack Gardner, USA, for his perspectives on the Fourth Industrial Revolution and global trends, and Rear Admiral (Ret.) David Titley, USN, for his insights on climate change.

Our book would not have been possible without the tireless support of our terrific research assistants: Kelly Bedard and Cameron Moubray at American University, and Joel Carter, John Trimble, and Paula Alvarez-Couceiro at the Johns Hopkins School of Advanced International Studies. In addition to providing many helpful suggestions, their efforts saved us countless hours of work, and made the documentation for the book monumentally less painful.

We are especially grateful to Jim Goldgeier, former dean of the School of International Service at American University, for encouraging us to publish this book as part of the Bridging the Gap series, and for his boundless encouragement during the years of effort required to bring it to fruition. We also thank Mara Karlin, Eliot Cohen, and Thayer McKell at SAIS for their patience and support as we completed this project while teaching in the Strategic Studies program.

We thank David McBride and his staff at Oxford University Press for their patience and flexibility with our elastic view of deadlines and publication targets. We also greatly appreciate the support of the editors of Oxford's Bridging the Gap series, Jim Goldgeier, Steve Weber, and Bruce Jentleson.

We would have been unable to undertake this project without the generous support of the Smith Richardson Foundation. We want to especially thank Marin Strmecki, Nadia Schadlow, and Kathy Lavery for all of their support and encouragement throughout the project.

Finally, we thank our family and friends for their patience, love, and unstinting support throughout this project, which took far longer than we anticipated. Dave would like to offer particular thanks to his wife, Susan, whose infinite patience throughout this long-running project was deeply appreciated.

Of course, despite all of these many sources of support and insights, any errors and omissions in the work are ours alone.

Notes

INTRODUCTION

1. US Department of Defense, "Secretary of Defense Speech, United States Military Academy (West Point, NY)," February 25, 2011.

2. This quotation has been variously attributed to people as different as Niels Bohr and Yogi Berra, but seems to have originated in Denmark. See "It's Difficult to Make Predictions, Especially about the Future," QuoteInvestigator.com, October 20, 2013, http://quoteinvestigator.com/2013/10/20/no-predict/.

3. Philip E. Tetlock and Dan Gardner, *Superforecasting: The Art and Science of Prediction*, New York: Crown Publishers, 2015, 4. The original study is Philip E. Tetlock, *Expert Political Judgment: How Good Is It? How Can We Know?*, Princeton, NJ: Princeton University Press, 2005.

4. Tetlock and Gardner, *Superforecasting*.

5. National Commission on Terrorist Attacks upon the United States, *The 9/11 Commission Report*, New York: Norton, 2004.

6. For example, the British Expeditionary Force (BEF) from 1939 to 1940 was generally well trained and reasonably well equipped, and knew that it would likely fight the Germans on the continent in the next war. Yet it was still overwhelmed by the unexpected German tactics and speed during the invasion of France and the Low Countries in May 1940. The result was a spectacularly quick defeat, capped by the mass evacuation of the BEF at Dunkirk.

7. Michael Howard, "Military Science in an Age of Peace," *RUSI Journal* 119, no. 1 (March 1974): 7.

8. Eisenhower continued: "There is a very great distinction [between plans and planning] because when you are planning for an emergency you must start with this one thing: the very definition of 'emergency' is that it is unexpected, therefore it is not going to happen the way you are planning. So, the first thing you do is

to take all the plans off the top shelf and throw them out the window and start once more. But if you haven't been planning you can't start to work, intelligently at least." President Dwight D. Eisenhower, remarks at the National Defense Executive Reserve Conference, November 14, 1957.

9. Nadia Schadlow, "Peace and War: The Space Between," *War on the Rocks*, August 18, 2014.

10. David Barno and Nora Bensahel, "Fighting and Winning in the 'Gray Zone,'" *War on the Rocks*, May 19, 2015. See also Nora Bensahel, "Darker Shades of Gray: Why Gray Zone Conflicts Will Become More Frequent and Complex," Foreign Policy Research Institute E-Notes, February 13, 2017.

11. A key question for the future, though well beyond the scope of this book, is whether future warfare will still be primarily a struggle defined by military means—or even involve the military at all. See David Barno and Nora Bensahel, "A New Generation of Unrestricted Warfare," *War on the Rocks*, April 19, 2016.

12. See, for example, Thomas L. Friedman, *Thank You for Being Late: An Optimist's Guide to Thriving in the Age of Accelerations*, New York: Farrar, Straus and Giroux, 2016; Richard Dobbs, James Manyika, and Jonathan Woetzel, *No Ordinary Disruption: The Four Global Forces Breaking All the Trends*, New York: PublicAffairs, 2015; National Intelligence Council, *Global Trends: Paradox of Progress*, January 9, 2017.

13. Dobbs, Manyika, and Woetzel, *No Ordinary Disruption*, 16–18.

14. Dobbs, Manyika, and Woetzel, *No Ordinary Disruption*, 43; Darrell Etherington, "Pokémon Go Estimated at over 75M Downloads Worldwide," *TechCrunch*, July 25, 2016.

15. By 2020, 5.4 billion people will have a cell phone, 5.3 billion people will have electricity, and only 3.5 billion will have running water. Roger Cheng, "By 2020, More People Will Own a Phone Than Have Electricity," *Cnet.com*, February 3, 2016.

16. The Apollo Guidance Computer (AGC) had 64KB of memory and 0.043 MHz of processing power. The original iPhone, released in 2007, had 4GB of storage and 412 MHz of processing power. The base model iPhone 7, released in 2016, had 32GB of storage and 2.34 GHz of processing power. See Chaim Gartenberg, "We Compared Hardware Specs for Every iPhone Ever Made," *TheVerge.com*, June 28, 2017.

17. Kim Tingley, "The Loyal Engineers Steering NASA's Voyager Probes across the Universe," *New York Times Magazine*, August 3, 2017.

18. For more on the Fourth Industrial Revolution, see Chapter 9.

CHAPTER 1

1. See James G. March, "Footnotes to Organizational Change," *Administrative Science Quarterly* 26, no. 4 (December 1981): 563.

2. Amy B. Zegart, *Spying Blind: The CIA, the FBI, and the Origins of 9/11*, Princeton, NJ: Princeton University Press, 2007, 16. See also Amy B. Zegart, "Agency Design and Evolution," in Robert F. Durant, ed., *The Oxford Handbook of American Bureaucracy*, Oxford: Oxford University Press, 2010, 217.

3. Michael T. Hannan and John Freeman, "Structural Inertia and Organizational Change," *American Sociological Review* 49, no. 2 (April 1984): 151.

4. Zegart, *Spying Blind*, 20–21.

5. Zegart, *Spying Blind*, 45; US Census Bureau, *Statistical Abstract of the United States: 2012*, Washington, DC: Government Printing Office, 2012, 506.

6. UCLA-LoPucki Bankruptcy Research Database, http://lopucki.law.ucla.edu/spreadsheet.htm, accessed June 2017.

7. Zegart argues that even the most poorly performing government agencies usually survive because interest groups and elected officials have vested interests in maintaining them. See *Spying Blind*, 46, 54.

8. Zegart argues, "The single most important reason the United States remained so vulnerable on September 11 was not the McDonald's wages paid to airport security workers, the Clinton administration's inability to capture or kill Osama bin Laden, or the Bush administration's failure to place terrorism higher on its priority list. It was the stunning inability of US intelligence agencies to adapt to the end of the Cold War." Zegart, *Spying Blind*, 3.

9. Williamson Murray, *Military Adaptation in War: With Fear of Change*, New York: Cambridge University Press, 2011, 1.

10. The other two are the failure to learn and the failure to anticipate. Eliot A. Cohen and John Gooch, *Military Misfortunes*, New York: Anchor Books, 2003, 26–7.

11. See, for example, Joint Chiefs of Staff, *Doctrine for the Armed Forces of the United States*, Joint Publication 1, March 25, 2013; and Headquarters, Department of the Army, *The Army*, Army Doctrine Publication 1, September 17, 2012.

12. Headquarters, Department of the Army, *Army Leadership*, Field Manual 6-22, October 2006, 5–7.

13. Headquarters, Department of the Army, *Army Leadership*, Army Doctrine Reference Publication 6-22, August 2012, 1–2.

14. Stephen Peter Rosen, *Winning the Next War: Innovation and the Modern Military*, Ithaca, NY: Cornell University Press, 1991, 2.

15. Max Weber, *From Max Weber: Essays in Sociology*, ed. H. H. Gerth and C. Wright Mills, New York: Oxford University Press, 1946, 214.

16. Weber, *From Max Weber*, 214–5.

17. Graham T. Allison, *Essence of Decision*, Boston: Little, Brown, 1971, 76–7; Murray, *Military Adaptation in War*, 18; and Barry R. Posen, "Foreword: Military Doctrine and the Management of Uncertainty," *Journal of Strategic Studies* 39, no. 2 (2016): 161.

18. Richard M. Cyert and James G. March, *A Behavioral Theory of the Firm*, Englewood Cliffs, NJ: Prentice-Hall, 1963, 102, 119.

19. Cyert and March, *Behavioral Theory*, 101.

20. Allison, *Essence of Decision*, 83.

21. Zegart, "Agency Design and Evolution," 224.

22. March, "Footnotes to Organizational Change," 572; Zegart, "Agency Design and Evolution," 220. See also Robert Komer's classic work on Vietnam, which details how bureaucratic inertia shaped the conduct of the war. R. W. Komer, *Bureaucracy Does Its Thing: Institutional Constraints on U.S.-GVN Performance in Vietnam*, R-967-ARPA, Santa Monica, CA: Rand Corporation, August 1972.

23. For more on incrementalism, see Charles E. Lindblom, "The Science of 'Muddling Through,'" *Public Administration Review* 19, no. 2 (Spring 1959): 79–88.

24. Allison, *Essence of Decision*, 88.

25. Zegart, *Spying Blind*, 53–4.

26. Hannan and Freeman, "Structural Inertia," 154.

27. Zegart, *Spying Blind*, 224.

28. Rosen, *Winning the Next War*, 2.

29. For a good overview of the problems caused by chance and friction, see Murray, *Military Adaptation in War*, 15–8.

30. Carl von Clausewitz, *On War*, ed. and trans. Michael Howard and Peter Paret, Princeton, NJ: Princeton University Press, 1976, 85.

31. Clausewitz, *On War*, 101.

32. Alan Clark, *Barbarossa: The Russian-German Conflict, 1941–1945*, New York: Perennial, 2002, 171–5.

33. Ian W. Toll, *Pacific Crucible: War at Sea in the Pacific, 1941–1942*, New York: Norton, 2012, 39–42.

34. Clausewitz, *On War*, 167.

35. Peter Paret, *Clausewitz and the State*, New York: Oxford University Press, 1976, 197.

36. Clausewitz, *On War*, 119.

37. Williamson Murray, *America and the Future of War*, Stanford, CA: Hoover Institution Press, 2017, 39.

38. Orlando Ward, "Foreword to Original Edition," in Russell A. Gugeler, *Combat Actions in Korea*, Washington, DC: Office of the Chief of Military History, US Army, 1970, iii.

39. Clausewitz, *On War*, 113–6; Paret, *Clausewitz and the State*, 197–8.

40. Carl von Clausewitz, *Principles of War*, ed. and trans. Hans W. Gatzke, Harrisburg, PA: Stackpole Company, 1960, 61.

41. Clausewitz, *On War*, 119.

42. Stephen Ambrose, *D-Day, June 6, 1944: The Climactic Battle of World War II*, New York: Simon & Schuster, 1994, 196–238.

43. Clausewitz, *On War*, 85.

44. Alan Beyerchen, "Clausewitz, Nonlinearity, and the Unpredictability of War," *International Security* 17, no. 3 (Winter 1992–1993): 73.

45. Beyerchen, "Clausewitz, Nonlinearity," 77.

46. Murray, *Military Adaptation in War*, 3. Emphasis added.

47. Clausewitz, *On War*, 75.

48. We discuss some of the ways in which the US military failed to adapt to the IED threat in Chapter 6.

49. Militaries also exist to deter wars, but this requires potential enemies to be convinced that they are ready and able to fight and win.

50. Ralph Ingersoll, *The Battle Is the Pay-off*, New York: Harcourt, Brace, 1943.

51. Michael Howard, "The Use and Abuse of Military History," in Michael Howard, *The Causes of War*, Cambridge, MA: Harvard University Press, 1983, 194.

52. Murray, *Military Adaptation in War*, 8–9.

53. Williamson Murray, "Does Military Culture Matter?," *Orbis* 43, no. 1 (Winter 1999): 28.

54. March, "Footnotes to Organizational Change," 572.

55. Williamson Murray, "Innovation: Past and Future," in Williamson Murray and Allan R. Millett, eds., *Military Innovation in the Interwar Period*, Cambridge; New York: Cambridge University Press, 1996, 313.

56. Timothy T. Lupfer, "The Challenge of Military Reform," in Asa A. Clark IV et al., eds., *The Defense Reform Debate*, Baltimore: Johns Hopkins University Press, 1984, 25.

57. Suzanne C. Nielsen, *An Army Transformed: The U.S. Army's Post-Vietnam Recovery and the Dynamics of Change in Military Organizations*, Carlisle, PA: Strategic Studies Institute, US Army War College, September 2010, 12.

58. Adam Grissom, "The Future of Military Innovation Studies," *Journal of Strategic Studies* 29, no. 5 (October 2006): 906.

59. Grissom, "Military Innovation Studies," 906.

60. See Rosen, *Winning the Next War*.

61. See Murray, *Military Adaptation in War*.

62. Theo Farrell and Terry Terriff, "The Sources of Military Change," in Theo Farrell and Terry Terriff, eds., *The Sources of Military Change*, Boulder, CO: Lynne Rienner Publishers, 2002, 6; Theo Farrell, "Improving in War: Military Adaptation and the British in Helmand Province, Afghanistan, 2006–2009," *Journal of Strategic Studies* 33, no. 4 (August 2010): 567–8.

63. Theo Farrell, "Introduction: Military Adaptation in War," in Theo Farrell, Frans Osinga, and James A. Russell, eds., *Military Adaptation in Afghanistan*, Stanford, CA: Stanford University Press, 2013, 7.

64. Dmitry (Dima) Adamsky and Kjell Inge Bjerga, "Conclusion: Military Innovation between Anticipation and Adaptation," in Dmitry (Dima) Adamsky and Kjell

Inge Bjerga, eds., *Contemporary Military Adaptation: Between Anticipation and Adaptation*, London: Routledge, 2012, 188.

65. Murray, *Military Adaptation in War*, 308.

66. Murray, *Military Adaptation in War*, 309–10. Emphasis added.

67. Grissom, "Military Innovation Studies," 905.

68. Murray, "Innovation," 2.

69. Murray, "Innovation," 2. Watts and Murray credit Andrew Marshall, then the director of the Office of Net Assessment within the Department of Defense, for first seeing this parallel. Barry Watts and Williamson Murray, "Military Innovation in Peacetime," in Murray and Millett, *Military Innovation*, 377. Marshall's office funded the research that led to the edited volume on innovation—just as it had funded Millett and Murray's previous work on military effectiveness, and subsequently funded Murray's work on adaptation.

70. Allan R. Millett and Williamson Murray, eds., *Military Effectiveness*, Boston: Allan & Unwin, 1988.

71. Murray and Millett, *Military Innovation*.

72. Williamson Murray and Allan R. Millett, "Introduction: Military Effectiveness Twenty Years Later," in Allan R. Millett and Williamson Murray, eds., *Military Effectiveness*, vol. 1: *The First World War*, new ed., Cambridge; New York: Cambridge University Press, 2010, xvi.

73. Murray and Millett, "Introduction: Military Effectiveness," xvii.

74. Murray, *Military Adaptation in War*.

75. Murray, *Military Adaptation in War*, 6.

76. Clausewitz, *On War*, 88.

77. See, for example, Frank E. Hartung, "The Sociology of Positivism," *Science & Society* 8, no. 4 (Fall 1944): 328–41.

78. For examples of such methodological approaches, see Stephen Van Evera, *Guide to Methods for Students of Political Science*, Ithaca, NY: Cornell University Press, 1997; John Gerring, *Social Science Methodology: A Criterial Framework*, Cambridge: Cambridge University Press, 2001.

79. Barry R. Posen, *The Sources of Military Doctrine: France, Britain, and Germany between the World Wars*, Ithaca, NY: Cornell University Press, 1984.

80. Rosen, *Winning the Next War*.

81. See the discussion of doctrine later in this chapter, as well as Grissom, "Military Innovation Studies," 908–10, 913–6.

82. Grissom, "Military Innovation Studies," 916–9. Grissom also identifies a fourth model of military innovation that focuses on interservice competition resulting from resource constraints. Most of that work, however, utilizes the historical approach, which is why we do not discuss it here.

83. Grissom, "Military Innovation Studies," 919–20.

84. Farrell, "Improving in War," 568–9. See also Farrell, "Introduction: Military Adaptation," 17–8.

85. Grissom, "Military Innovation Studies," 930.

86. See, for example, James A. Russell, *Innovation, Transformation, and War: Counterinsurgency in Anbar and Ninewa Provinces, Iraq, 2005–2007*, Stanford, CA: Stanford University Press, 2011; Sergio Catignani, "'Getting COIN' at the Tactical Level in Afghanistan: Reassessing Counter-insurgency Adaptation in the British Army," *Journal of Strategic Studies* 35, no. 4 (August 2012): 513–39; Nina Kollars, "Organising Adaptation in War," *Survival* 57, no. 6 (December 2015–January 2016): 111–26.

87. Nina A. Kollars, "War's Horizon: Soldier-Led Adaptation in Iraq and Vietnam," *Journal of Strategic Studies* 38, no. 4 (June 2015): 531.

88. Sergio Catignani, "Coping with Knowledge: Organizational Learning in the British Army?," *Journal of Strategic Studies* 37, no. 1 (January 2014): 30–64; Kollars, "War's Horizon"; Kollars, "Organising Adaptation in War."

89. Our definitions of the levels of war are drawn from Edward N. Luttwak, "The Operational Level of War," *International Security* 5, no. 3 (Winter 1980–1981): 61–79; Joint Chiefs of Staff, *Joint Operations*, Joint Publication 3-0, January 17, 2017, I-12–I-14.

90. J. F. C. Fuller, *The Foundations of the Science of War*, London: Hutchinson, 1926, 35.

91. Fuller, *Foundations*, 35.

92. Posen, "Foreword," 172. See also Harald Høiback, "What Is Doctrine?," *Journal of Strategic Studies* 34, no. 6 (December 2011): 897.

93. Posen, *Sources of Military Doctrine*.

94. Deborah D. Avant, *Political Institutions and Military Change: Lessons from Peripheral Wars*, Ithaca, NY: Cornell University Press, 1994.

95. Elizabeth Kier, *Imagining War*, Princeton, NJ: Princeton University Press, 1997.

96. Kimberly Marten Zisk, *Engaging the Enemy*, Princeton, NJ: Princeton University Press, 1993.

97. Avant, *Political Institutions*, 3; Elizabeth Kier, "Culture and Military Doctrine: France between the Wars," *International Security* 19, no. 4 (Spring 1995): 68.

98. In 2016, Posen wrote, "Doctrine exists at almost every level of military activity, from the lowly infantry company to the nuclear forces of a Cold War super-power. But it is high-level doctrine, which encompasses all of a state's military power, that has perhaps most captured the interest of political scientists, historians, and military theorists. There is no agreed term for this subject. . . . In past work, I simply called this body of principles 'military doctrine.'" Posen, "Foreword," 159.

99. Action is a key element of how *The Oxford Companion to Military History* defines doctrine: "An approved set of principles and methods, intended to provide large military organizations with a common outlook and a uniform basis for action." Richard Holmes, ed., *The Oxford Companion to Military History*, Oxford: Oxford University Press, 2001.

100. General Gordon R. Sullivan, US Army, "Doctrine: A Guide to the Future," *Military Review* 72, no. 2 (February 1992): 3.

101. Michael Evans, "Forward from the Past: The Development of Australian Army Doctrine, 1972–Present," Study Paper No. 301, Land Warfare Studies Centre, September 1999, 2.

102. See also Posen, "Foreword," 164–6.

103. Holmes, *Oxford Companion to Military History*.

104. John Spencer, "What Is Army Doctrine?," Modern War Institute, March 21, 2016.

105. Høiback, "What Is Doctrine?," 890.

106. Fuller, *Foundations*, 254.

107. Fuller, *Foundations*, 254.

108. Paul Johnston, "Doctrine Is Not Enough: The Effect of Doctrine on the Behavior of Armies," *Parameters* 30, no. 3 (Autumn 2000): 30–1.

109. Author count of doctrinal publications listed at "Doctrine and Training" section of the Publications tab at http://www.apd.army.mil/default.aspx.

110. Fuller, *Foundations*, 254.

111. Fuller, *Foundations*, 254.

112. Frank Hoffman, "Changing Tires on the Fly: The Marines and Postconflict Stability Ops," Foreign Policy Research Institute E-Notes, September 2, 2006; Jim Michaels, *A Chance in Hell*, New York: St. Martin's Press, 2010, 38.

113. Stephen Biddle, *Military Power*, Princeton, NJ: Princeton University Press, 2004, 4–5.

114. For more on technological determinism, see Grissom, "Military Innovation Studies," 926; Donald MacKenzie and Judy Wajcman, *The Social Shaping of Technology*, 2nd ed., Buckingham, England: Open University Press, 1999, 3–5.

115. Farrell and Terriff, "Sources of Military Change," 12–3. Barry Posen stands out as a notable exception to this generalization; he argues that "the influence of technology is seldom direct, and is usually filtered through organizational biases and statesmen's perceptions of the international political system." See Posen, *Sources of Military Doctrine*, 236.

116. See, for example, MacKenzie and Wajcman, *Social Shaping of Technology*; Nelly Oudshoorn and Trevor Pinch, eds., *How Users Matter: The Co-Construction of Users and Technologies*, Cambridge, MA: MIT Press, 2003.

117. Grissom, "Military Innovation Studies," 926–7; Farrell and Terriff, "Sources of Military Change," 13.

118. Oudshoorn and Pinch, *How Users Matter*, 1.

119. See, for example, Alastair Iain Johnston, "Thinking about Strategic Culture," *International Security* 19, no. 4 (Spring 1995): 32–64; Colin Gray, "Strategic Culture as Context: The First Generation of Theory Strikes Back," *Review of International Studies* 25, no. 1 (January 1999): 49–69.

120. One of the first books to systematically address the different service cultures remains one of the best today: Carl H. Builder, *The Masks of War: American Military Styles in Strategy and Analysis*, Baltimore: Johns Hopkins University Press, 1989.

121. See the many excellent examples discussed in Theo Farrell, "Culture and Military Power," *Review of International Studies* 24, no. 3 (July 1998): 407–16. See also Risa A. Brooks and Elizabeth Stanley-Mitchell, eds., *Creating Military Power*, Stanford, CA: Stanford University Press, 2007.

CHAPTER 2

1. Michael Howard, "Military Science in an Age of Peace," *RUSI Journal* 119, no. 1 (March 1974): 7.

2. Timothy T. Lupfer, *The Dynamics of Doctrine: The Changes in German Tactical Doctrine during the First World War*, Fort Leavenworth, KS: US Army Command and General Staff College, 1981, 43–4.

3. Lupfer, *The Dynamics of Doctrine*, 46–7.

4. One prominent example of the seriousness of interwar German thought on this concept is *Achtung-Panzer*, published in 1937 by Heinz Guderian. Major General Guderian was a forceful prewar advocate for using tanks in mass, and was in many ways the architect of the German panzer division. He subsequently led an armored corps during the invasion of Poland and the 19th Army Corps attacking through the Ardennes during the battle for France. See Alistair Horne, *To Lose a Battle: France 1940*, London: Macmillan, 1969, 47–53. For an English translation of the original book, see Heinz Guderian, *Achtung-Panzer! The Development of Tank Warfare*, trans. Christopher Duffy, London: Cassell, 1999.

5. See, for example, Major General Werner Widder, "Auftragstaktik and Innere Führung: Trademarks of German Leadership," *Military Review* 82, no. 5 (September–October 2002): 3–9. These principles were the forerunners of the current US Army concept of mission command, as discussed in Chapter 10.

6. Williamson Murray, *Military Adaptation in War: With Fear of Change*, New York: Cambridge University Press, 2011, 126–7.

7. Murray also notes that the unusual degree to which German senior officers were willing to criticize their own performance "was in stark contrast to the author's own experience with the United States Air Force, in which he served as a junior officer in the mid- to late 1960s." Murray, *Military Adaptation in War*, 124.

8. Germany was at war with France and Great Britain during this period, but the level of activity between October 1939 and April 1940 was so low that the period was referred to as the "Phony War," or "Sitzkrieg." Little combat occurred beyond sporadic air operations. Horne, *To Lose a Battle*, 101–9.

9. Robert M. Citino, *The Path to Blitzkrieg: Doctrine and Training in the German Army, 1920–1939*, Boulder, CO: Lynne Rienner Publishers, 1999, 92–4.

10. Pascal Marie Henri Lucas, *The Evolution of Tactical Ideas in France and Germany during the War of 1914–1918*, trans. P. V. Kieffer, Paris: Berger-Leorault, 1925, 147. Lucas quotes a senior French commander on the challenges of adapting to the open warfare in the German offensive breakthrough in the spring of 1918.

11. Williamson Murray, "Armored Warfare: The British, French, and German Experiences," in Williamson Murray and Allan R. Millett, eds., *Military Innovation in the Interwar Period*, Cambridge; New York: Cambridge University Press, 1996, 31–2.

12. Martin S. Alexander, "In Defence of the Maginot Line: Security Policy, Domestic Politics and the Economic Depression in France," in Robert Boyce, ed., *French Foreign and Defence Policy, 1918–1940: The Decline and Fall of a Great Power*, New York: Routledge, 1998, 167–8.

13. Murray, "Armored Warfare," 31.

14. France suffered 1.3 million military deaths, 4.2 million wounded, and 537,000 missing or made prisoners in World War I. Taken together, those numbers account for more than 70 percent of the 8.4 million men mobilized. William L. Shirer, *The Collapse of the Third Republic: An Inquiry into the Fall of France*, New York: Simon & Schuster, 1969, 133.

15. Robert A. Doughty, *The Seeds of Disaster: The Development of French Army Doctrine, 1919–1939*, Hamden, CT: Archon Books, 1985, 29–35.

16. Even in these disparate armies, aside from the German military, there was little consensus on how to organize and employ tanks in anything other than infantry-supporting roles. John T. Hendrix, "The Interwar Army and Mechanization: The American Approach," *Journal of Strategic Studies* 16, no. 1 (March 1993): 75–108.

17. As one French soldier recalled, "We should have been perfectly prepared to spend whole days potting at one another from entrenched positions . . . it would have been well within our capacity to stand firm in face of an assault through a curtain of wire . . . or to have gone over the top courageously . . . in short, we could have played our part without difficulty in operations beautifully planned by our own staff and the enemy's, if only they had been in accordance with the well-digested lessons learned at peace-time maneuvers." Marc Bloch, *Strange Defeat: A Statement of Evidence Written in 1940*, trans. Gerard Hopkins, New York: Octagon Books, 1968, 48–9. Doughty also argues that "the French military recognized that proposing rigid solutions to every problem would kill initiative, but there is little doubt that it opted for a rigid doctrine." See *The Seeds of Disaster*, 34–7.

18. Meir Finkel, *On Flexibility: Recovery from Technological and Doctrinal Surprise on the Battlefield*, Stanford, CA: Stanford University Press, 2011, 205–6, 211–20.

19. Martin S. Alexander, *The Republic in Danger: General Maurice Gamelin and the Politics of French Defence, 1933–1940*, New York: Cambridge University Press, 1992, 357.

20. Robert A. Doughty, *Breaking Point: Sedan and the Fall of France, 1940*, Hamden, CT: Archon Books, 1990, 70–1.

21. Doughty, *Breaking Point*, 344–7; Doughty, *The Seeds of Disaster*, 98–9.

22. Horne, *To Lose a Battle*, 335–41.

23. Barry Watts and Williamson Murray, "Military Innovation in Peacetime," in Murray and Millett, *Military Innovation*, 372–3.

24. Although the IDF includes Israel's army, air force, and small navy, we use the terms "IDF" and "Israeli army" interchangeably throughout this section for simplicity.

25. One example of an erroneous lesson learned was the belief that tanks could successfully operate independently and without infantry support. David Rodman, "Combined Arms Warfare: The Israeli Experience in the 1973 Yom Kippur War," *Defence Studies* 15, no. 2 (April 2015): 162.

26. Finkel, *On Flexibility*, 150.

27. Murray, *Military Adaptation in War*, 272–3.

28. Finkel, *On Flexibility*, 65.

29. Half-tracks are lightly armored trucks combining wheels in the front for steering with a tracked assembly for power and traction under the rear cargo area. The IDF used sizable numbers of US World War II surplus M3 half-tracks during this period for armored infantry mobility.

30. Anthony Cordesman and Abraham Wagner, *The Lessons of Modern War*, vol. 1: *The Arab-Israeli Conflicts, 1973–1989*, Boulder, CO: Westview Press, 1990, 52–64.

31. Finkel, *On Flexibility*, 65–7; Cordesman and Wagner, *Lessons of Modern War*, 52–4.

32. Murray, *Military Adaptation in War*, 269–70.

33. The Egyptian operational plan envisioned seizing up to 25 kilometers of the Sinai, conducting an operational pause to construct defenses in order to defeat IDF counterattacks, and subsequently negotiating for a settlement from a position of advantage relative to prewar positions. Michael Eisenstadt and Kenneth Pollack, "Armies of Snow and Armies of Sand: The Impact of Soviet Military Doctrine on Arab Armies," *Middle East Journal* 55, no. 4 (Autumn 2001): 563–4. See also Dani Asher, *The Egyptian Strategy for the Yom Kippur War: An Analysis*, trans. Moshe Tlamim, Jefferson, NC: McFarland, 2009, 145.

34. Fielded in the late 1960s by both the United States and Soviet Union, these wire-guided projectiles allowed infantry soldiers to attack tanks from as far as three kilometers away, guiding their deadly projectile by wire-link commands into their far-distant armored target. The ATGM was a breakthrough in the long-standing battle between tank and antitank weapons because of its range, precision, lethality, and portability. Moscow provided thousands of these man-portable Sagger

ATGMs to Egypt and supplemented them with tens of thousands of shoulder-fired unguided antitank rockets such as the RPG-7.

35. Eisenstadt and Pollack, "Armies of Snow," 564–5.

36. Eisenstadt and Pollack, "Armies of Snow," 565.

37. Eisenstadt and Pollack, "Armies of Snow," 565.

38. Hamdy Sobhy Abouseada, "The Crossing of the Suez Canal, October 6, 1973," Strategy Research Project, US Army War College, April 2000, 9–10.

39. Colonel T. N. Dupuy, USA, Ret., *Elusive Victory: The Arab Israeli Wars, 1947–1974*, Fairfax, VA: Hero Books, 1984, 419.

40. Murray, *Military Adaptation in War*, 296.

41. Edgar O'Ballance, *No Victor, No Vanquished: The Yom Kippur War*, San Rafael, CA: Presidio Press, 1978, 115–7, 161–2.

42. Major Edwin L. Kennedy, "Failure of Israeli Armored Tactical Doctrine, Sinai, 6–8 October 1973," *Armor* 99, no. 6 (November–December 1990): 30–1.

43. Over 200 Egyptian tanks and 250 other armored vehicles were destroyed by IDF tank gunnery and, in the south, air power. Eisenstadt and Pollack, "Armies of Snow," 566.

44. This attack was evidently directly ordered by Egyptian president Anwar Sadat to relieve intense Israeli military pressure on Damascus, capital of his Syrian ally Hafez al Assad. Saad el Shazly, *The Crossing of the Suez*, San Francisco: American Mideast Research, 2003, 245–8.

45. The US military studied the lessons of this battle in great detail, which shaped its subsequent doctrine of AirLand Battle. See Chapter 5.

46. Murray, *Military Adaptation in War*, 296.

CHAPTER 3

1. For more on the fog and friction of war, see Chapter 1.

2. Examples include France fighting on its own soil in World War I, or Israel battling along its narrow borders during most of its major wars.

3. For more on why bureaucracies, and especially military bureaucracies, resist change, see Chapter 1.

4. Lida Mayo, *The Ordnance Department: On Beachhead and Battlefront*, Washington, DC: Office of the Chief of Military History, US Army, 1968, 322–38.

5. Steven J. Zaloga, *Sherman Medium Tank, 1942–1945*, Oxford: Osprey Publishing, 1978, 34–9; Steven Zaloga, *Armored Thunderbolt: The U.S. Army Sherman in World War II*, Mechanicsburg, PA: Stackpole Books, 2008, 338–44. The countering argument, advanced both during and after the war by defenders of the M4, was that the Sherman, although not the best tank of the war, could be produced in such large numbers and with great enough mechanical reliability that it was good enough to enable the Allies to overwhelm the wide array of

German defenses. See Zaloga, *Armored Thunderbolt*, 327–30; Mayo, *Ordnance Department*, 336–8.

6. Gordon A. Harrison, *Cross-Channel Attack*, Washington, DC: Center of Military History, US Army, 1993, 46–83, 158–98.

7. Captain Michael D. Doubler, *Busting the Bocage: American Combined Arms Operations in France, 6 June–31 July 1944*, Fort Leavenworth, KS: Combat Studies Institute, US Army Command and General Staff College, November 1988, 14–5.

8. Martin Blumenson, *Breakout and Pursuit*, Washington, DC: Center of Military History, US Army, 1993, 12.

9. Doubler, *Busting the Bocage*, 21.

10. According to a survey taken after Normandy, out of a sample of 100 First Army junior officers, none reported having any prior knowledge of the hedgerows or training to handle them. Doubler, *Busting the Bocage*, 21.

11. Matthew A. Boal, "Field Expedient Armor Modifications to US Armored Vehicles," master's thesis, US Army Command and General Staff College, June 16, 2006, 28; Steven Zaloga, *Armored Attack, 1944: U.S. Army Tank Combat in the European Theater from D-Day to the Battle of the Bulge*, Mechanicsburg, PA: Stackpole Books, 2011, 128.

12. Doubler, *Busting the Bocage*, 31–2. See also Blumenson, *Breakout and Pursuit*, 187–8.

13. Blumenson, *Breakout and Pursuit*, 206–7. Both the Greendozer and "salad fork" breaching devices were developed by the 747th Tank Battalion. "After Action Report, 747th Tank Battalion, June thru December 1944, January thru April 1945," 2, available through the Ike Skelton Combined Arms Research Digital Library, http://cgsc.cdmhost.com/cdm/ref/collection/p4013coll8/id/3499.

14. Doubler, *Busting the Bocage*, 33–4.

15. Major David M. Russen, "Combat History: 102nd Cavalry Reconnaissance Squadron (MECZ), World War II (D Day—June 6, 1944 through VE Day—May 8, 1945)," 4, available at http://njcavalryandarmorassociation.org/History%20 of%20the%20102nd.pdf.

16. Doubler, *Busting the Bocage*, 35.

17. Mayo, *Ordnance Department*, 254.

18. "After Action Report, 735th Tank Bn, July–Nov 44, Feb–Mar 45," 2, available at http://cgsc.cdmhost.com/cdm/singleitem/collection/p4013coll8/id/3596/rec/8. See also Doubler, *Busting the Bocage*, 34.

19. Blumenson, *Breakout and Pursuit*, 207. Tank destroyer units and self-propelled gun units also requested the devices, but tanks were given top priority.

20. The Greendozer was a device constructed from railroad beams welded to the front of a tank that helped it cut through hedgerows. See Zaloga, *Armored Thunderbolt*, 157, 159–60; "After Action Report, 747th Tank Battalion, June thru December 1944, January thru April 1945."

21. Doubler, *Busting the Bocage*, 35.

22. Some of the bombs fell short onto friendly lines, inflicted hundreds of American casualties, including General Lesley McNair, the visiting chief of Army Ground Forces.

23. Blumenson, *Breakout and Pursuit*, 232.

24. Boal, "Field Expedient Armor Modifications," 31.

25. For simplicity, when we refer to German tank guns, we include German antitank guns as well since the Germans typically used the same high-velocity cannons in both weapons systems. Both presented nearly identical and highly lethal threats to Allied tanks.

26. Mayo, *Ordnance Department*, 327.

27. Boal, "Field Expedient Armor Modifications," 22.

28. Zaloga, *Armored Thunderbolt*, 281.

29. Zaloga, *Armored Thunderbolt*, 283–4.

30. Zaloga, *Armored Thunderbolt*, 284.

31. Zaloga, *Armored Thunderbolt*, 284.

32. Zaloga, *Armored Thunderbolt*, 279–86.

33. Boal, "Field Expedient Armor Modifications," 23.

34. Zaloga, *Armored Thunderbolt*, 283.

35. Another 1,100 M4s were destroyed in the Italian campaign. Zaloga, *Armored Thunderbolt*, 339–45.

36. Hubert C. Johnson, *Breakthrough! Tactics, Technology, and the Search for Victory on the Western Front in World War I*, Novato, CA: Presidio Press, 1994, 1–27.

37. The British military also concurrently and largely independently developed tracked mechanical vehicle designs in the first years of World War I, aimed at breaking through heavily defended trench lines and restoring mobility to the battlefield. Johnson, *Breakthrough*, 127–32.

38. Roger Pierre Laroussinie, quoted in Olivier Lahaie, "The Development of Tank Warfare on the Western Front," in Alaric Searle, ed., *Genesis, Employment, Aftermath: First World War Tanks and the New Warfare, 1900–1945*, Solihull, West Midlands, England: Helion, 2015, 58.

39. Robert A. Doughty, *Pyrrhic Victory: French Strategy and Operations in the Great War*, Cambridge, MA: Belknap Press of Harvard University Press, 2005, 337.

40. Lahaie, "Development of Tank Warfare," 58.

41. Johnson, *Breakthrough*, 130.

42. Johnson, *Breakthrough*, 131; Lahaie, "Development of Tank Warfare," 59.

43. Elizabeth Greenhalgh, *The French Army and the First World War*, Cambridge: Cambridge University Press, 2014, 121.

44. Greenhalgh, *French Army*, 228.

45. Elizabeth Greenhalgh, "Technology Development in Coalition: The Case of the First World War Tank," *International History Review* 22, no. 4 (December 2000): 810.

46. Lahaie, "Development of Tank Warfare," 78–9.

47. Greenhalgh, *French Army*, 227; Tim Gale, "'A Charming Toy': The Surprisingly Long Life of the Renault Light Tank, 1917–1940," in Searle, *Genesis, Employment, Aftermath*, 195.

48. Tim Gale, *The French Army's Tank Force and Armoured Warfare in the Great War: The* Artillerie Spéciale, Burlington, VT: Ashgate, 2013, 109, 123.

49. The tanks' failure in this initial outing nearly terminated the program. Only strong advocacy by chief of the General Staff and commander-in-chief Philippe Petain rescued it from cancelation. Lahaie, "Development of Tank Warfare," 73.

50. Greenhalgh, *French Army*, 241.

51. Doughty, *Pyrrhic Victory*, 370.

52. Heinz Guderian, quoted in Gale, "A Charming Toy," 203.

53. Gale, "A Charming Toy," 205.

54. M. P. M. Finch, "*Outre-Mer* and *Métropole*: French Officers' Reflections on the Use of the Tank in the 1920s," *War in History* 15, no. 3 (July 2008): 294.

55. Finch, "*Outre-Mer* and *Métropole*," 294. See also Greenhalgh, *French Army*, 389.

56. Zaloga, *Armored Thunderbolt*, 94–7.

57. Tank destroyers began the war as wheeled or half-tracked vehicles carrying 37mm and 75mm antitank guns, but evolved into fully tracked vehicles carrying a higher-velocity cannon designed primarily to defeat enemy tanks. They suffered from light armored protection and an open crew compartment vulnerable to artillery fire, and had no better mobility than the medium tank upon whose chassis many were built. By the end of the war, their value was largely discounted by armored force combat leaders, who found them wholly inadequate to mobile armored warfare against German panzers. David E. Johnson, *Fast Tanks and Heavy Bombers*, Ithaca, NY: Cornell University Press, 1998, 149–52; Christopher R. Gabel, *Seek, Strike, and Destroy: U.S. Army Tank Destroyer Doctrine in World War II*, Leavenworth Papers No. 12, Fort Leavenworth, KS: US Army Command and General Staff College, 1985, 20, 27–32.

58. Zaloga, *Armored Thunderbolt*, 49–50.

59. These ranged from the capable Panzer III and IV, which were common in 1942 and 1943 in North Africa, to later long-barreled versions of the Panzer IV, to the highly lethal Panzer V Panther and Panzer VI Tiger and King Tiger. The latter models were often encountered by the Allies in 1944 in Normandy and the Battle of the Bulge, and later in 1945 during the advance into Germany. Zaloga, *Armored Thunderbolt*, 94–8.

60. Zaloga, *Armored Thunderbolt*, 86–9.

61. In the fall of 1943, McNair wrote the following in a note to Devers: "The M4 tank, particularly the M4A3, has been widely hailed as the best tank on the battlefield today. There are indications that the enemy concurs in this view. Apparently, the M4 is an ideal combination of mobility, dependability, speed, protection, and

firepower. Other than this particular request—which represents the British view—there has been no call from any theater for a 90mm tank gun. There appears to be no fear on the part of our forces of the German Mark VI (Tiger) tank. . . There can be no basis for the T26 [prototype] tank other than the conception of a tank versus tank duel—which is believed unsound and unnecessary. Both British and American battle experience has demonstrated that the antitank gun in suitable numbers and disposed properly is the master of the tank. Any attempt to armor and gun tanks so as to outmatch antitank guns is foredoomed to failure. . . There is no indication that the 76mm antitank gun is inadequate against the German Mark VI (Tiger) tank." Zaloga, *Armored Thunderbolt*, 123–4.

62. Zaloga, *Armored Thunderbolt*, 125.

63. Zaloga, *Armored Thunderbolt*, 105.

64. Mayo, *Ordnance Department*, 323.

65. Johnson, *Fast Tanks*, 195.

66. Mayo, *Ordnance Department*, 326–7.

67. Johnson, *Fast Tanks*, 196.

68. Johnson, *Fast Tanks*, 196.

69. Johnson, *Fast Tanks*, 197–9.

70. The T26E1 that Marshall ordered in December 1943 ultimately became the M26 Pershing once put into production. Continued bureaucratic infighting and some final technical difficulties delayed the beginning of M26 production for almost a year longer. Zaloga, *Armored Thunderbolt*, 125.

71. David T. Zabeki, ed., *World War II in Europe: An Encyclopedia*, New York: Routledge, 2015, 1122, 1136.

72. Zaloga, *Armored Thunderbolt*, 81, 85, 94–7.

CHAPTER 4

1. Headquarters, Department of the Army, *Army Leadership*, Army Doctrine Publication 6-22, August 1, 2012, 1.

2. Headquarters, Department of the Army, *Leader Development*, Field Manual 6-22, June 2015, 5-7.

3. Headquarters, Department of the Army, *Operations*, Field Manual 3-0, October 2017, 2-56.

4. Headquarters, Department of the Army, *Leader Development*, 1-1.

5. Headquarters, Department of the Army, *Army Leadership*, Army Doctrine Reference Publication 6-22, August 1, 2012, 9-5. See also Headquarters, Department of the Army, *Mission Command*, Army Doctrine Reference Publication 6-0, May 2012, 1-2.

6. Headquarters, Department of the Army, *Army Leadership*, Army Doctrine Reference Publication 6-22, especially 5-1 and 5-2, and 9-4 through 9-6; Headquarters, Department of the Army, *Leader Development*, 5-7 and 5-8.

7. Tactical leadership occurs at roughly the same level as tactical adaptation in technology, as discussed in Chapter 3.

8. For more on the levels of war, see Chapter 1.

9. As we discuss in Chapter 9, Daniel Kahneman and Amos Tversky were the first scholars to study these systematic decision-making errors. For a clearly written overview of their complex work, see Michael Lewis, *The Undoing Project*, New York: Norton, 2017.

10. For more on hubris as an occupational hazard, see David Barno and Nora Bensahel, "What We Saw in *War Machine*," *War on the Rocks*, June 13, 2017.

11. Upon being promoted to the rank of one-star general, Barno was given the following tongue-in-cheek warning: "From now on, you will never again have a bad meal, and you will never again hear the truth."

12. Captain John Abizaid later became General John Abizaid, who commanded US Central Command from 2003 to 2007. See Chapters 5 through 8.

13. Captain David Barno commanded C Company of the First Ranger Battalion in Grenada, alongside Abizaid, who commanded A Company in the same battalion. Unless otherwise cited, this section is based on Barno's personal recollections.

14. Mark Adkin, *Urgent Fury: The Battle for Grenada*, New York: Lexington Books, 1989, 197–9.

15. Edgar F. Raines Jr., *The Rucksack War: U.S. Operational Logistics in Grenada, 1983*, Washington, DC: Center of Military History, US Army, 2010, 242.

16. US Atlantic Command, based in Norfolk, Virginia, directed the operation. It changed the parachute drop time in order to provide more time in which to land a SEAL team from a nearby navy destroyer to gather last-minute intelligence before the airdrop. Raines, *Rucksack War*, 179–80; Adkin, *Urgent Fury*, 196.

17. Raines, *Rucksack War*, 243–5.

18. Adkin, *Urgent Fury*, 211.

19. Adkin, *Urgent Fury*, 213.

20. Ronald H. Cole, *Operation Urgent Fury: The Planning and Execution of Joint Operations in Grenada, 12 October–2 November 1983*, Washington, DC: Joint History Office, Office of the Chairman of the Joint Chiefs of Staff, 1997, 41–2.

21. Adkin, *Urgent Fury*, 215.

22. Adkin, *Urgent Fury*, 215.

23. Harold G. Moore and Joseph L. Galloway, *We Were Soldiers Once . . . and Young*, New York: Open Road Media, 2012, 202, 224.

24. Merle L. Pribbenow, "The Fog of War: The Vietnamese View of the Ia Drang Battle," *Military Review* 81, no. 1 (January–February 2001): 95.

25. Dan Reed, "Shootout at LZ Albany: The 7th Cavalry's Second Fight for Survival at the Battle of Ia Drang," *Vietnam* 28, no. 4 (December 2015): 33.

26. Dallas J. Henry, "The Battle at LZ Albany," *Infantry* 103, no. 2 (April–June 2014): 21.

27. Moore and Galloway, *We Were Soldiers Once*, 209.

28. Henry, "Battle at LZ Albany," 21.

29. Captain Steven M. Leonard, OD, "Steel Curtain: The Guns on the Ia Drang," *Field Artillery*, July–August 1998, 20.

30. Moore and Galloway, *We Were Soldiers Once*, 212.

31. Henry, "Battle at LZ Albany," 22.

32. Moore and Galloway, *We Were Soldiers Once*, 216.

33. Moore and Galloway, *We Were Soldiers Once*, 215; Pribbenow, "Fog of War," 95.

34. Pribbenow, "Fog of War," 95.

35. Reed, "Shootout at LZ Albany," 35–6.

36. Pribbenow, "Fog of War," 96.

37. Reed, "Shootout at LZ Albany," 34.

38. Joseph L. Galloway, "Vietnam Story," *U.S. News & World Report*, October 29, 1990.

39. Pribbenow, "Fog of War," 96.

40. One member of 2/7 Cav, PFC Toby Braveboy, walked out of the jungle four days later and was rescued by a helicopter. Moore and Galloway, *We Were Soldiers Once*, 218, 254–5.

41. Moore and Galloway, *We Were Soldiers Once*, 233–5.

42. Moore did not personally make this decision, but he had the ability to countermand the call, which he did not do. Moore and Galloway, *We Were Soldiers Once*, 153–5.

43. Moore and Galloway, *We Were Soldiers Once*, 153–5.

44. Field Marshal Viscount Slim, *Defeat into Victory*, New York: Cooper Square Press, 2000, 8.

45. Slim, *Defeat into Victory*, 119.

46. Slim, *Defeat into Victory*, 119.

47. T. R. Moreman, *The Jungle, the Japanese and the British Commonwealth Armies at War, 1941–45*, New York: Frank Cass, 2005, 53–4.

48. Moreman, *The Jungle*, 38–9.

49. Ian Grant and Kazuo Tamayama, *Burma 1942: The Japanese Invasion*, Chichester, England: Zampi Press, 1999, 319–20.

50. Slim, *Defeat into Victory*, 105.

51. Grant and Tamayama, *Burma, 1942*, 320.

52. Brian Bond, "Slim and Fourteenth Army in Burma," in Brian Bond and Kyoichi Tachikawa, eds., *British and Japanese Military Leadership in the Far Eastern War, 1941–45*, New York: Routledge, 2004, 44.

53. Bond, "Slim and Fourteenth Army," 43; Slim, *Defeat into Victory*, 120–1.

54. Bond, "Slim and Fourteenth Army," 46; Ronald Lewin, *Slim: The Standardbearer*, London: Cooper, 1976, 139.

55. Bond, "Slim and Fourteenth Army," 61–3, 98.

56. Lewin, *Slim*, 115.

57. These included less reliance on motor vehicles and roads, the confidence to operate for extended periods while isolated and supplied only by air, and the ability to use small-unit day and night patrols in the jungle. Moreman, *The Jungle*, 99.

58. Moreman, *The Jungle*, 99, 168–9; Grant and Tamayama, *Burma, 1942*, 299; Lewin, *Slim*, 175.

59. Slim, *Defeat into Victory*, 179–80.

60. Lewin, *Slim*, 174, 217–9.

61. Vicki Constantine Croke, *Elephant Company: The Inspiring Story of an Unlikely Hero and the Animals Who Helped Him Save Lives in World War II*, New York: Random House, 2014, 255–76.

62. Bond, "Slim and Fourteenth Army," 44, 50–1.

63. Lewin, *Slim*, 136–8.

64. On at least one occasion, Slim gave his pep talk mostly in the wrong language. Slim, *Defeat into Victory*, 186.

65. Williamson Murray and Allan R. Millett, *A War to Be Won: Fighting the Second World War*, Cambridge, MA: Belknap Press of Harvard University Press, 2000, 492.

66. Lewin, *Slim*, 136–7.

67. Lewin, *Slim*, 191; Slim, *Defeat into Victory*, 180.

68. The British Fourth Corps—encircled by the Japanese at Imphal—depended completely on aerial resupply from April 9 until it was relieved on June 22. Between April 18 and June 30, British and American planes airlifted in 19,000 reinforcements, 13,000 tons of cargo, and 835,000 gallons of petrol, and evacuated 13,000 casualties and 43,000 noncombatants. Moreman, *The Jungle*, 131–2.

69. Slim's campaign included creating a corps of elephants to build bridges and an initiative to recruit refugees, guerrillas, and other local irregulars to attack the Japanese behind their lines. Croke, *Elephant Company*, 255–76.

70. Lewis Sorley, *Westmoreland: The General Who Lost Vietnam*, Boston: Houghton Mifflin Harcourt, 2011, 16–24.

71. The commanders who served in the US Army's airborne divisions in World War II went on to great acclaim and created a powerful internal community of influence in the army called the "airborne mafia." Westmoreland became part of this group after commanding the 504th Parachute Infantry Regiment after the war in the 82nd Airborne Division and subsequently the independent 187th Airborne regimental combat team during combat in Korea. Brian McAllister Linn, *Elvis's Army: Cold War GIs and the Atomic Battlefield*, Cambridge, MA: Harvard University Press, 2016, 40.

72. Mark Perry, *Four Stars*, Boston: Houghton Mifflin, 1989, 136. The president considered three other army generals to take command in Vietnam: Harold K. Johnson, Creighton Abrams, and Bruce Palmer. All later went on the record as saying that winning in Vietnam required a counterinsurgency approach that differed dramatically from the attrition-based strategy that Westmoreland put in place. Lewis Sorley, *A Better War: The Unexamined Victories and Final Tragedies of America's Last Years in Vietnam*, New York: Harcourt Brace, 1999, 1.

73. Andrew F. Krepinevich Jr., *The Army and Vietnam*, Baltimore: Johns Hopkins University Press, 1986, 197.

74. Carl von Clausewitz, *On War*, ed. and trans. Michael Howard and Peter Paret, Princeton, NJ: Princeton University Press, 1976, 88.

75. Krepinevich describes this model of warfare as the Army Concept, which we discuss further in Chapter 5. See *The Army and Vietnam*, 4–7.

76. In April 1967 Westmoreland told the president that the "crossover point" had been reached at which the Viet Cong and North Vietnamese Army would be unable to replace their losses in all but two provinces. Krepinevich, *The Army and Vietnam*, 187.

77. Krepinevich, *The Army and Vietnam*, 259.

78. According to Krepinevich, a May 1967 army study found that "approximately 25% of all operations are security or minor pacification missions, and approximately 75% are search and destroy." Krepinevich, *The Army and Vietnam*, 182.

79. Sorley, *A Better War*, 4.

80. Anonymous army officer quoted in John A. Nagl, *Learning to Eat Soup with a Knife: Counterinsurgency Lessons from Malaya and Vietnam*, Westport, CT: Praeger, 2002, 152–3.

81. Krepinevich, *The Army and Vietnam*, 179.

82. Shelby L. Stanton, *The Rise and Fall of an American Army: U.S. Ground Forces in Vietnam, 1965–1973*, Novato, CA: Presidio, 1985, 65; Howard K. Butler, *Army Aviation and Logistics in Vietnam*, St. Louis, MO: US Army Aviation Command Historical Office, 1985, 357–9.

83. Nagl, *Learning to Eat Soup*, 177–80. As one high-level example, the army's own 1966 study called "Program for the Pacification and Long Term Development of South Vietnam" noted that the situation had "seriously deteriorated," and recommended that pacification be given top priority. Inside MACV, junior officers were not shy about disputing the large-unit, search-and-destroy value of the attrition approach. Yet another example of new ideas within the command were the Marine Corps' combined action platoons. Krepinevich, *The Army and Vietnam*, 172–7, 181–2; Nagl, *Learning to Eat Soup*, 177.

84. In Krepinevich's words: "The use of massive firepower as a crutch in lieu of an innovative counterinsurgency strategy alienated the population and provided the enemy with an excellent source of propaganda." Krepinevich, *The Army and Vietnam*, 198–9.

85. Krepinevich, *The Army and Vietnam*, 202–3.

86. Krepinevich, *The Army and Vietnam*, 187.

87. Krepinevich, *The Army and Vietnam*, 190. There are several modern revisionist accounts of the Vietnam War that argue that Westmoreland deserves less blame for the war's ultimate outcome, was ill served by his analytics, and actually chose the only reasonable strategy. See, for example, Gregory A. Daddis, *No Sure Victory: Measuring U.S. Army Effectiveness and Progress in the Vietnam War*, New York: Oxford University Press, 2011, 111–7; Dale Andrade, "Westmoreland Was Right: Learning the Wrong Lessons from the Vietnam War," *Small Wars &*

Insurgencies 19, no. 2 (June 2008): 145–81. We find these accounts less persuasive than the majority of studies that criticize Westmoreland's decision-making, particularly when evaluating his adaptability.

88. Sorley, *A Better War*, 7–9; Sorley, *Westmoreland*, 180–2.

89. MACV's August 1968 Long Range Planning Task Group concluded shortly after Westmoreland's departure: "All of our U.S. combat accomplishments have made no significant, positive difference to the rural Vietnamese—for there is still no real security in the countryside...Destruction of NVA and VC units and individuals—that is, the 'kill VC' syndrome—has become an end in itself—an end that has at times become self-defeating...The Viet Cong thrive in an environment of insecurity. It is essential for them to demonstrate that the GVN is not capable of providing security to its citizens. *And, they have succeeded*." Krepinevich, *The Army and Vietnam*, 254.

90. See Lewis, *The Undoing Project*, as well as Amos Tversky and Daniel Kahneman, "Availability: A Heuristic for Judging Frequency and Probability," *Cognitive Psychology* 5, no. 2 (September 1973): 207–32.

CHAPTER 5

1. Quoted in Conrad C. Crane, *Avoiding Vietnam: The U.S. Army's Response to Defeat in Southeast Asia*, Carlisle, PA: Strategic Studies Institute, US Army War College, September 2002, 4. Emphasis added.

2. This is also one of the key themes of Chapter 8.

3. The US Marine Corps was involved in the process of rewriting the manual, and it was copublished by the marines as well. Yet, as discussed later, active-duty and retired army officers led most of the initiative.

4. Another, and likely more important, factor was the Sunni Awakening, where Sunni tribal leaders decided to realign themselves with the United States in order to defeat their previous ally, al-Qaeda in Iraq. See Chapter 7.

5. See, for example, Gian P. Gentile, "A Strategy of Tactics: Population-Centric COIN and the Army," *Parameters* 39, no. 3 (August 2009): 5–17; Colonel Gian Gentile, *Wrong Turn: America's Deadly Embrace of Counterinsurgency*, New York: New Press, 2013; Douglas Porch, *Counterinsurgency: Exposing the Myths of the New Way of War*, New York: Cambridge University Press, 2013. For a response to these criticisms, see David H. Ucko, "Critics Gone Wild: Counterinsurgency as the Root of All Evil," *Small Wars & Insurgencies* 25, no. 1 (2014): 161–79.

6. See, for example, Henry Nuzum, *Shades of CORDS in the Kush: The False Hope of "Unity of Effort" in American Counterinsurgency*, Carlisle, PA: Strategic Studies Institute, US Army War College, April 2010; US House of Representatives, Committee on Armed Services, Subcommittee on Oversight & Investigations, "Agency Stovepipes vs. Strategic Agility: Lessons We Need to Learn from Provincial Reconstruction Teams in Iraq and Afghanistan," September 2008;

Project on National Security Reform, *Forging a New Shield*, Arlington, VA: Project on National Security Reform, November 2008.

7. Weigley argues that this strategy extends as far back as the Revolutionary War. See Russell F. Weigley, *The American Way of War: A History of United States Military Strategy and Policy*, Bloomington: University of Indiana Press, 1973.

8. Andrew F. Krepinevich, Jr., *The Army and Vietnam*, Baltimore: Johns Hopkins University Press, 1986, 5. Although he refers to conventional war as mid-intensity conflict—presumably in comparison to a nuclear conflict—this book refers to it as high-intensity conflict.

9. See David H. Ucko, *The New Counterinsurgency Era: Transforming the U.S. Military for Modern Wars*, Washington, DC: Georgetown University Press, 2009, Chapter 1 and 25–30.

10. Krepinevich, *The Army and Vietnam*, 6–7.

11. Krepinevich, *The Army and Vietnam*, 7.

12. For extensive details, see Krepinevich, *The Army and Vietnam*, and John A. Nagl, *Learning to Eat Soup with a Knife: Counterinsurgency Lessons from Malaya and Vietnam*, Westport, CT: Praeger, 2002.

13. Krepinevich, *The Army and Vietnam*, 164–260; Nagl, *Learning to Eat Soup*, 151–87.

14. Thomas E. Ricks, *The Generals*, New York: Penguin, 2012, 326–32; Kenneth J. Campbell, "Once Burned, Twice Cautious: Explaining the Weinberger-Powell Doctrine," *Armed Forces & Society* 24, no. 3 (Spring 1998): 358–62; David Fitzgerald, *Learning to Forget: US Army Counterinsurgency Doctrine and Practice from Vietnam to Iraq*, Stanford, CA: Stanford University Press, 2013, 40–1.

15. Ricks, *The Generals*, 336–9.

16. David W. Barno, "Military Adaptation in Complex Operations," *Prism* 1, no. 1 (December 2009): 29.

17. Crane, *Avoiding Vietnam*, 16–7.

18. For more on the 1973 war, see Chapter 2.

19. Major Paul H. Herbert, *Deciding What Has to Be Done: General William E. DePuy and the 1976 Edition of FM 100-5, Operations*, Leavenworth Papers Number 16, Fort Leavenworth, KS: Combat Studies Institute, US Army Command and General Staff College, 1988, 29–31; Ricks, *The Generals*, 337–40; Benjamin M. Jensen, *Forging the Sword: Doctrinal Change in the U.S. Army*, Stanford, CA: Stanford University Press, 2016, 33–43.

20. Jeffrey W. Long, *The Evolution of U.S. Army Doctrine: From Active Defense to AirLand Battle and Beyond*, Fort Leavenworth, KS: US Army Command and General Staff College, 1991, 31. For more on Active Defense, see John L. Romjue, "From Active Defense to AirLand Battle: The Development of Army Doctrine, 1973–1982," Fort Monroe, VA: United States Training and Doctrine Command Historical Office, June 1984, 6–11; Herbert, *Deciding What Has to Be Done*, 79–85.

21. Herbert, *Deciding What Has to Be Done*, 1; Lieutenant Colonel Donald B. Vought, "Preparing for the Wrong War?," *Military Review* 57, no. 5 (May 1977): 28–9; Krepinevich, *The Army and Vietnam*, 260.

22. Ricks, *The Generals*, 337.

23. Richard Duncan Downie, *Learning from Conflict: The U.S. Military in Vietnam, El Salvador, and the Drug War*, Westport, CT: Praeger, 1998, Chapters 3 and 4.

24. This publication was Field Manual 31-22, *U.S. Army Counterinsurgency Forces*. Downie, *Learning from Conflict*, 53.

25. These two publications were Field Manual 100-20, *Low Intensity Conflict*, and Field Manual 31-16, *Counterguerilla Operations*, respectively. Krepinevich, *The Army and Vietnam*, 272; Andrew F. Krepinevich Jr., "Recovery from Defeat: The U.S. Army and Vietnam," in George J. Andreopolis and Harold E. Selesky, eds., *The Aftermath of Defeat: Societies, Armed Forces, and the Challenge of Recovery*, New Haven, CT: Yale University Press, 1994, 137.

26. For example, in a book chapter published in 1994, Krepinevich noted that the 1987 version of Field Manual 90-4, *Air Assault Operations*, contains only 11 lines of text about counterinsurgency throughout its 85 pages, and that the 1981 version of Field Manual 7-30, *Infantry, Airborne, and Air Assault Brigade Operations*, does not discuss unconventional operations at all. Krepinevich, "Recovery from Defeat," 137.

27. Peter R. Mansoor, *Baghdad at Sunrise: A Brigade Commander's War in Iraq*, New Haven, CT: Yale University Press, 2008, 345.

28. Ricks, *The Generals*, 337; John P. Lovell, "Vietnam and the U.S. Army: Learning to Cope with Failure," in George K. Osborn et al., eds., *Democracy, Strategy, and Vietnam: Implications for American Policymaking*, Lexington, MA: Lexington Books, 1987, 132; Krepinevich, "Recovery from Defeat," 134.

29. Colonel John D. Waghelstein, "Post-Vietnam Counterinsurgency Doctrine," *Military Review*, 65, no. 5 (May 1985): 42–4.

30. Barno, "Military Adaptation," 29.

31. Michael J. Mazarr, *Light Forces and the Future of U.S. Military Strategy*, McLean, VA: Brassey's, 1990, 35–7. Two existing infantry divisions, the Seventh and 25th, would be converted to light-infantry divisions, and two new divisions, the Sixth and 10th Mountain, would be added to the force structure, increasing the total of active-duty divisions at the time from 16 to 18 by 1988.

32. The army still had 14 divisions that were primarily focused on a conventional fight with the Soviets, and the majority of those were far larger than the new light-infantry divisions. Mazarr, *Light Forces*, 11.

33. Lovell, "Vietnam and the U.S. Army," 132; see also Colonel David J. Barratto, "Special Forces in the 1980s: A Strategic Reorientation," *Military Review* 63, no. 3 (March 1983): 8.

34. Downie, *Learning from Conflict*, 63–81.

35. The Strategic Studies Institute at the Army War College published Summers's work in 1981; it was reprinted as a book the following year. See Harry G. Summers Jr., *On Strategy: A Critical Analysis of the Vietnam War*, Novato, CA: Presidio Press, 1982.

36. Crane, *Avoiding Vietnam*, 6–7; Krepinevich, *The Army and Vietnam*, 262; Krepinevich, "Recovery from Defeat," 135; Fitzgerald, *Learning to Forget*, 53–9.

37. Crane suggests that this exchange may not be historically accurate, but it was nonetheless extremely influential. Summers, *On Strategy*, 1; Crane, *Avoiding Vietnam*, 7–10.

38. Downie, *Learning from Conflict*, 74; Crane, *Avoiding Vietnam*, 9.

39. It was criticized by many in the army for focusing too much on defense and not enough on offense; for ignoring the psychological dimensions of warfare; and for focusing too narrowly on Europe. Herbert, *Deciding What Has to Be Done*, 96–8.

40. Crane, *Avoiding Vietnam*, 8.

41. The term "fight outnumbered and win" originated in the 1976 version of Field Manual 100-5. It is most frequently associated with the 1982 version, however, even though that phrase does not appear in that document.

42. For more on AirLand Battle, see Romjue, "From Active Defense to AirLand Battle"; Richard Lock-Pullan, "How to Rethink War: Conceptual Innovation and AirLand Battle Doctrine," *Journal of Strategic Studies* 28, no. 4 (August 2005): 679–702.

43. Barno, "Military Adaptation," 29.

44. For more on the ground campaign in Operation Desert Storm, see Michael R. Gordon and General Bernard E. Trainer, *The Generals' War: The Inside Story of the Conflict in the Gulf*, Boston: Little, Brown, 1995.

45. Quoted in Crane, *Avoiding Vietnam*, 1.

46. Quoted in Nagl, *Learning to Eat Soup*, 211–2.

47. Downie, *Learning from Conflict*, 59; Janine Davidson, *Lifting the Fog of Peace: How Americans Learned to Fight Modern War*, Ann Arbor: University of Michigan Press, 2010, 139–40.

48. Ucko, *New Counterinsurgency Era*, 48–51.

49. Conrad C. Crane, *Cassandra in Oz: Counterinsurgency and Future War*, Annapolis, MD: Naval Institute Press, 2016, 45.

50. For more on the war planning, see Michael R. Gordon and General Bernard E. Trainor, *Cobra II: The Inside Story of the Invasion and Occupation of Iraq*, New York: Pantheon Books, 2006.

51. For details on the prewar planning process and the assumptions it contained, see Nora Bensahel et al., *After Saddam: Prewar Planning and the Occupation of Iraq*, MG-642-A, Santa Monica, CA: Rand Corporation, 2008.

52. Mansoor, *Baghdad at Sunrise*, 345–6; Austin Long, *Doctrine of Eternal Recurrence: The U.S. Military and Counterinsurgency Doctrine, 1960–1970 and 2003–2006*, Rand Counterinsurgency Study Paper 6, Santa Monica, CA: Rand Corporation, 2008, 22; Ucko, *New Counterinsurgency Era*, 69.

53. Wallace had commanded Fifth Corps during major combat operations in Iraq and took command of the Combined Arms Center in June 2003. See Gordon and Trainor, *Cobra II*.

54. Fred Kaplan, *The Insurgents: David Petraeus and the Plot to Change the American Way of War*, New York: Simon & Schuster, 2013, 133.

55. Kaplan, *The Insurgents*, 133.

56. Kaplan, *The Insurgents*, 134.

57. Kaplan, *The Insurgents*, 136–7. For additional critiques of the manual, see Long, *Doctrine of Eternal Recurrence*, 21–2; Ucko, *New Counterinsurgency Era*, 66.

58. Headquarters, Department of the Army, *Counterinsurgency Operations*, Field Manual-Interim 3-07.22, October 2004.

59. Kaplan, *The Insurgents*, 3; Ucko, *New Counterinsurgency Era*, 78.

60. Kaplan, *The Insurgents*, 82–4.

61. Kaplan, *The Insurgents*, 85–90.

62. Nagl, *Learning to Eat Soup*.

63. Peter Maass, "Professor Nagl's War," *New York Times Magazine*, January 11, 2004.

64. Kaplan, *The Insurgents*, 106–7.

65. The number of counterinsurgency-related articles published in *Parameters*, the journal of the US Army War College, also increased, from three in 2004 to 11 in 2005. Ucko, *New Counterinsurgency Era*, 77.

66. Ucko, *New Counterinsurgency Era*, 76–7; Kaplan, *The Insurgents*, 97–107.

67. Kalev I. Sepp, "Best Practices in Counterinsurgency," *Military Review* 85, no. 3 (May–June 2005): 8–12. This issue also contains articles by Conrad Crane, who, as discussed subsequently, led the writing team for the new counterinsurgency manual, and Montgomery McFate, an anthropologist who helped write the new manual's chapter on intelligence.

68. Kaplan, *The Insurgents*, 108–16. Quotation from 113.

69. Kaplan, *The Insurgents*, 14–31. The article was published as John R. Galvin, "Uncomfortable Wars: Toward a New Paradigm," *Parameters* 16, no. 4 (Winter 1986): 2–8.

70. Ucko, *New Counterinsurgency Era*, 75; Kaplan, *The Insurgents*, 71–6.

71. David Cloud and Greg Jaffe, *The Fourth Star: Four Generals and the Epic Struggle for the U.S. Army*, New York: Crown Publishers, 2009, 216–7; John A. Nagl, "Constructing the Legacy of Field Manual 3-24," *Joint Force Quarterly* 58 (3rd Quarter 2010): 118; Ucko, *New Counterinsurgency Era*, 75; Kaplan, *The Insurgents*, 128–9.

72. This was also the advice Petraeus received from Eliot Cohen, whom he had asked to review the draft doctrine. Crane, *Cassandra in Oz*, 46.

73. Octavian Manea, "Reflections on the 'Counterinsurgency Decade': Small Wars Journal Interview with General David H. Petraeus," *Small Wars Journal*, September 1, 2013.

74. Cloud and Jaffe, *Fourth Star*, 217–8; Kaplan, *The Insurgents*, 137.

75. Those who had attended the Basin Harbor conference included T. X. Hammes, from the National Defense University; Steven Metz, from the Army War College; and Janine Davidson, from SAIC. Other notable participants included Frank Hoffman, from the Marine Warfighting Lab; Michèle Flournoy, from the Center for Strategic and International Studies; and Lieutenant Colonel Richard Lacquement, from the Office of the Deputy Assistant Secretary of State for Stability Operations. Lacquement and Nagl later drafted parts of the new manual; all of those named attended the February 2006 conference discussed later. Kilcullen, however, did not attend the October 2005 conference; he had been invited but returned to Australia to help with counterterrorism policy after the 2005 bombings in Bali. Kaplan, *The Insurgents*, 138–41; Crane, *Cassandra in Oz*, 43.

76. Kaplan, *The Insurgents*, 144; Crane, *Cassandra in Oz*, 45.

77. Kaplan, *The Insurgents*, 144–6.

78. Crane, *Cassandra in Oz*, 48. Lieutenant General William Caldwell IV, who succeeded Petraeus as the commander of the Combined Arms Center, later recalled, "Traditionally, when we write Army doctrine, it's done in-house. The Army has a very deliberate set procedure . . . [the new counterinsurgency manual] was really the first deviation from the way Army manuals are written, done in 2006 in a much more open and collaborative manner, many [in] academia and others being brought into the process." Quoted in Nagl, "Constructing the Legacy," 120.

79. Crane, *Cassandra in Oz*, 58.

80. Crane, *Cassandra in Oz*, 58.

81. Kaplan, *The Insurgents*, 147–9; Crane, *Cassandra in Oz*, 53.

82. Nagl, "Constructing the Legacy," 118.

83. These drew on a number of sources, including David Galula's classic *Counterinsurgency Warfare: Theory and Practice*; Kalev Sepp's "Best Practices in Counterinsurgency"; an article written by Petraeus that was later published in *Military Review*; parts of Nagl's *Learning to Eat Soup*; and his own article on Phase 4 operations, which had been published in the May–June 2005 issue of *Military Review*. Kaplan, *The Insurgents*, 149–50.

84. Kaplan, *The Insurgents*, 150; Crane, *Cassandra in Oz*, 54–5.

85. Eliot Cohen, Lieutenant Colonel Conrad Crane, US Army, Retired, Lieutenant Colonel Jan Horvath, US Army, and Lieutenant Colonel John Nagl, US Army, "Principles, Imperatives, and Paradoxes of Counterinsurgency," *Military Review* 86, no. 2 (March–April 2006): 49–53. The authors outlined and drafted much of the article on the second day of the February 2006 conference described later. The article was also posted on *Small Wars Journal*, a blog read by many members of the military, to ensure it reached as many soldiers as possible. See Nagl, "Constructing the Legacy," 118; Kaplan, *The Insurgents*, 164–5; Crane, *Cassandra in Oz*, 55.

86. Crane lists the chapter authors in *Cassandra in Oz*, 53–4.

87. Crane, *Cassandra in Oz*, 56.

88. Crane, *Cassandra in Oz*, 59–63.

89. Kaplan, *The Insurgents*, 146–8; Crane, *Cassandra in Oz*, 64–7.

90. Figures 8.2, 8.3, and 8.4 show the increase in the number of insurgent attacks, Afghan casualties, and US troop fatalities by this time.

91. The journalists were invited as participants, not reporters; the meeting was off the record. Kaplan, *The Insurgents*, 148–9, 153.

92. Crane, *Cassandra in Oz*, 1–5.

93. From March 2003 through the end of January 2006, 2,243 US troops died in Iraq. See Michael E. O'Hanlon and Ian Livingston, "Iraq Index: Tracking Variables of Reconstruction & Security in Post-Saddam Iraq," Brookings Institution, January 31, 2011, 12.

94. Crane, *Cassandra in Oz*, 68.

95. For good summaries of these debates, see Kaplan, *The Insurgents*, 156–64; Crane, *Cassandra in Oz*, 70–8.

96. Crane, *Cassandra in Oz*, 79.

97. This set of revisions took longer than anticipated; Petraeus and Crane both thought that they would go faster and enable the final version to be published in late summer or early fall. Crane, *Cassandra in Oz*, 78.

98. Crane, *Cassandra in Oz*, 79–89.

99. Crane, *Cassandra in Oz*, 90.

100. Kaplan, *The Insurgents*, 216.

101. Nagl, "Constructing the Legacy," 118.

102. John A. Nagl, "Learning and Adapting to Win," *Joint Force Quarterly* 58 (3rd Quarter 2010), 123.

103. Given the army's arcane system for categorizing doctrine, it is significant that the army designated the manual as 3-24 instead of continuing the interim manual's designation as 3-07.22. The leading number 3 signifies that the doctrine focuses on operations; 07 is the category for stability operations, and the subcategory 22 suggests that counterinsurgency was simply one aspect of stability operations. Crane argued that counterinsurgency was far more than that and merited its own operational category. Petraeus agreed, and 24 became the new operational category for counterinsurgency. Crane, *Cassandra in Oz*, 50–1.

104. Stories about the new manual ran on the front page of several prestigious newspapers, including the *New York Times* and the *Washington Post*, and were even featured in international publications like *The Economist*. In what was and will likely continue to be the most unusual bit of publicity for any official military publication, Nagl was interviewed by comedian Jon Stewart on the wildly popular *Daily Show*. Cloud and Jaffe, *Fourth Star*, 220; Crane, *Cassandra in Oz*, 123–4.

105. Cloud and Jaffe, *Fourth Star*, 220; Crane, *Cassandra in Oz*, 123–4.

106. James A. Baker and Lee H. Hamilton, Co-Chairs, *The Iraq Study Group Report*, New York: Vintage Books, 2006, 6.

107. See Figures 8.3 and 8.4 for details.

108. See Bensahel et al., *After Saddam*.

109. Headquarters, Department of the Army / Headquarters, Marine Corps Combat Development Command, *Counterinsurgency*, Field Manual 3-24 / Marine Corps Warfighting Publication 3-33.5, December 15, 2006, 1-26 to 1-28.

110. Ucko, *New Counterinsurgency Era*, 111; Kaplan, *The Insurgents*, 151.

111. Headquarters, Department of the Army / Headquarters, Marine Corps Combat Development Command, *Counterinsurgency*, 1-27 to 1-28.

112. The US military officially defines civil affairs operations as "actions planned, coordinated, executed, and assessed to enhance awareness of, and manage the interaction with, the civil component of the operational environment; [actions that] identify and mitigate underlying causes of instability within civil society; and/or involve the application of functional specialty skills normally the responsibility of civil government." Joint Chiefs of Staff, *DOD Dictionary of Military and Associated Terms*, November 2018, 35.

113. Headquarters, Department of the Army, *Civil Affairs Operations*, Field Manual 41-10, February 2000, 1-2, 2-28, 4-6.

114. Joint Chiefs of Staff, *Joint Doctrine for Civil-Military Operations*, Joint Publication 3-57, June 21, 1995; Joint Chiefs of Staff, *Interagency Coordination during Joint Operations*, vol. 1, Joint Publication 3-08, October 9, 1996; Headquarters, Department of the Army, *Civil Affairs Operations*, Field Manual 41-10, February 2000; Joint Chiefs of Staff, *Joint Doctrine for Civil-Military Operations*, Joint Publication 3-57, February 8, 2001.

115. Joint Chiefs of Staff, *Joint Doctrine for Civil-Military Operations*, June 21, 1995, IV-3.

116. Headquarters, Department of the Army, *Civil Affairs Operations*, Field Manual 41-10, January 11, 1993, 5-9.

117. Joint Chiefs of Staff, *Joint Doctrine for Civil-Military Operations*, June 21, 1995, IV-3.

118. Headquarters, Department of the Army, *Civil Affairs Operations*, February 2000, 4-5, 4-6; Joint Chiefs of Staff, *Joint Doctrine for Civil-Military Operations*, February 8, 2001, ix, II-17 to II-21.

119. The two largest US military operations during the 1990s occurred in Bosnia and Kosovo. But since both operations were commanded by NATO, they relied on NATO doctrine for civil-military cooperation (CIMIC) instead of US doctrine. See, for example, James J. Landon, "CIMIC: Civil Military Cooperation," in Larry Wentz, ed., *Lessons from Bosnia: The IFOR Experience*, Washington, DC: Institute for National Strategic Studies, National Defense University / CCRP, 1997, 119–38; Thomas R. Mockaitis, *Civil-Military Cooperation in Peace*

Operations: The Case of Kosovo, Carlisle, PA: Strategic Studies Institute, US Army War College, October 2004.

120. Chris Seiple, *The U.S. Military / NGO Relationship in Humanitarian Interventions*, Carlisle, PA: Peacekeeping Institute, Center for Strategic Leadership, US Army War College, 1996, 21, 24, 32, 38–40.

121. Seiple, *U.S. Military / NGO Relationship*, 40–1.

122. Seiple, *U.S. Military / NGO Relationship*, 42–3.

123. This meeting included not only NGOs and IOs, but also representatives from the Bangladeshi government and from the British, Pakistani, and Japanese helicopter units that were participating in the military operation. Seiple, *U.S. Military / NGO Relationship*, 83.

124. Seiple, *U.S. Military / NGO Relationship*.

125. Civil-military tensions in the CMOC in Somalia were much higher than northern Iraq or Bangladesh, partly because of the scope and size of the operation but also because of broader political disagreements between the United States and the United Nations. Seiple, *U.S. Military / NGO Relationship*, Chapter 4. For more on the political tensions, see Nora Bensahel, "Humanitarian Relief and Nation Building in Somalia," in Robert J. Art and Patrick M. Cronin, eds., *The United States and Coercive Diplomacy*, Washington, DC: United States Institute of Peace Press, 2003, 21–56.

126. Seiple, *U.S. Military / NGO Relationship*, Chapter 5.

127. Major Aaron L. Wilkins, "The Civil Military Operations Center (CMOC) in Operation *Uphold Democracy* (Haiti)," paper presented to the Research Department of the Air Command and Staff College in partial fulfilment of graduation requirements, March 1997.

128. Joint Chiefs of Staff, *Joint Doctrine for Civil-Military Operations*, June 21, 1995, IV-4 to IV-5.

129. Joint Chiefs of Staff, *Interagency Coordination during Joint Operations*, vol. 1, III-16 to III-18.

130. Headquarters, Department of the Army, *Civil Affairs Operations*, February 2000, 5-10 to 5-12, Appendix H.

131. Joint Chiefs of Staff, *Joint Doctrine for Civil-Military Operations*, February 8, 2001, IV-10 to IV-14.

132. Donald P. Wright et al., *A Different Kind of War: Operation ENDURING FREEDOM, October 2001–September 2005*, Fort Leavenworth, KS: Combat Studies Institute Press, US Army Combined Arms Center, 2010, 194, 197. See also Olga Oliker et al., *Aid during Conflict: Interaction between Civilian Assistance Providers in Afghanistan, September 2001–June 2002*, MG-212, Santa Monica, CA: Rand Corporation, 2004, 20–2.

133. Several senior US counterterrorism officials did warn that al-Qaeda posed a grave threat to the United States, but their arguments for taking forceful action rarely

prevailed. See National Commission on Terrorist Attacks upon the United States, *The 9/11 Commission Report*, New York: Norton, 2004.

134. See Oliker et al., *Aid during Conflict*.

135. See the transcripts of the first and second presidential debates between George W. Bush and Al Gore, available at debates.org; for the views of Condoleezza Rice, see Dale Russakoff, "Lessons of Might and Right," *Washington Post*, September 9, 2001.

136. These phases were first described by William Flavin in "Civil-Military Operations: Afghanistan," US Army Peacekeeping and Stability Operations Institute, March 23, 2004.

137. The four-phase planning construct was reaffirmed, somewhat ironically as it turns out, in joint doctrine that was published the day before the September 11 attacks. See Joint Chiefs of Staff, *Doctrine for Joint Operations*, Joint Publication 3-0, September 10, 2001, III-18 to III-21.

138. For more on the role of special forces during these phases of the war, see Chapter 7.

139. Phase 4 did emphasize longer-term cooperation with coalition partners to prevent terrorism from reemerging inside Afghanistan, but said little about how this was to be achieved. For more on each phase, see Wright et al., *Different Kind of War*, 46–7.

140. Flavin, "Civil-Military Operations," ix.

141. Flavin, "Civil-Military Operations," 73; Wright et al., *Different Kind of War*, 46–7.

142. White House, Office of the Press Secretary, "Presidential Address to the Nation," Treaty Room, October 7, 2001.

143. Wright et al., *Different Kind of War*, 82–3.

144. Oliker et al., *Aid during Conflict*, 43, 48.

145. Oliker et al., *Aid during Conflict*, 49, 66–7.

146. Flavin, "Civil-Military Operations," 18.

147. The CJCMOTF commander later recalled that General Tommy Franks, the CENTCOM commander, "told me directly, with his finger in my face, that I would not get involved in nation building." Wright et al., *Different Kind of War*, 194.

148. Oliker et al., *Aid during Conflict*, 50–1.

149. It was initially commanded by the commander of the 122nd Rear Operations Cell, but Kratzer was named when it became clear that the CJCMOTF needed to be led by a general officer. Flavin, "Civil-Military Operations," 18.

150. The exercise was part of the annual Bright Star series of exercises with Egypt. Flavin, "Civil-Military Operations," 18.

151. These units were the 352nd Civil Affairs Command, the 511th Military Police Company, and the 489th and 96th Civil Affairs battalions. Flavin, "Civil-Military Operations," 18; Oliker et al., *Aid during Conflict*, 50; Wright et al., *Different Kind of War*, 193.

152. Flavin, "Civil-Military Operations," 18; Oliker et al., *Aid during Conflict*, 66; Wright et al., *Different Kind of War*, 193; Brian Neumann, Lisa Mundy, and Jon Mikolashek, *Operation ENDURING FREEDOM: The United States Army in Afghanistan, March 2002–April 2005*, Washington, DC: US Army Center for Military History, [2013?], 32.

153. Flavin, "Civil-Military Operations," xiv, 28, 73.

154. Neumann et al., *Operation ENDURING FREEDOM*, 32.

155. Flavin, "Civil-Military Operations," 22, 51; Oliker et al., *Aid during Conflict*, 43, 67; Wright et al., *Different Kind of War*, 82–3, 194; Neumann et al., *Operation ENDURING FREEDOM*, 32.

156. Oliker et al., *Aid during Conflict*, 86; Wright et al., *Different Kind of War*, 82–3, 194–5.

157. Flavin, "Civil-Military Operations," 48; Wright et al., *Different Kind of War*, 194.

158. Flavin, "Civil-Military Operations," 19; Wright et al., *Different Kind of War*, 194.

159. Flavin, "Civil-Military Operations," 36.

160. Oliker et al., *Aid during Conflict*, 72.

161. Flavin, "Civil-Military Operations," x, 20.

162. Flavin, "Civil-Military Operations," 50.

163. Oliker et al., *Aid during Conflict*, 74; Neumann et al., *Operation ENDURING FREEDOM*, 33.

164. Wright et al., *Different Kind of War*, 194–5.

165. Oliker et al., *Aid during Conflict*, 70; Wright et al., *Different Kind of War*, 196–7.

166. Flavin, "Civil-Military Operations," 41, 47–8.

167. Wright et al., *Different Kind of War*, 197; Oliker et al., *Aid during Conflict*, 49, 91.

168. CHLC personnel initially stopped wearing their uniforms as a force protection measure. Many local leaders continued to request that they dress in civilian clothes, partially for security reasons but also because they wanted to appear as though they were in charge of their areas rather than as dependent on leaders in Kabul and the US forces that supported them. Flavin, "Civil-Military Operations," 49.

169. Flavin, "Civil-Military Operations," 49–50; Wright et al., *Different Kind of War*, 198.

170. Flavin, "Civil-Military Operations," 21.

171. Wright et al., *Different Kind of War*, 4. The Afghan Interim Administration was established by the December 2001 Bonn Agreement, which also named Hamid Karzai as its chairman. See James F. Dobbins, *Foreign Service: Five Decades on the Frontlines of American Diplomacy*, Santa Monica, CA: Rand Corporation; Washington, DC: Brookings Institution Press, 2017, Chapter 24.

172. Flavin, "Civil-Military Operations," xiv, 22; Wright et al., *Different Kind of War*, 210–1.

173. He also took formal control of the OMC from the US embassy, so he would also oversee efforts to train the Afghan security forces. Wright et al., *Different Kind of War*, 211.

174. Flavin, "Civil-Military Operations," 22; Wright et al., *Different Kind of War*, 223.

175. Oliker et al., *Aid during Conflict*, 89.

176. Wright et al., *Different Kind of War*, 229.

177. Wright et al., *Different Kind of War*, 223–5; Neumann et al., *Operation ENDURING FREEDOM*, 34.

178. Wright et al., *Different Kind of War*, 226.

179. Barbara J. Stapleton, "A British Agencies Afghanistan Group Briefing Paper on the Development of Joint Regional Teams in Afghanistan," British Agencies Afghanistan Group, January 2003, 26; Michael J. McNerney, "Stabilization and Reconstruction in Afghanistan: Are PRTs a Model or a Muddle?," *Parameters* 35, no. 4 (Winter 2005–2006): 32.

180. Stapleton, "British Agencies Afghanistan Group," 5; Wright et al., *Different Kind of War*, 226.

181. Stapleton, "British Agencies Afghanistan Group," 5, 11.

182. Stout had helped develop the concept of the CJCMOTF, and initially went to Afghanistan to propose a study of that organization at the one-year mark. When he briefed the director of CJTF-180 on his planned study, he noted that there was no official political-military strategy at that point, and that he had with him an unsigned draft of such a strategy that the State Department had developed earlier in the year. His knowledge of that draft and his previous experience led Lieutenant General McNeill and his staff to keep Stout in Afghanistan and make him responsible for turning the JRT concept into reality. Though the pol-mil strategy draft was never approved, it did shape the ways in which JRTs were developed. Wright et al., *Different Kind of War*, 226–7.

183. Stapleton, "British Agencies Afghanistan Group," 19; Wright et al., *Different Kind of War*, 228.

184. Cameron S. Sellers, "Provincial Reconstruction Teams: Improving Effectiveness," master's thesis, Naval Postgraduate School, September 2007, 10, 33.

185. Stapleton, "British Agencies Afghanistan Group," 16–9, 22, 26. Presidential and parliamentary elections were initially scheduled for July 2004, in accordance with the terms of the Bonn Agreement. However, technical challenges and security concerns delayed the presidential election until October, and the parliamentary elections were not held until September 2005.

186. Stapleton, "British Agencies Afghanistan Group," 6–7, 16.

187. Stapleton, "British Agencies Afghanistan Group," 6, 16; McNerney, "Stabilization and Reconstruction," 36.

188. Wright et al., *Different Kind of War*, 228.

189. Wright et al., *Different Kind of War*, 226; Neumann et al., *Operation ENDURING FREEDOM*, 34.

190. Their concerns and recommendations for improvement are described in detail in Stapleton, "British Agencies Afghanistan Group."

191. Wright et al., *Different Kind of War*, 227.

192. Wright et al., *Different Kind of War*, 228.

193. Wright et al., *Different Kind of War*, 227–8, 256–7.

194. McNerney, "Stabilization and Reconstruction," 36.

195. The PRTs reported to CJTF-180 through the CJCMOTF, just as the CHLCs had. Neumann et al., *Operation ENDURING FREEDOM*, 36.

196. McNerney, "Stabilization and Reconstruction," especially 36–41.

197. Robert M. Perito, "The U.S. Experience with Provincial Reconstruction Teams in Afghanistan," Special Report 152, United States Institute of Peace, October 2005, 3; Wright et al., *Different Kind of War*, 257.

198. For more on the internationalization of PRTs in Afghanistan, see Perito, "U.S. Experience," especially 2–3; McNerney, "Stabilization and Reconstruction," especially 38–9.

199. Wright et al., *Different Kind of War*, 4.

200. See Chapter 4.

201. In order to maintain as much objectivity as possible, Bensahel wrote this section based on both primary and secondary sources and, where needed, on Barno's earlier interviews and publications. Information not otherwise cited comes from Barno's personal recollection.

202. David W. Barno, "Theater Command in Afghanistan: Taking Charge of 'The Other War' in 2003–2005," in Michael A. Sheehan and Erich Marquardt, eds., *Counterterrorism & Irregular Warfare*, United States Military Academy Combating Terrorism Center, West Point, NY, 2019.

203. Wright et al., *Different Kind of War*, 244. See also Lieutenant General David W. Barno, US Army (Retired), "Fighting 'The Other War'": Counterinsurgency Strategy in Afghanistan, 2003–2005," *Military Review* 87, no. 5 (September–October 2007), note 8.

204. Nuzum, *Shades of CORDS*, 19–20; Barno, "Theater Command in Afghanistan."

205. Wright et al., *Different Kind of War*, 245, 255.

206. These included T. E. Lawrence's *Seven Pillars of Wisdom*, the Marine Corps' *Small Wars Manual*, and John Nagl's *Learning to Eat Soup*. Kaplan, *The Insurgents*, 320.

207. The chart that CFC-A used to brief this concept is reproduced in Wright et al., *Different Kind of War*, 246, and Barno, "Fighting 'The Other War,'" 35.

208. Wright et al., *Different Kind of War*, 245.

209. The other three pillars involved defeating terrorism and denying sanctuary, enabling Afghan security structures, and engaging regional states. See Barno, "Fighting 'The Other War,'" 37–41.

210. Barno, "Fighting 'The Other War,'" 39.
211. Neumann et al., *Operation ENDURING FREEDOM*, 48; Barno, "Fighting 'The Other War,'" 33.
212. Pillar 3 was modeled after the New York Police Department's successful efforts to reduce crime in the 1990s. Neumann et al., *Operation ENDURING FREEDOM*, 48; Barno, "Fighting 'The Other War,'" 39–40.
213. McNerney, "Stabilization and Reconstruction," 38.
214. Barno, "Fighting 'The Other War,'" 40.
215. Wright et al., *Different Kind of War*, 255.
216. Neumann et al., *Operation ENDURING FREEDOM*, 58; Barno, "Fighting 'The Other War,'" 40. Barno also notes that he took a calculated risk in sending the PRTs into more dangerous areas. Although the PRTs did not include enough military personnel to defend themselves against a significant attack, Taliban and other insurgents knew that US combat aircraft would arrive within 20 minutes of any distress call, which served as a powerful deterrent.
217. Barno, "Fighting 'The Other War,'" 40; Nuzum, *Shades of CORDS*, 22.
218. Wright et al., *Different Kind of War*, 255. The United Kingdom continued to lead the PRT in Mazar-e Sharif, and by this time New Zealand led the PRT in Bamian and Germany led the PRT in Konduz.
219. Barno, "Fighting 'The Other War,'" 36. Regional commands in the west and north were established in September 2004 and in 2005, respectively. Wright et al., *Different Kind of War*, 294.
220. Wright et al., *Different Kind of War*, 255, 278, 286, 292–7, 326; Neumann et al., *Operation ENDURING FREEDOM*, 58.
221. For a detailed description of the general PRT model, see Perito, "U.S. Experience."
222. Wright et al., *Different Kind of War*, 258–60, 293; McNerney, "Stabilization and Reconstruction," 41.
223. North Atlantic Treaty Organization, *ISAF Provincial Reconstruction Team (PRT) Handbook*, 4th ed. [n.d.], 105–11.
224. NATO as an organization took command of ISAF in August 2003, replacing a rotating system of lead nations. The United Nations authorized ISAF to expand beyond Kabul in October 2003, and the alliance developed a plan to do so in four stages. NATO expanded ISAF to the north in October 2004, to the west in September 2005, to the south in July 2006, and to the east in October 2006. North Atlantic Treaty Organization, *ISAF Provincial Reconstruction Team (PRT) Handbook*, 91–6; Andrew R. Hoehn and Sarah Harting, *Risking NATO: Testing the Limits of the Alliance in Afghanistan*, MG-974, Santa Monica, CA: Rand Corporation, 2010, 25–9.
225. PRTs in Iraq were structured quite differently from the ones in Afghanistan, however. See Robert M. Perito, "Provincial Reconstruction Teams in Iraq," Special

Report 185, United States Institute of Peace, March 2007, especially 1–3; Ucko, *New Counterinsurgency Era*, 163.

226. See Perito, "U.S. Experience"; McNerney, "Stabilization and Reconstruction"; "Provincial Reconstruction Teams in Afghanistan: An Interagency Assessment," Office of the Coordinator for Stabilization and Reconstruction, Department of State; Joint Center for Operational Analysis, United States Joint Forces Command; and Bureau of Policy and Program Coordination, United States Agency for International Development, April 5, 2006; US House of Representatives, Committee on Armed Services, Subcommittee on Oversight & Investigations, "Agency Stovepipes."

227. Neumann et al., *Operation ENDURING FREEDOM*, 37, 59.

228. See Headquarters, Department of the Army, *Civil Affairs Operations*, Field Manual 3-05.40, September 2006; Joint Chiefs of Staff, *Civil-Military Operations*, Joint Publication 3-57, July 8, 2008, especially II-29 to II-33.

CHAPTER 6

1. Marjorie Censer, "Army Report: Military Has Spent $32 Billion since '95 on Abandoned Weapons Programs," *Washington Post*, May 27, 2011. See also Russell Rumbaugh, "What We Bought: Defense Procurement from FY01 to FY10," Stimson Center, October 2011.

2. Close air support (CAS) is "air action by fixed-wing and rotary-wing aircraft against hostile targets that are in close proximity to friendly forces and requires detailed integration of each air mission with the fire and movement of those forces." See Joint Chiefs of Staff, *Close Air Support*, Joint Publication 3-09.3, November 25, 2014, xi. The army prefers to use the term "close combat attack" when its helicopters conduct CAS missions, to signify that the aircraft are conducting aerial maneuvers in support of ground forces. We use the joint term throughout this section.

3. A total of 4,482 US troops died during the Iraq War. See Michael E. O'Hanlon and Ian Livingston, "Iraq Index: Tracking Variables of Reconstruction & Security in Post-Saddam Iraq," Brookings Institution, January 31, 2011, 8.

4. Humvees are officially called High Mobility Multipurpose Wheeled Vehicles (HMMWVs).

5. Kollars notes that this term "eventually became a particular point of embarrassment for DoD." Nina Kollars, "Military Innovation's Dialectic: Gun Trucks and Rapid Acquisition," *Security Studies 23*, no. 4 (October-December 2014): 800–1.

6. John Barry, "'Hillbilly Armor,'" *Newsweek*, December 19, 2004; Thom Shanker and Eric Schmitt, "Armor Scarce for Heavy Trucks Transporting U.S. Cargo in Iraq," *New York Times*, December 10, 2004.

7. Mark Thompson, "How Safe Are Our Troops?," *Time*, December 13, 2004.

8. Christopher J. Lamb, Matthew J. Schmidt, and Berit G. Fitzsimmons, "MRAPs, Irregular Warfare, and Pentagon Reform," *Joint Force Quarterly* 55 (4th Quarter 2009): 77. General John Abizaid served in Grenada as Captain John Abizaid. See Chapter 4.

9. The secret Manhattan Project developed nuclear weapons during World War II. Bryan Bender, "Panel on Iraq Bombings Grows to $3B Effort," *Boston Globe*, June 25, 2006.

10. Bender, "Panel on Iraq Bombings."

11. O'Hanlon and Livingston, "Iraq Index," 8.

12. Michael Moran, "Frantically, the Army Tries to Armor Humvees," *MSNBC.com*, April 15, 2004.

13. Moran, "Army Tries to Armor"; also see Chapter 3.

14. Steve Liewer, "Sandbags Become Makeshift Vehicle Armor," *Stars and Stripes*, April 26, 2004.

15. CSM Joseph C. Paccioretti, "The Pioneers," paper written for the US Army Sergeants Major Academy, May 18, 2008, available at http://cdm15040.contentdm.oclc.org/cdm/ref/collection/p15040coll2/id/4882.

16. Liewer, "Sandbags."

17. Robert M. Gates, *Duty: Memoirs of a Secretary at War*, New York: Alfred A. Knopf, 2014, 120.

18. Kollars, "Military Innovation's Dialectic," 801–2; Jacques S. Gansler, William Lucyshyn, and William Varettoni, "Acquisition of Mine-Resistant, Ambush-Protected (MRAP) Vehicles: A Case Study," Center for Public Policy and Private Enterprise, School of Public Policy, University of Maryland, March 2010; Richard Currey, "Waiting for Justice: The Saga of Army Lt. Julian Goodrum, PTSD, Hillbilly Armor, and Whistle-Blowing," *VVA Veteran*, March–April 2005.

19. Kollars, "Military Innovation's Dialectic," 801.

20. Barry, "Hillbilly Armor."

21. Seabees are the navy's construction battalions, or CBs. See http://www.navy.mil/navydata/personnel/seabees/seabee1.html.

22. "Soldiers Must Rely on 'Hillbilly Armor' for Protection," *ABC News*, December 8, 2004.

23. Seth Robson, "2nd ID Vehicles Are Upgraded to Survive Rough Traveling in Iraq," *Stars and Stripes*, August 20, 2004.

24. Robson, "2nd ID Vehicles."

25. Nina A. Kollars, "War's Horizon: Soldier-Led Adaptation in Iraq and Vietnam," *Journal of Strategic Studies* 38, no. 4 (2015): 546; Kollars, "Military Innovation's Dialectic," 808.

26. Dreadnoughts were armored battleships from the early 20th century. Juliana Gittler, "'Skunk Werks' Armor Shop Helps Soldiers through Better Protection for U.S. Vehicles," *Stars and Stripes*, October 31, 2004.

27. Kollars, "War's Horizon," 546. Emphasis added.

28. Kollars, "Military Innovation's Dialectic," 806–7.

29. Kollars, "War's Horizon," 547.

30. US Department of Defense, Office of the Assistant Secretary of Defense, "Special Defense Department Briefing on Uparmoring HMMWV," December 15, 2004. For more on early army efforts to address the IED threat, see Colonel Joel D. Rayburn and Colonel Frank K. Sobchak, eds., *The U.S. Army in the Iraq War*, vol. 1: *Invasion, Insurgency, Civil War, 2003–2006*, Carlisle, PA: Strategic Studies Institute and US Army War College Press, January 2019, 236–7.

31. Shanker and Schmitt, "Armor Scarce."

32. Peter Eisler, Blake Morrison, and Tom Vanden Brook, "Pentagon Balked at Pleas from Officers in Field for Safer Vehicles," *USA Today*, July 16, 2007; Kollars, "Military Innovation's Dialectic," 806.

33. Barry, "Hillbilly Armor"; US Department of Defense, "Special Defense Department Briefing on Uparmoring HMMWV."

34. Barry, "Hillbilly Armor"; Sharon K. Weiner, "Organizational Interests versus Battlefield Needs: The U.S. Military and Mine-Resistant Ambush Protected Vehicles in Iraq," *Polity* 42, no. 4 (October 2010): 467.

35. John Hendren, "Unarmored Military Vehicles to Be Restricted," *Los Angeles Times*, February 13, 2005.

36. General Peter Schoomaker, testimony to the US House of Representatives, Committee on Armed Services, November 17, 2004.

37. Barry, "Hillbilly Armor"; Currey, "Waiting for Justice."

38. General Peter Schoomaker, testimony to the US House of Representatives, Committee on Armed Services, November 17, 2004.

39. Barry, "Hillbilly Armor."

40. Eric Schmitt, "Iraq-Bound Troops Confront Rumsfeld over Lack of Armor," *New York Times*, December 8, 2004.

41. William Branigin, "Bush, Rumsfeld Pledge to Protect Troops," *Washington Post*, December 9, 2004.

42. Branigin, "Bush, Rumsfeld Pledge."

43. US Department of Defense, "Special Defense Department Briefing on Uparmoring HMMWV."

44. See, for example, "Soldiers Must Rely on 'Hillbilly Armor' for Protection"; Shanker and Schmitt, "Armor Scarce"; Thompson, "How Safe Are Our Troops?"; and Barry, "Hillbilly Armor."

45. Eisler et al., "Pentagon Balked."

46. Bender, "Panel on Iraq Bombings."

47. The vehicles had other safety features as well. See Captain Wayne A. Sinclair, "Answering the Landmine," *Marine Corps Gazette* 80, no. 7 (July 1996): 38.

48. Sinclair, "Answering the Landmine," 37.

49. Eisler et al., "Pentagon Balked"; Weiner, "Organizational Interests," 478.

50. Eisler et al., "Pentagon Balked."

51. Eisler et al., "Pentagon Balked."

52. Eisler et al., "Pentagon Balked"; Lamb et al., "MRAPs," 77.

53. O'Hanlon and Livingston, "Iraq Index," 8.

54. Eisler et al., "Pentagon Balked."

55. Gates, *Duty*, 121.

56. Eisler et al., "Pentagon Balked."

57. The vehicle that was eventually procured—the Iraqi Light Armored Vehicle—was almost the same as an MRAP. Eisler et al., "Pentagon Balked."

58. The officer who drafted the request had learned about the MRAP from Wayne Sinclair, the author of the 1996 article published in *Marine Corps Gazette*. Eisler et al., "Pentagon Balked."

59. Lamb et al., "MRAPs," 79.

60. Gansler et al., "Acquisition."

61. Sharon Weinberger and Noah Shachtman, "Military Dragged Feet on Bomb-Proof Vehicles (Updated Again)," *Wired*, May 22, 2007.

62. Eisler et al., "Pentagon Balked."

63. Weinberger and Shachtman, "Military Dragged Feet"; Lamb et al., "MRAPs," 79.

64. Gansler et al., "Acquisition," 15.

65. This meeting included five three-star generals, four two-star generals, and seven one-star generals. Eisler et al., "Pentagon Balked."

66. Eisler et al., "Pentagon Balked."

67. Lamb et al., "MRAPs," 79.

68. Eisler et al., "Pentagon Balked."

69. Eisler et al., "Pentagon Balked"; Andrew Feickert, "Joint Light Tactical Vehicle (JLTV): Background and Issues for Congress," RS22942, Congressional Research Service, updated June 24, 2019, 2, 6; Todd South, "Here's When the Marine Corps' Newest Tactical Vehicle Hits the Rest of the Fleet," *Marine Corps Times*, August 16, 2019.

70. Hagee later said that this decision was not intended to affect the MRAP request. Gansler et al., "Acquisition," 7, 15.

71. JIEDDO evolved out of the original Army IED Defeat Task Force, discussed in the previous section. See Andrew Smith, *Improvised Explosive Devices in Iraq, 2003–09: A Case Study of Operational Surprise and Institutional Response*, Carlisle, PA: Strategic Studies Institute, US Army War College, April 2011, 14–5.

72. US Department of Defense, *Joint Improvised Explosive Device Defeat Organization Annual Report 2010*, 11.

73. Bender, "Panel on Iraq Bombings"; Lamb et al., "MRAPs," 78–9. Bensahel also observed this when she worked on projects sponsored by JIEDDO while at the Rand Corporation.

74. John T. Bennett, "What Next for U.S. Joint Anti-IED Efforts?," *Defense News*, September 17, 2007.

75. Eisler et al., "Pentagon Balked"; Lamb et al., "MRAPs," 80; Gansler et al., "Acquisition," 8–9.

76. US Department of Defense, Office of the Under Secretary of Defense for Acquisition, Technology, and Logistics, *Report of the Defense Science Board Task Force on the Fulfillment of Urgent Operational Needs*, July 2009, 3, 9; Kollars, "Military Innovation's Dialectic," 809–10.

77. Weiner, "Organizational Interests," 478.

78. Gansler et al., "Acquisition," 9, 15.

79. Eisler et al., "Pentagon Balked."

80. Between 2003 and 2007, Congress gave the Pentagon more than $521 billion in supplemental funding for wartime needs like the MRAP, above and beyond the almost $2 trillion it received through the regular budgeting process for long-term requirements like the JLTV. All figures in 2009 dollars. US Department of Defense, *Fiscal Year 2009 Supplemental Request*, April 2009, 1.

81. Eisler et al., "Pentagon Balked."

82. Gansler et al., "Acquisition," 16.

83. Weiner, "Organizational Interests," 465.

84. O'Hanlon and Livingston, "Iraq Index," 8.

85. Taylor quoted in Lamb et al., "MRAPs," 80.

86. General Richard Cody, Vice Chief of Staff, US Army, testimony to the US House of Representatives, Committee on Armed Services, Readiness Subcommittee, March 13, 2007.

87. General Robert Magnus, Assistant Commandant, US Marine Corps, testimony to the US House of Representatives, Committee on Armed Services, Readiness Subcommittee, March 13, 2007.

88. Gates, *Duty*, 119; Tom Vanden Brook, "New Vehicles Protect Marines in 300 Attacks in Iraq Province," *USA Today*, April 19, 2007.

89. Vanden Brook, "New Vehicles Protect Marines."

90. The Marine Corps had ordered 3,700 MRAPs, and the US Army had ordered 2,300. Gates, *Duty*, 122.

91. Gates, *Duty*, 122.

92. Gates, *Duty*, 122. Emphasis added.

93. Gates gave the MRAP program a DX rating, which gave it legal priority over both military and civilian programs for access to needed materials and equipment, including specialty steel, tires, and axles. It is not a common rating; in November 2007, for example, only six other programs had received it. Gates, *Duty*, 123; Gansler et al., "Acquisition," 36–7.

94. Gates, *Duty*, 123.

95. Lamb et al., "MRAPs," 80; Gansler et al., "Acquisition," 16.

96. Gates, *Duty*, 124.

97. Eisler et al., "Pentagon Balked."

98. According to Gates, "I learned the background story the same way I heard about the vehicle in the first place: from the newspaper. Two and a half months after my first briefing, I read in *USA Today* that the Pentagon had first tested MRAPs in 2000 and that the Marine Corps had requested its first twenty-seven of them in December 2003 for explosive disposal teams." Gates, *Duty*, 121.

99. *Congressional Record*, Volume 153 (2007), Part 14 (Senate), 19742–8. The article mentioned in the previous citation estimated that if MRAPs had been fielded after the first formal request in February 2005, as many as 742 troops could have been saved. Eisler et al., "Pentagon Balked."

100. *Congressional Record*, Volume 153 (2007), Part 14 (Senate), 19743.

101. Lamb et al., "MRAPs," 80; Gansler et al., "Acquisition," 16.

102. Lamb et al., "MRAPs," 81.

103. Gates, *Duty*, 124.

104. Gates, *Duty*, 125.

105. These included Abrams tanks, Bradley fighting vehicles, and Stryker armored vehicles. Gates, *Duty*, 124.

106. Gates, *Duty*, 125.

107. Carter Malkasian, *Illusions of Victory: The Anbar Awakening and the Rise of the Islamic State*, New York: Oxford University Press, 2017.

108. For more on the surge, see Chapter 8.

109. Of the 126 US soldiers who died in Iraq in May 2007, 82 were killed by IEDs. O'Hanlon and Livingston, "Iraq Index," 8.

110. In July 2007, Senator Joe Biden (D-DE) forcefully rebuked this concern during a floor speech: "Many believed that Congress would not support funding the MRAP while also fielding better armored Humvees. I do not know of a single wartime funding request that Congress has denied. There have been some items added to the supplemental bills that were clearly not urgent or war related, but nothing directly linked to current operations was refused. Nonetheless, it appears that the military did not believe that our support for needed equipment was real. Even today, I hear that leaders are concerned that they must cut multiple existing programs to pay for this growing MRAP requirement. There may be programs that we could all agree are not as vital for a wartime Army, but I do not want that debate and concern to slow lifesaving equipment." *Congressional Record*, Volume 153 (2007), Part 14 (Senate), 19742–8.

111. Kollars, "Military Innovation's Dialectic," 811. Emphasis added.

112. This section draws heavily on Jonathan Bernstein, *AH-64 Apache Units of Operations Enduring Freedom and Iraqi Freedom*, Osprey Combat Aircraft 57, Oxford: Osprey Publishing, 2005; Chris Bishop, *Apache AH-64 Boeing (McDonnell Douglas), 1976–2005*, Oxford: Osprey Publishing, 2005.

113. The other four weapons systems were the M1 Abrams main battle tank, the M2/
    3 Bradley infantry fighting vehicle, the UH-60 Blackhawk utility/assault helicop-
    ter, and the Patriot air defense missile system. See Robert H. Scales Jr., *Certain
    Victory: The US Army in the Gulf War*, Washington, DC: Office of the Chief of
    Staff, United States Army, 1993, 19–20.

114. For more on AirLand Battle, see Chapter 5.

115. Major Darren W. Buss, "Evolution of Army Attack Aviation: A Chaotic Coupled
    Pendulums Analogy," School of Advanced Military Studies, US Army Command
    and General Staff College, Fort Leavenworth, KS, 2013, 10–1.

116. This suite was called the Target Acquisition and Designation Sight / Pilot's Night
    Vision System, or TADS/PNVS. Bishop, *Apache AH-64 Boeing*, 8–12.

117. "Fire and forget" meant that the crew could fire the missile when it found
    the target and then quickly leave the area. The missile would continue to
    maneuver independently to seek out and destroy its original target, even if the
    target moved.

118. Apaches were first briefly used in combat during the opening hours of the
    December 1989 invasion of Panama. Bishop, *Apache AH-64 Boeing*, 16–7.

119. US General Accounting Office, "Operation Desert Storm: Apache Helicopter
    Was Considered Effective in Combat, but Reliability Problems Persist," GAO/
    NSIAD-92-146, April 1992, 3.

120. Bernstein, *AH-64 Apache Units*, 8.

121. The army also believed that the 1991 Gulf War had validated the AirLand Battle
    doctrine, and it deeply shaped the perspectives of the junior officers who became
    the army's most senior generals during the wars in Iraq and Afghanistan. See
    Chapter 5.

122. These included limits on flying at altitude, concerns about taking casualties,
    and restrictive rules of engagement. Two Apaches also crashed during train-
    ing in nearby Albania, which further eroded support for using them. See
    John Gordon IV, Bruce Nardulli, and Walter L. Perry, "The Operational
    Challenges of Task Force Hawk," *Joint Force Quarterly* 29 (Autumn–Winter
    2001–2002): 52–7.

123. Bishop, *Apache AH-64 Boeing*, 11–2. These included the ability to simultaneously
    track up to 128 targets and engage up to 16 at once with its Hellfire missiles, and
    share data with other helicopters. Bernstein, *AH-64 Apache Units*, 9.

124. Donald P. Wright et al., *A Different Kind of War: Operation ENDURING
    FREEDOM, October 2001–September 2005*, Fort Leavenworth, KS: Combat
    Studies Institute Press, US Army Combined Arms Center, 2010, 127–9.

125. Bishop, *Apache AH-64 Boeing*, 13–4.

126. The flechette warhead contained 1,180 hardened steel flechettes, which are small,
    nail-like projectiles designed to be used against personnel or soft targets. Bishop,
    Apache AH-64 Boeing, 14.

127. Wright et al., *A Different Kind of War*, 140. Due to troop constraints imposed by Washington, the initial US units deployed without any of their usual supporting artillery.

128. Ground fire support controllers use the warning term "danger close" when calling in fires in situations where friendly forces are "within close proximity of the target." Joint Chiefs of Staff, *DOD Dictionary of Military and Associated Terms*, Joint Publication 1-02, November 8, 2010, as amended through February 15, 2016, 59. This can often include fires within 200 meters of friendly troops, but the danger distance varies by the type of munition being used. In infantry firefights, ranges this close to the enemy were relatively common.

129. For more on Operation Anaconda, see Walter L. Perry and David Kassing, *Toppling the Taliban: Air-Ground Operations in Afghanistan, October 2001–June 2002*, RR-381, Santa Monica, CA: Rand Corporation, 2015, 98–104; Sean Naylor, *Not a Good Day to Die: The Untold Story of Operation Anaconda*, New York: Berkley Books, 2005.

130. Bishop, *Apache AH-64 Boeing*, 47. Standard firing positions for Apaches would employ "stand-off" from their targets, often hovering behind concealment and employing their weapons at maximum range from the enemy. This afforded the aircraft and crew much better protection and chances for survival after delivering an attack.

131. Naylor, *Not a Good Day to Die*, 225.

132. Bishop, *Apache AH-64 Boeing*, 34.

133. Buss, "Evolution," 39–40.

134. Robert Draper, "Boondoggle Goes Boom," *New Republic*, June 19, 2013.

135. Steven Brill, "Donald Trump, Palantir, and the Crazy Battle to Clean Up a Multi-Billion-Dollar Procurement Swamp," *Fortune*, April 1, 2017.

136. US Government Accountability Office, "Defense Major Automated Information Systems: Cost and Schedule Commitments Need to Be Established Earlier," GAO-15-282, February 2015, 62.

137. Draper, "Boondoggle Goes Boom"; Sara Carter, "The Software You've Likely Never Heard About That Could Have Saved Soldiers' Lives—So Why Did the Army Stop It from Being Used?," *The Blaze*, April 21, 2014.

138. See, for example, Operation Anaconda in the Shah-i-Kot Valley, discussed earlier in this chapter.

139. See data available in Jason H. Campbell and Jeremy Shapiro, "Afghanistan Index: Tracking Variables of Reconstruction & Security in Post-9/11 Afghanistan," Brookings Institution, November 18, 2008.

140. Rob Evans, "Afghanistan War Logs: How the IED Became the Taliban's Weapon of Choice," *The Guardian*, July 25, 2010; Spencer Ackerman, "Taliban Suicide Attacks Have Held Steady for Most of the War," *Wired*, February 24, 2012.

141. Draper, "Boondoggle Goes Boom"; Noah Shachtman, "Brain, Damaged: Army Says Its Software Mind Is 'Not Survivable,'" *Wired*, August 8, 2012; Noah Shachtman, "No Spy Software Scandal Here, Army Claims," *Wired*, November 30, 2012; Brill, "Donald Trump."

142. Representative Duncan D. Hunter (R-CA), whose opposition to DCGS-A is discussed later, summarized what troops had told him about the army's system as follows: "'DCGS doesn't work. It doesn't work at all. We don't use it. We use PowerPoint and Google Maps.'" Sara Carter, "'Matter of Life and Limb': The Congressman Who's Going to Battle with the Army over a Software Program," *The Blaze*, April 22, 2014. See also Draper, "Boondoggle Goes Boom."

143. Palantir was originally founded with venture capital from In-Q-Tel, a nonprofit investment arm of the CIA. Draper, "Boondoggle Goes Boom"; Brill, "Donald Trump."

144. Doug Philippone, a former army ranger who was hired by Palantir in 2008, later described his reaction to seeing the software for the first time: "My jaw dropped, and I said, 'I wish I'd had this in Ramadi.'" Brill, "Donald Trump."

145. Draper, "Boondoggle Goes Boom"; Brill, "Donald Trump."

146. Brill, "Donald Trump."

147. Brill, "Donald Trump."

148. For more on the Rapid Equipping Force, see Chapter 10.

149. Carter, "Software."

150. Draper, "Boondoggle Goes Boom."

151. Major General Michael T. Flynn, USA, Captain Matt Pottinger, USMC, and Paul D. Batchelor, DIA, "Fixing Intel: A Road Map for Making Intelligence Relevant in Afghanistan," Center for a New American Security, January 2010.

152. Brill, "Donald Trump."

153. The full text of Flynn's JUON request is available at https://www.politico.com/pdf/PPM223_mg_flynn_juons.pdf.

154. Flynn JUON request.

155. Brill, "Donald Trump." This example is part of a larger pattern of overpromising what DCGS-A would be able to do, since the program was perpetually behind schedule. The army consistently denied that was the case, but according to someone who worked on DCGS-A, it was able to keep the program technically on schedule by extending the timelines for some requirements. Speaking about the army, this person claimed, "They'll say, 'OK, we delivered Increment One on time.' But what they delivered has only forty percent of the requirement that it was supposed to have when it got started. The rest of the requirement gets slipped into Increment Two." Draper, "Boondoggle Goes Boom."

156. The CTTSO falls within the office of the Assistant Secretary of Defense for Special Operations / Low-Intensity Conflict. See https://www.tswg.gov/?q=vendors_about.

157. Noah Shachtman, "Spy Chief Called Silicon Valley Stooge in Army Software Civil War," *Wired*, August 1, 2012.

158. Brill, "Donald Trump."

159. In a 2017 interview, Katharina McFarland, who was in charge of army acquisition from 2012 until early 2017, continued to defend DCGS-A. When asked why commanders in the field had requested Palantir, she replied, "So, they had salesmen in the field who pushed a few people. That's not something we respond to." Brill, "Donald Trump."

160. Brill, "Donald Trump"; Shachtman, "Spy Chief."

161. Patrick Tucker, "The War over Soon-to-Be-Outdated Army Intelligence Systems," *Defense One*, July 5, 2016.

162. Tucker, "Army Intelligence Systems"; Brill, "Donald Trump."

163. Charles Hoskinson, "Computer Bug Hurts Army Ops," *Politico*, June 29, 2011. The full text of the letter is available at https://www.politico.com/pdf/PPM156_july_2010_hac-d_letter.pdf.

164. Hoskinson, "Computer Bug."

165. Hoskinson, "Computer Bug."

166. Hoskinson, "Computer Bug."

167. As a result of the senators' letter, CENTCOM hosted a meeting of Palantir representatives and several army officials involved in the DCGS-A program. The army officials claimed that the Palantir software did far less than DCGS-A, while Palantir answered each one of their objections. An internal CENTCOM summary of the meeting reportedly stated, "Lots of bad blood here—and one side or the other was definitely lying." Draper, "Boondoggle Goes Boom."

168. Army Test and Evaluation Command, "Palantir Operational Assessment Report," April 2012, 1, available at https://www.wired.com/images_blogs/dangerroom/2012/08/1830_001.pdf.

169. Army Test and Evaluation Command, "Palantir Operational Assessment Report," April 2012, 4–5.

170. Army Test and Evaluation Command, "Palantir Operational Assessment Report," April 2012, 6.

171. The report was unclassified but designated "For Official Use Only," which exempted it from release to the public and the requirements of the Freedom of Information Act. This designation is commonly applied to these types of reports.

172. The text of this order is available at https://www.wired.com/images_blogs/dangerroom/2012/08/1829_001.pdf. Since the issuer's name has been blacked out, it is not clear whether this message came from Legere herself or whether she ordered it to be issued. A bid protest that Palantir filed with the US Court of Federal Claims on June 30, 2016, alleges that Legere pressured Major General Gino Dellarocco, the commanding general of ATEC, to revise the report and ordered

him to revoke the original one. That bid protest is available at https://admin. govexec.com/media/di_1_-_complaint.pdf.

173. There were a few other small changes, such as a new statement noting that the description of the system's capabilities came from Palantir rather than reflecting an independent ATEC assessment. Army Test and Evaluation Command, "Palantir Operational Assessment Report," May 2012, available at https://www. wired.com/images_blogs/dangerroom/2012/08/1831_001.pdf.

174. Carter, "Matter of Life and Limb."

175. A later army investigation concluded that the changes made to the report, though strange, "were not attributable to anyone attempting to improperly advance" a specific agenda. A spokesman for Representative Hunter responded, however, that "the Army's report in no way satisfies why the ATEC changed its findings, or why ground combat units were denied these critical capabilities." Shachtman, "No Spy Software Scandal."

176. Carter, "Matter of Life and Limb."

177. Shachtman, "Brain, Damaged."

178. Shachtman, "Brain, Damaged."

179. Shachtman, "Brain, Damaged."

180. Shachtman, "Spy Chief."

181. Ken Dilanian, "Top Army Brass Defend Troubled Intelligence System," Associated Press, July 10, 2014.

182. Carter, "Matter of Life and Limb." Duncan D. Hunter is the son of Duncan L. Hunter, the Republican congressman and chairman of the House Armed Services Committee who pushed for more up-armoring kits, as discussed earlier.

183. Brill, "Donald Trump."

184. Lolita C. Baldor, "Congress Eyes Probe into Army Program," *San Diego Union-Tribune*, July 24, 2012.

185. Rumors abounded for years within the army that Hunter had a direct financial stake in Palantir, but Hunter's staff repeatedly denied this was the case. One news story noted that Hunter's campaign had received no contributions from Palantir, but did receive more than $50,000 from three companies involved with DCGS-A (Lockheed Martin, Northrop Grumman, and General Dynamics). Ashly McGlone, "Hunter, IEDs, and Campaign Contributions," *Morning Call*, August 3, 2012.

186. Stephanie Gaskell, "House Panel Probes Army IED Review," *Politico*, August 1, 2012. The text of the letter is available at https://web.archive.org/web/20141014143203/http://oversight.house.gov/wp-content/uploads/2012/08/2012-08-01-DEI-Chaffetz-to-Panetta-re-ATEC-assess-of-Palantir.pdf.

187. Shachtman, "No Spy Software Scandal."

188. Rowan Scarborough, "Rep. Duncan Hunter Takes Army to Task over Requests for IED Finder," *Washington Times*, March 28, 2013.

189. Austin Wright, "Hunter Battles Army on Intel," *Politico*, May 1, 2013.

190. Draper, "Boondoggle Goes Boom."

191. "Special Forces, Marines Embrace Palantir Software," *Military.com*, July 1, 2013.

192. Brill, "Donald Trump."

193. Draper, "Boondoggle Goes Boom."

194. Rowan Scarborough, "Problems with Army's Battlefield Intel System Unresolved after Two Years," *Washington Times*, May 1, 2014.

195. The army requested $267 million for DCGS-A in its budget request for fiscal year 2014, but the defense appropriations bill provided only $110 million. Rowan Scarborough, "Army Mulls Funding for Controversial Intel Network," *Washington Times*, February 14, 2014.

196. In 2014, 55 US troops were killed in Afghanistan, which was approximately 11 percent of the 499 US troops killed at the peak of the violence in 2010. That number dropped to 22 in 2015, 14 in 2016, and 11 from January through September 2017. Ian S. Livingston and Michael O'Hanlon, "Afghanistan Index: Tracking Variables of Reconstruction & Security in Post-9/11 Afghanistan," Brookings Institution, September 29, 2017.

197. Brill, "Donald Trump."

198. Brill, "Donald Trump."

199. Jen Judson, "Army Releases RFP for DCGS-A Increment 2," *Army Times*, December 23, 2015.

200. The bid protest is available at https://admin.govexec.com/media/di_1_-_complaint.pdf.

201. The GAO denial is available at https://www.gao.gov/assets/680/677570.pdf.

202. Tucker, "Army Intelligence Systems"; Brill, "Donald Trump."

203. Brill, "Donald Trump."

204. Steven Brill reports that, upon cross-examination, one of the army's key witnesses admitted that "he had never considered whether Palantir could meet the needs of the program by making the modifications it always makes to its basic product when working with any new customer. Instead, the Army's expert said, he had been told to opine not on whether Palantir could fulfill the Army's need but on whether Palantir could fulfill the specified requirements—the ones, such as the accounting software, that Palantir claimed were written to exclude the company. He also conceded that he had 'insufficient information' to know whether Palantir's platform could be tailored to meet the Army's needs."

205. Somewhere in this lengthy process, the army stopped referring to the upgrade as Increment 2, and instead referred to it as Increment 1, Capability Drop 1. Jen Judson, "The Army Turns to a Former Legal Opponent to Fix Its Intel Analysis System," *C4ISRNET*, March 9, 2018.

206. Shane Harris, "Palantir Wins Competition to Build Army Intelligence System," *Washington Post*, March 26, 2019; Jen Judson, "Palantir—Who Successfully Sued the Army—Has Won a Major Army Contract," *Defense News*, March 29, 2019.

CHAPTER 7

1. Austin Long, "The Anbar Awakening," *Survival* 50, no. 2 (April–May 2008): 77; James A. Russell, *Innovation, Transformation, and War: Counterinsurgency in Anbar and Ninewa Provinces, Iraq, 2005–2007*, Stanford, CA: Stanford University Press, 2011, 19.

2. As David Kilcullen later wrote, "Most tribal Iraqis I have spoken with consider [AQI's] brand of 'Islam' utterly foreign to their traditional and syncretic version of the faith." See David Kilcullen, "Anatomy of a Tribal Revolt," *Small Wars Journal*, August 30, 2007.

3. Kilcullen, "Anatomy."

4. John A. McCary, "The Anbar Awakening: An Alliance of Incentives," *Washington Quarterly* 32, no. 1 (Winter 2008–2009): 44; Major Niel Smith, US Army, and Colonel Sean MacFarland, US Army, "Anbar Awakens: The Tipping Point," *Military Review* 88, no. 2 (March–April 2008): 47; Russell, *Innovation, Transformation, and War*, 19.

5. For more on both battles, see Bing West, *No True Glory: A Frontline Account of the Battle for Fallujah*, New York: Bantam Books, 2005. Thomas E. Ricks discusses the second battle in *Fiasco*, New York: Penguin, 2006, 398–406.

6. Smith and MacFarland, "Anbar Awakens," 42.

7. Chief Warrant Officer-4 Timothy S. McWilliams and Lieutenant Colonel Kurtis P. Wheeler, *Al-Anbar Awakening*, vol. 1: *American Perspectives: U.S. Marines and Counterinsurgency in Iraq, 2004–2009*, Quantico, VA: Marine Corps University Press, 2009, 142.

8. Russell, *Innovation, Transformation, and War*, 96–7; Carter Malkasian, "Did the Coalition Need More Forces in Iraq? Evidence from Al Anbar," *Joint Force Quarterly* 46 (3rd Quarter 2007), especially 121.

9. Russell, *Innovation, Transformation, and War*, 19, 95.

10. Most of their revenues (and therefore their influence) came from smuggling, banditry, and other quasi-legitimate businesses that AQI either took over directly or disrupted through ongoing violence. McCary, "The Anbar Awakening," 44; Long, "The Anbar Awakening," 77; Kilcullen, "Anatomy."

11. Long, "The Anbar Awakening," 77; Kilcullen, "Anatomy."

12. Marc Lynch, "Explaining the Awakening: Engagement, Publicity, and the Transformation of Iraqi Sunni Political Attitudes," *Security Studies* 20, no. 1 (January-March 2011): 43. See also Kilcullen, "Anatomy."

13. Stephen Biddle, Jeffrey A. Friedman, and Jacob N. Shapiro, "Testing the Surge: Why Did Violence Decline in Iraq in 2007?," *International Security* 37, no. 1 (Summer 2012), especially 18–9.

14. Jeannie L. Johnson, *The Marines, Counterinsurgency, and Strategic Culture: Lessons Learned and Lost in America's Wars*, Washington, DC: Georgetown University Press, 2018, 242.

15. Johnson, *The Marines*, 242; Jim Michaels, *A Chance in Hell*, New York: St. Martin's Press, 2010, 90–1.

16. Long, "The Anbar Awakening," 78, 80.

17. Biddle et al., "Testing the Surge," 20; Smith and MacFarland, "Anbar Awakens," 42.

18. See Russell, *Innovation, Transformation, and War*, 65–7; Michael R. Gordon and General Bernard E. Trainor, *The Endgame: The Inside Story of the Struggle for Iraq, from George W. Bush to Barack Obama*, New York: Pantheon Books, 2012, 171–2; LtCol Julian D. Alford and Maj. Edwin O. Rueda, "Winning in Iraq," *Marine Corps Gazette* 90, no. 6 (June 2006): 29–30.

19. According to Major Ben Connable, "If you looked at the color-tone map of all the cities that were good, bad, and ugly, Baghdad was yellow. Ramadi was orange going toward red, as in the deeper the color, the worse it gets. So Ramadi, according to MNF-I [Multi-National Force–Iraq, the four-star theater command], MNC-I [Multi-National Corps–Iraq, its three-star subordinate command] intelligence experts, was the worst city in the country, in that time frame." McWilliams and Wheeler, *American Perspectives*, 131–2.

20. Kimberly Kagan, "The Anbar Awakening: Displacing al Qaeda from Its Stronghold in Western Iraq," Institute for the Study of War and WeeklyStandard.com, August 21, 2006–March 30, 2007, 5.

21. Smith and MacFarland, "Anbar Awakens," 42–4.

22. Thomas E. Ricks, *The Gamble*, New York: Penguin, 2009, 47–8. The entire memo is reproduced as Appendix A.

23. MacFarland's brigade was not a pure army unit; it included several smaller marine units, and its staff included personnel from all four services. See Colonel Sean MacFarland, "Addendum: Anbar Awakens," *Military Review* 88, no. 3 (May–June 2008): 2–3; McWilliams and Wheeler, *American Perspectives*, 177, 184–5.

24. Joshua Hutcheson, "'Ready First' Brigade Takes Responsibility for Tal Afar Area," *Stars and Stripes*, February 22, 2006.

25. George Packer, "The Lesson of Tal Afar," *New Yorker*, April 10, 2006.

26. Ricks, *The Gamble*, 60. This description was supported by an internal army review of more than three dozen US units that operated in Iraq in 2005, which concluded that McMaster's unit had conducted counterinsurgency most successfully. Ricks, *Fiasco*, 420; see also 419–24.

27. Packer, "Lesson of Tal Afar."

28. Packer, "Lesson of Tal Afar."

29. Packer, "Lesson of Tal Afar."

30. Russell, *Innovation, Transformation, and War*, 112.

31. McWilliams and Wheeler, *American Perspectives*, 131; Michaels, *A Chance in Hell*, 52. Casey's approach is discussed extensively in Chapter 8.

32. Michaels, *A Chance in Hell*, 53.

33. Russell, *Innovation, Transformation, and War*, 112.

34. Russell, *Innovation, Transformation, and War*, 113–4; Michaels, *A Chance in Hell*, 53–4.

35. Smith and MacFarland, "Anbar Awakens," 41.

36. This task had been started by the previous unit in Ramadi, the Second Brigade of the 28th Infantry Division, from the Pennsylvania National Guard. McWilliams and Wheeler, *American Perspectives*, 178; Russell, *Innovation, Transformation, and War*, 113.

37. McWilliams and Wheeler, *American Perspectives*, 178; Michaels, *A Chance in Hell*, 60.

38. Smith and MacFarland, "Anbar Awakens," 45.

39. For more details on how they built COPs, see Russell, *Innovation, Transformation, and War*, 117–32; Ricks, *The Gamble*, 64–5; Michaels, *A Chance in Hell*, 60–1, 67–74.

40. Ricks, *The Gamble*, 65–6; Smith and MacFarland, "Anbar Awakens," 45.

41. McWilliams and Wheeler, *American Perspectives*, 179–80.

42. Smith and MacFarland, "Anbar Awakens," 45–6.

43. As Major Eric Remoy explained, "Our fear was that the desire to cease operations would come from outside the brigade. Our concern was VIPs would come out and see what we're doing in Ramadi and the casualties we were taking and say—without understanding what we were doing and what our end state was—and tell us to pull back on the reins." Michaels, *A Chance in Hell*, 80–1.

44. Smith and MacFarland, "Anbar Awakens," 46.

45. Russell, *Innovation, Transformation, and War*, 114; Michaels, *A Chance in Hell*, 121–4; Ricks, *The Gamble*, 220–1.

46. MacFarland addressed this concern by saying, "I'm a product of Catholic schools, and I was taught that every saint has a past and every sinner can have a future." Ricks, *The Gamble*, 67; see also "A Profile of Iraq's Anbar Province," *Talk of the Nation*, National Public Radio, September 20, 2007.

47. David Ucko, "Militias, Tribes and Insurgents: The Challenge of Political Reintegration in Iraq," *Conflict, Security & Development* 8, no. 3 (October 2008): 360.

48. Russell, *Innovation, Transformation, and War*, 114–5.

49. McCary, "The Anbar Awakening," 44, 51; Smith and MacFarland, "Anbar Awakens," 43–6; McWilliams and Wheeler, *American Perspectives*, 134; Lynch, "Explaining the Awakening," especially 38–9.

50. For detailed examples of how this worked, see the interview with Major Daniel Zappa in McWilliams and Wheeler, *American Perspectives*, 201–10.

51. Patriquin was the author of the infamous PowerPoint briefing called "How to Win in Iraq," featuring stick-figure illustrations and a tone reminiscent of *Sesame Street*. Its simplicity made it go viral throughout the US military. The briefing is available at https://abcnews.go.com/images/US/how_to_win_in_anbar_v4.pdf. Patriquin was killed by an IED on December 6, 2006. Smith and MacFarland, "Anbar Awakens," 46–50; Ricks, *The Gamble*, 68.

52. McWilliams and Wheeler, *American Perspectives*, 183.

53. "A Profile of Iraq's Anbar Province."

54. Smith and MacFarland, "Anbar Awakens," 44, 47; McWilliams and Wheeler, *American Perspectives*, 181; Michaels, *A Chance in Hell*, 112, 140–1.

55. Smith and MacFarland, "Anbar Awakens," 43; Ricks, *The Gamble*, 64.

56. Smith and MacFarland, "Anbar Awakens," 47; Russell, *Innovation, Transformation, and War*, 115; Michaels, *A Chance in Hell*, 114–5.

57. Smith and MacFarland, "Anbar Awakens," 47; Russell, *Innovation, Transformation, and War*, 115; Joshua Partlow, "Sheikhs Help Curb Violence in Iraq's West, U.S. Says," *Washington Post*, January 27, 2007.

58. Partlow, "Sheikhs Help Curb Violence."

59. Sattar was a smuggler and highway bandit. He had provided support to AQI in 2003 and 2004, but he and his family began to clash with the group as it started to threaten their sources of illicit revenue. AQI had killed his father and two of his brothers, so he had some personal motives to fight the group. Long, "The Anbar Awakening," 80; Mark Kukis, "Turning Iraq's Tribes against Al-Qaeda," *Time*, December 26, 2006.

60. Russell, *Innovation, Transformation, and War*, 115; Smith and MacFarland, "Anbar Awakens," 47–8.

61. MacFarland had seen the group's proclamation several days earlier, since Sattar shared it with one of the brigade's lieutenant commanders. One of its 11 planks called for declaring a state of emergency and taking over the provincial government, but MacFarland convinced them to change that plank during the September 9 meeting. Otherwise, MacFarland said, "Ten of them [the planks] I would have written for them almost exactly the same way they wrote them." Ricks, *The Gamble*, 67; see also Michaels, *A Chance in Hell*, 95–8, 104–7; Smith and MacFarland, "Anbar Awakens," 48.

62. Michaels, *A Chance in Hell*, 105.

63. Khalid Al-Ansary and Ali Adeeb, "Most Tribes in Anbar Agree to Unite against Insurgents," *New York Times*, September 18, 2006.

64. Kukis, "Turning Iraq's Tribes."

65. Smith and MacFarland, "Anbar Awakens," 44. Former insurgents who were able to pass the screening tests were allowed to join, however. Ucko, "Militias, Tribes and Insurgents," 360.

66. Smith and MacFarland, "Anbar Awakens," 49.

67. Both Michaels and Smith and MacFarland write that this sheikh was considering joining the Awakening; Ricks states that the tribe intended to stay neutral, but the checkpoints angered AQI because they fell along the main route between Fallujah and Ramadi. Michaels, *A Chance in Hell*, 168; Smith and MacFarland, "Anbar Awakens," 49; Ricks, *The Gamble*, 69.

68. Smith and MacFarland, "Anbar Awakens," 49; Michaels, *A Chance in Hell*, 170.

69. Before making this decision, Ferry consulted the senior battalion commander and the brigade's executive officer, who both said it was his decision. Michaels, *A Chance in Hell*, 167–71; Ricks, *The Gamble*, 69.

70. Ricks provides a vignette about tactical adaptation in the middle of a firefight: "A drone reconnaissance aircraft was sent to circle over the fight. Patriquin [the captain responsible for tribal outreach] called the sheikh of the tribe. 'Hey, look,' he said. 'We can't tell who is who. Could you have your guys wave towels over their heads so we can identify friend from foe?' That done, Marine F-18 warplanes rolled into bomb those without towels, and then arriving US Army tanks began to fire on fleeing al Qaeda vehicles." Ricks, *The Gamble*, 69; see also Michaels, *A Chance in Hell*, 171.

71. Smith and MacFarland, "Anbar Awakens," 50; Michaels, *A Chance in Hell*, 171–4.

72. Smith and MacFarland, "Anbar Awakens," 50.

73. Michaels, *A Chance in Hell*, 175.

74. Smith and MacFarland, "Anbar Awakens," 50–1; Michaels, *A Chance in Hell*, 175; Ricks, *The Gamble*, 69.

75. Russell, *Innovation, Transformation, and War*, 132; Partlow, "Sheikhs Help Curb Violence."

76. Smith and MacFarland, "Anbar Awakens," 44; Russell, *Innovation, Transformation, and War*, 132.

77. McWilliams and Wheeler, *American Perspectives*, 180.

78. McWilliams and Wheeler, *American Perspectives*, 146, 180.

79. Partlow, "Sheikhs Help Curb Violence."

80. In order to provide continuity, MacFarland gave three of his battalions to the brigade replacing his, and those battalions gradually rotated out later in 2007. Monte Morin, "'Raider Brigade' Takes Over Ramadi," *Stars and Stripes*, February 19, 2007; "A Profile of Iraq's Anbar Province."

81. Smith and MacFarland, "Anbar Awakens," 52, state that 85 members of the brigade were killed; Morin, "'Raider Brigade,'" reports that the figure was 86.

82. Writing in August 2007, Kilcullen, "Anatomy," noted that the provincial and national governments had also developed an interest in sustaining the movement: "The Government of Iraq sees benefits in terms of grass-roots political reconciliation and reduced violence, and is keen to take control of, and credit for, the process. Provincial governments also see the benefits of self-securing districts, freeing up police and military forces for other tasks."

83. Russell, *Innovation, Transformation, and War*, 57.

84. Long, "The Anbar Awakening," 84; Michaels, *A Chance in Hell*, 220–9.

85. Ricks, *The Gamble*, 203; Michael E. O'Hanlon and Jason Campbell, "Iraq Index: Tracking Variables of Reconstruction & Security in Post-Saddam Iraq," Brookings Institution, January 28, 2008, 13; and Michael E. O'Hanlon and Jason Campbell, "Iraq Index: Tracking Variables of Reconstruction & Security in Post-Saddam Iraq," Brookings Institution, July 28, 2008, 11.

86. Biddle et al. argue that the Awakening was necessary for the reduction in violence, but not sufficient on its own—the same argument that they make about the effects of the surge. Instead, they argue, "A synergistic interaction between the surge and the Awakening is the best explanation for why violence declined in Iraq in 2007. Without the surge, the Anbar Awakening would not have spread far or fast enough. And without the surge, sectarian violence would likely have continued for a long time to come... Yet the surge, though necessary, was insufficient to explain 2007's sudden reversal in fortunes. Without the Awakening to thin the insurgents' ranks and unveil the holdouts to U.S. troops, the violence would probably have remained very high until well after the surge had been withdrawn and well after U.S. voters had lost patience with the war . . . a synergistic interaction between them created something new that neither could have achieved alone." Biddle et al., "Testing the Surge," 10–1; see also 36.

87. See, for example, Johnson, *The Marines*, 243–4; McWilliams and Wheeler, *American Perspectives*, 137.

88. Russell, *Innovation, Transformation, and War*, 133.

89. This section draws heavily on Doug Stanton, *Horse Soldiers: The Extraordinary Story of a Band of U.S. Soldiers Who Rode to Victory in Afghanistan*, New York: Scribner, 2009. Sensitive to protecting their identities at the time, Stanton refers to Captain Mark Nutsch as Mitch Nelson and his deputy Chief Warrant Officer Bob Pennington as Cal Spencer; we use both individuals' real names.

90. Each of the army's five active-duty special forces groups had, and still has, a specific geographic area of responsibility and is aligned with the combatant command in that region. First Group is aligned with US Pacific Command (now called US Indo-Pacific Command); Third Group is aligned with US Africa Command; Fifth Group is aligned with US Central Command; Seventh Group is aligned with US Southern Command; and 10th Group is aligned with US European Command. Stanton, *Horse Soldiers*, 27.

91. Tom Clancy with John Gresham, *Special Forces: A Guided Tour of U.S. Army Special Forces*, New York: Berkley Books, 2001, 148–9.

92. Drew Brooks, "The True Story of '12 Strong,' Horse Soldiers Were First to Enter Afghanistan," *Fayetteville Observer*, January 21, 2018.

93. Stanton, *Horse Soldiers*, 38.

94. Stanton, *Horse Soldiers*, 38–9.

95. Brooks, "True Story of 12 Strong."

96. Brooks, "True Story of 12 Strong."

97. Stanton, *Horse Soldiers*, 27–8.

98. Stanton, *Horse Soldiers*, 32–3. For more on the role of the CIA, see Gary Schroen, *First In: An Insider's Account of How the CIA Spearheaded the War on Terror in Afghanistan*, New York, Ballantine Books, 2005; Brian Glyn Williams, "General Dostum and the Mazar i Sharif Campaign: New Light on the Role of Northern

Alliance Warlords in Operation Enduring Freedom," *Small Wars & Insurgencies* 21, no. 4 (December 2010): 615–6; Rod Paschall, "The 27-Day War," *MHQ: The Quarterly Journal of Military History* 24, no. 3 (Spring 2012), especially 56–7, 61.

99. Williams, "General Dostum," 617; Paschall, "The 27-Day War," 56.

100. Fifth Group became the core of Joint Special Forces Operations North, also known as Task Force Dagger. It was commanded by Colonel John Mulholland and included elements of the 160th Special Operations Aviation Regiment and special tactics personnel from Air Force Special Operations Command. Richard W. Stewart, *Operation ENDURING FREEDOM: The United States Army in Afghanistan, October 2001–March 2002*, CMH Pub 70-83-1, US Army Center for Military History, [2004], 8; Charles H. Briscoe et al., *Weapon of Choice: U.S. Army Special Operations Forces in Afghanistan*, Fort Leavenworth, KS: Combat Studies Institute Press, 2003, 52–4.

101. One group member called the publisher of *The Bear Went over the Mountain*, a book about the Soviet invasion and occupation of Afghanistan, and asked for 600 copies. The book was out of print, but the publisher raced to get it reprinted and shipped to the group. Stanton, *Horse Soldiers*, 34–5, 47; Briscoe et al., *Weapon of Choice*, 53.

102. Briscoe et al., *Weapon of Choice*, 127–8.

103. Stanton, *Horse Soldiers*, 43. ODA 595 also became known as Tiger 02, but we continue to refer to it by its number.

104. Stanton, *Horse Soldiers*, 38–9.

105. Afghanistan includes several different major ethnic groups—including the Pashtun, Tajik, Uzbek, Turkmen, and Hazara—who often compete for power and influence in the country.

106. Paschall, "The 27-Day War," 59; Stanton, *Horse Soldiers*, 66–7. For more on Dostum's background, see Williams, "General Dostum," 611–5.

107. Stanton, *Horse Soldiers*, 105; Bob Woodward, *Bush at War*, New York: Simon & Schuster, 2002, 274.

108. The first strikes against Afghanistan occurred on the night of October 7, 2001. For more on conventional military operations in Afghanistan, see Walter L. Perry and David Kassing, *Toppling the Taliban: Air-Ground Operations in Afghanistan, October 2001–June 2002*, RR-381, Santa Monica, CA: Rand Corporation, 2015.

109. Stewart, *Operation ENDURING FREEDOM*, 10–1; Briscoe et al., *Weapon of Choice*, 96, 117–22.

110. ODA 555 flew in the same night, but deployed further south. Its mission was to work with General Bismullah Khan and support his forces near Kabul. Paschall, "The 27-Day War," 59–60; Stanton, *Horse Soldiers*, 69–77, 79–83; Stewart, *Operation ENDURING FREEDOM*, 10–1.

111. Howard Altman, "Saddled with Challenges," *Tampa Bay Times*, January 6, 2018.

112. Williams, "General Dostum," 617; Stanton, *Horse Soldiers*, 99–106.

113. Stanton, *Horse Soldiers*, 73, 100–20; Williams, "General Dostum," 617–8; Briscoe et al., *Weapon of Choice*, 122–3.

114. Stanton, *Horse Soldiers*, 52–6; Briscoe et al., *Weapon of Choice*, 125.

115. Stanton, *Horse Soldiers*, 120–5.

116. Stanton, *Horse Soldiers*, 120–1.

117. Stanton, *Horse Soldiers*, 73, 120–1; Stewart, *Operation ENDURING FREEDOM*, 11–3; Briscoe et al., *Weapon of Choice*, 123.

118. Stanton, *Horse Soldiers*, 116, 126.

119. Pennington later recalled, "We thought we were going to go in on foot, link up with the indigenous forces and maybe ride around in something like a Toyota Hilux." Altman, "Saddled with Challenges."

120. Stanton, *Horse Soldiers*, 123–6; Williams, "General Dostum," 618.

121. Stephen Biddle, *Afghanistan and the Future of Warfare: Implications for Army and Defense Policy*, Carlisle, PA: Strategic Studies Institute, US Army War College, November 2002, 9–10. See also Briscoe et al., *Weapon of Choice*, 125.

122. Altman, "Saddled with Challenges"; Greg Myre, "'12 Strong': When the Afghan War Looked Like a Quick, Stirring Victory," *Morning Edition*, National Public Radio, January 18, 2018.

123. Stanton, *Horse Soldiers*, 125, 134–5.

124. Nutsch requested either McClellan or Australian saddles, which were light enough to use on the small Afghan horses. Yet one of the two later airdrops contained much heavier Western saddles, which Nutsch had specifically said that he did not need. Briscoe et al., *Weapon of Choice*, 127–8.

125. Paschall, "The 27-Day War," 61.

126. Stanton, *Horse Soldiers*, 138–44.

127. Stanton, *Horse Soldiers*, 138–41.

128. Briscoe et al., *Weapon of Choice*, 134. See also Perry and Kassing, *Toppling the Taliban*, 51–2; Stanton, *Horse Soldiers*, 138.

129. Stewart, *Operation ENDURING FREEDOM*, 11; Briscoe et al., *Weapon of Choice*, 126–7.

130. Perry and Kassing, *Toppling the Taliban*, 52; Williams, "General Dostum," 618–9; Stewart, *Operation ENDURING FREEDOM*, 11.

131. Stanton, *Horse Soldiers*, 141–6, 151–2.

132. Stanton, *Horse Soldiers*, 152–5.

133. Stanton, *Horse Soldiers*, 155.

134. Stanton, *Horse Soldiers*, 156.

135. Stanton, *Horse Soldiers*, 160–8.

136. Nutsch sent two other soldiers north with Marchal, who remained in a camp to call in strikes on the targets that Marchal identified. Stanton refers to Marchal by the pseudonym Sam Diller. See *Horse Soldiers*, 155, 170, 180, 185.

137. Stanton, *Horse Soldiers*, 185.

138. Williams, "General Dostum," 620.

139. Paschall, "The 27-Day War," 61.

140. Williams, "General Dostum," 619.

141. Stanton, *Horse Soldiers*, 189–90. For the harrowing details of the flights that dropped off the two air controllers and logistical supplies, see Briscoe et al., *Weapon of Choice*, 135–8.

142. The military calls this the "designated mean point of impact." Stanton, *Horse Soldiers*, 159–60.

143. Paschall, "The 27-Day War," 61; Stanton, *Horse Soldiers*, 186–7.

144. Stanton, *Horse Soldiers*, 213–6.

145. Stanton, *Horse Soldiers*, 64.

146. Williams, "General Dostum," 625.

147. Williams, "General Dostum," 624.

148. Stanton, *Horse Soldiers*, 235–42; Stewart, *Operation ENDURING FREEDOM*, 13.

149. Williams, "General Dostum," 623.

150. Stanton, *Horse Soldiers*, 242–53, 256–7; Williams, "General Dostum," 626–7.

151. Williams, "General Dostum," 627.

152. Briscoe et al., *Weapon of Choice*, 94.

153. Williams, "General Dostum," 627; Paschall, "The 27-Day War," 64.

154. Williams, "General Dostum," 629.

CHAPTER 8

1. For more on the levels of war, see Chapter 2.

2. Sanchez took command from Lieutenant General William Wallace, who had commanded the army's Fifth Corps during major combat operations. His new headquarters was called Combined Joint Task Force-7. Thomas E. Ricks, *Fiasco*, New York: Penguin, 2006, 172–6.

3. US Department of Defense, Press Operations, "Secretary Rumsfeld Media Availability with Jay Garner," June 18, 2003.

4. Carter Malkasian, "Counterinsurgency in Iraq, May 2003–January 2007," in Daniel Marston and Carter Malkasian, eds., *Counterinsurgency in Modern Warfare*, Oxford: Osprey Publishing, 2008, 243; Brian Burton and John Nagl, "Learning as We Go: The US Army Adapts to Counterinsurgency in Iraq, July 2004–December 2006," *Small Wars & Insurgencies* 19, no. 3 (September 2008): 304; George Packer, "The Lesson of Tal Afar," *New Yorker*, April 10, 2006; Ricks, *Fiasco*, 392; Colonel Joel D. Rayburn and Colonel Frank K. Sobchak, eds., *The U.S. Army in the Iraq War*, vol. 1: *Invasion, Insurgency, Civil War, 2003–2006*, Carlisle, PA: Strategic Studies Institute and US Army War College Press, January 2019, 303.

5. The major exception was the approach of Major General David Petraeus, who adopted many elements of the counterinsurgency strategy that he later integrated into the counterinsurgency manual discussed in Chapter 5. David Cloud and Greg Jaffe, *The Fourth Star: Four Generals and the Epic Struggle for the U.S. Army*, New York: Crown Publishers, 2009, 117–22, 129–33, 138–42; Malkasian, "Counterinsurgency in Iraq," 243; Packer, "Lesson of Tal Afar."

6. Malkasian, "Counterinsurgency in Iraq," 243.

7. The battle of Fallujah ended because of political pressure from the Iraqi government. Yet Malkasian argues that the operation was a broader failure because US civilian officials pressed the military to undertake the operation without building any Iraqi political support before the operation began, and because military leaders did not sufficiently mitigate the effects of civilian casualties. See Carter Malkasian, "Signaling Resolve, Democratization, and the First Battle of Fallujah," *Journal of Strategic Studies* 29, no. 3 (June 2006): 423–52. See also Rayburn and Sobchak, *U.S. Army*, vol. 1, 281–2.

8. Malkasian, "Counterinsurgency in Iraq," 244–6.

9. As late as November 2005, Rumsfeld stated that describing the violence in Iraq as an insurgency gives those involved "a greater legitimacy than they seem to merit." He went on to say, "Why do you—why would you call Zarqawi and his people insurgents against a legitimate Iraqi government with their own constitution? It just—do they have broad popular support in that country? No . . . I think you can have a legitimate insurgency in a country that has popular support and has a cohesiveness and has a legitimate gripe. These people don't have a legitimate gripe. They've got a peaceful way to change that government through the constitution, through the elections. These people aren't trying to promote something other than disorder and to take over that country and turn it into a caliphate, and then spread it around the world. This is a group of people who don't merit the word 'insurgency,' I think." US Department of Defense, Press Operations, "News Briefing with Secretary of Defense Donald Rumsfeld and Gen. Peter Pace," November 29, 2005.

10. General John Abizaid served in Grenada as Captain John Abizaid, whom we discuss as an example of successful tactical leadership adaptation in Chapter 4.

11. Abizaid reportedly referred to those conducting violence in Iraq as "guerrillas" as early as July 2003, and both Sanchez and Abizaid used the term "insurgency" while testifying to Congress in May 2004. See Sydney J. Freedberg Jr., "Abizaid of Arabia," *Atlantic Monthly* 292, no. 5 (December 2003): 32–6; US Senate, Committee on Armed Services, "Allegations of Mistreatment of Iraqi Prisoners," May 19, 2004.

12. George W. Casey Jr., *Strategic Reflections: Operation* Iraqi Freedom *July 2004–February 2007*, Washington, DC: National Defense University Press, October 2012, Chapters 1 and 2.

13. Cloud and Jaffe, *The Fourth Star*, 161.

14. Ricks, *Fiasco*, 392–3; Burton and Nagl, "Learning as We Go," 306.

15. Bob Woodward, *The War Within: A Secret White House History, 2006–2008*, New York: Simon & Schuster, 2008, 5.

16. General George Casey, testimony to the US Senate, Committee on Armed Services, February 1, 2007. See also Casey, *Strategic Reflections*, 13; Rayburn and Sobchak, *U.S. Army*, vol. 1, 319–20.

17. Casey, *Strategic Reflections*, 58.

18. Thomas E. Ricks, *The Gamble*, New York: Penguin, 2009, 61; Cloud and Jaffe, *The Fourth Star*, 197; Woodward, *The War Within*, 5; James A. Russell, *Innovation, Transformation, and War: Counterinsurgency in Anbar and Ninewa Provinces, Iraq, 2005–2007*, Stanford, CA: Stanford University Press, 2011, 5; Burton and Nagl, "Learning as We Go," 305.

19. Russell, *Innovation, Transformation, and War*, 7.

20. For Casey's full description of his campaign plan, see *Strategic Reflections*, 25–33.

21. According to Sean Naylor, Casey "had no authority to give orders to JSOC forces." See *Relentless Strike: The Secret History of Joint Special Operations Command*, New York: St. Martin's Press, 2015, 270, 297.

22. Casey, *Strategic Reflections*, 70–1.

23. Senator John McCain, for example, used this phrase at least twice to describe operations in Iraq in 2006. William Safire, "Whack-a-Mole," *New York Times Magazine*, October 29, 2006. See also Peter R. Mansoor, *Baghdad at Sunrise: A Brigade Commander's War in Iraq*, New Haven, CT: Yale University Press, 2008, 344; Russell, *Innovation, Transformation, and War*, 56, 229.

24. Malkasian, "Counterinsurgency in Iraq," 248.

25. Ricks, *The Gamble*, 12–3; Russell, *Innovation, Transformation, and War*, 6, 56.

26. Rayburn and Sobchak, *U.S. Army*, vol. 1, 328.

27. Woodward, *The War Within*, 5; Burton and Nagl, "Learning as We Go," 313; Bensahel, personal observations.

28. See Ben Brody, "Bones, Fobbits and Green Beans: The Definitive Glossary of Modern U.S. Military Slang," *Salon.com*, December 4, 2013.

29. Casey directed his commanders to, in his words, "focus on getting the ISF [Iraqi Security Forces] to the point where they could plan and conduct security operations at the platoon/police station level with limited coalition support by January [2005]," when parliamentary elections were scheduled. Casey, *Strategic Reflections*, 32.

30. Malkasian, "Counterinsurgency in Iraq," 250–1; Cloud and Jaffe, *The Fourth Star*, 192; Rayburn and Sobchak, *U.S. Army*, vol. 1, 327.

31. In the fall of 2004, for example, US and Iraqi estimates of the number of trained police differed by approximately 60,000. Ricks, *Fiasco*, 395.

32. These included desertion, illness, injuries, and time off to take pay home (since they were often paid in cash). Woodward, *The War Within*, 63.

33. Malkasian, "Counterinsurgency in Iraq," 247–8.

34. Malkasian, "Counterinsurgency in Iraq," 250–1.

35. White House, Office of the Press Secretary, "President Addresses Nation, Discusses Iraq, War on Terror," Fort Bragg, North Carolina, June 28, 2005. Bush repeated this phrase again in late November when releasing the National Strategy for Victory in Iraq. See White House, Office of the Press Secretary, "President Outlines Strategy for Victory in Iraq," US Naval Academy, Annapolis, Maryland, November 30, 2005.

36. "President Addresses Nation, Discusses Iraq, War on Terror"; Malkasian, "Counterinsurgency in Iraq," 251.

37. "President Addresses Nation, Discusses Iraq, War on Terror"; Malkasian, "Counterinsurgency in Iraq," 251; Cloud and Jaffe, *The Fourth Star*, 190.

38. Casey, *Strategic Reflections*, 71–2.

39. Burton and Nagl, "Learning as We Go," 315; Russell, *Innovation, Transformation, and War*, 6, 107–8.

40. Casey wanted the army to send individual officers rather than units to serve on the training teams, because he was concerned that units would do too much of the work for their Iraqi counterparts, and felt that his command could effectively shape the approach of these individuals in theater to prevent them from doing so. However, the ad hoc nature of these training teams because one of their many limitations. Rayburn and Sobchak, *U.S. Army*, vol. 1, 385.

41. Cloud and Jaffe, *The Fourth Star*, 192–3; Malkasian, "Counterinsurgency in Iraq," 253.

42. Russell, *Innovation, Transformation, and War*, 6.

43. US Department of Defense, *Measuring Stability and Security in Iraq*, October 2005, 28.

44. Office of the Special Inspector General for Iraqi Reconstruction, "Interim Analysis of Iraqi Security Force Information Provided by the Department of Defense Report, *Measuring Stability and Security in Iraq*," SIGIR-08-015, April 25, 2008. See also Olga Oliker, *Iraqi Security Forces: Defining Challenges and Assessing Progress*, CT-277, Santa Monica, CA: Rand Corporation, March 2007, 1–2; Office of the Special Investigator General for Iraqi Reconstruction, "Challenges in Obtaining Reliable and Useful Data on Iraqi Security Forces Continue," SIGIR-09-002, October 21, 2008.

45. Rayburn and Sobchak, *U.S. Army*, vol. 1, 490.

46. Rayburn and Sobchak, *U.S. Army*, vol. 1, 491.

47. Casey and CENTCOM agreed that one of these two brigades would deploy instead to Kuwait, where it would serve as an operational reserve. Rayburn and Sobchak, *U.S. Army*, vol. 1, 512.

48. Rayburn and Sobchak, *U.S. Army*, vol. 1, 530.

49. Colonel Bill Hix, the chief strategist on Casey's staff, conducted the survey with Kalev Sepp. Both men later played key roles in drafting the new counterinsurgency

doctrine in 2006, as discussed in Chapter 5. See Rayburn and Sobchak, *U.S. Army*, vol. 1, 445–7.

50. Casey, *Strategic Reflections*, 73–4, 164; Russell, *Innovation, Transformation, and War*, 5; Fred Kaplan, *The Insurgents: David Petraeus and the Plot to Change the American Way of War*, New York: Simon & Schuster, 2013, 167–8, 173–4; Rayburn and Sobchak, *U.S. Army*, vol. 1, 456–7.

51. Casey reportedly said, "Because the Army won't change itself, I am going to change it here in Iraq." Cloud and Jaffe, *The Fourth Star*, 204–5; see also Russell, *Innovation, Transformation, and War*, 5–6.

52. David H. Ucko, *The New Counterinsurgency Era: Transforming the U.S. Military for Modern Wars*, Washington, DC: Georgetown University Press, 2009, 76; Janine Davidson, *Lifting the Fog of Peace: How Americans Learned to Fight Modern War*, Ann Arbor: University of Michigan Press, 2010, 182; Ricks, *Fiasco*, 414; Cloud and Jaffe, *The Fourth Star*, 204–5; Russell, *Innovation, Transformation, and War*, 6.

53. Casey, *Strategic Reflections*, 73.

54. Michael R. Gordon and General Bernard E. Trainor, *The Endgame: The Inside Story of the Struggle for Iraq, from George W. Bush to Barack Obama*, New York: Pantheon Books, 2012, 173; Casey, *Strategic Reflections*, 85.

55. Gordon and Trainor, *The Endgame*, 173.

56. Rayburn and Sobchak argue that Casey and his staff relied on data and indicators that were "deceptive or simply poor." See *U.S. Army*, vol. 1, 510–7.

57. Bensahel first wrote the analysis presented subsequently in the spring of 2006, using data that were widely available at the time. It was published as Nora Bensahel, "Securing Iraq: The Mismatch of Demand and Supply," in Markus E. Bouillon, David M. Malone, and Ben Rowswell, eds., *Iraq: Preventing a New Generation of Conflict*, Boulder, CO: Lynne Rienner, 2007, especially 229–31.

58. One of these brigades ended up staying in Kuwait as a reserve force instead of returning home. Casey, *Strategic Reflections*, 83.

59. Ricks, *The Gamble*, 31; Cloud and Jaffe, *The Fourth Star*, 222.

60. Peter D. Feaver, "Anatomy of the Surge," *Commentary* 125, no. 4 (April 2008): 26; Peter D. Feaver, "The Right to Be Right: Civil-Military Relations and the Iraq Surge Decision," *International Security* 35, no. 4 (Spring 2011): 100; Cloud and Jaffe, *The Fourth Star*, 211–2.

61. Malkasian, "Counterinsurgency in Iraq," 255; Feaver, "Anatomy of the Surge," 26; Burton and Nagl, "Learning as We Go," 317.

62. Cloud and Jaffe, *The Fourth Star*, 223.

63. Rayburn and Sobchak, *U.S. Army*, vol. 1, 535–7, 542–5.

64. Cloud and Jaffe, *The Fourth Star*, 223.

65. Ricks, *The Gamble*, 38.

66. By June, Casey no longer seemed to be hedging. During a major meeting on Iraq policy at Camp David that month, Casey writes that he told the president that

"in the aftermath of the Samarra bombing, the fundamental nature of the conflict had changed from an insurgency against the coalition to a struggle among Iraq's ethnic and sectarian groups for political and economic power in Iraq." See *Strategic Reflections*, 92, 104.

67. Casey held 25 videoconferences with the president, the secretary of defense, and the National Security Council between the February 22 bombing and the end of May. The Samarra bombing and ensuing sectarian violence were only discussed during two of those meetings, while accelerating the withdrawal of US forces was discussed during three meetings. Rayburn and Sobchak, *U.S. Army*, vol. 1, 539.

68. Rayburn and Sobchak, *U.S. Army*, vol. 1, 559.

69. Rayburn and Sobchak, *U.S. Army*, vol. 1, 533–9, 655; Casey also rejects the term "civil war" throughout his memoirs. See *Strategic Reflections*.

70. Rayburn and Sobchak, *U.S. Army*, vol. 1, 559.

71. Rayburn and Sobchak, *U.S. Army*, vol. 1, 545.

72. Casey, *Strategic Reflections*, 108.

73. Rayburn and Sobchak, *U.S. Army*, vol. 1, 583.

74. Woodward, *The War Within*, 59; Russell, *Innovation, Transformation, and War*, 6.

75. Rayburn and Sobchak, *U.S. Army*, vol. 1, 578, 656.

76. Woodward, *The War Within*, 71–2. The number of US troops in Iraq temporarily increased around the January and December 2005 elections, but had otherwise remained somewhere between 130,000 and 140,000 since Casey took command. Michael E. O'Hanlon and Nina Kamp, "Iraq Index: Tracking Variables of Reconstruction & Security in Post-Saddam Iraq," Brookings Institution, April 27, 2006, 20.

77. Naylor, *Relentless Strike*, 244, 279–89; Casey, *Strategic Reflections*, 103–4.

78. Michael E. O'Hanlon and Ian Livingston, "Iraq Index: Tracking Variables of Reconstruction & Security in Post-Saddam Iraq," Brookings Institution, January 31, 2011, 4, 5, 12.

79. Frederick W. Kagan, "Choosing Victory: A Plan for Success in Iraq—Phase I Report," A Report of the Iraq Planning Group at the American Enterprise Institute, January 2007, 21.

80. Ricks, *The Gamble*, 49. See also Feaver, "Anatomy of the Surge," 26.

81. Casey, *Strategic Reflections*, 112.

82. Casey reluctantly extended the deployment of the 172nd Stryker Brigade in order to provide the US reinforcements, which showed the limits of his withdrawal strategy as well. Cloud and Jaffe, *The Fourth Star*, 232.

83. Casey, *Strategic Reflections*, 119.

84. Solomon Moore and Julian Barnes, "Many Iraqi Troops Are No-Shows in Baghdad," *Los Angeles Times*, September 23, 2006.

85. Kagan, "Choosing Victory," 21. Emphasis added.

86. Malkasian, "Counterinsurgency in Iraq," 256.

87. Casey, *Strategic Reflections*, 114. Emphasis added; in this case, home station means the United States.

88. Casey, *Strategic Reflections*, 115.

89. Cloud and Jaffe, *The Fourth Star*, 241–2.

90. Cloud and Jaffe, *The Fourth Star*, 241–2.

91. Casey writes, "We projected having all 10 of the current Iraqi divisions in the lead and under the operational control of the Iraqi Ground Forces Command by the spring. We projected having all Iraqi provinces responsible for their security by the fall. We also projected completing the planned development of the ISF by the end of 2007, less the national police retraining and the completion of the prime minister's initiative to expand the armed forces and police." Casey, *Strategic Reflections*, 124.

92. These numbers include data from July 2004, when Casey took command, to October 2006. O'Hanlon and Livingston, "Iraq Index," 4, 12.

93. In his memoirs, Casey writes that while he met with administration officials in October 2006, "There were 'chicken-egg' discussions about whether the security situation had to improve before the political track could begin. I strongly argued that both tracks needed to move forward simultaneously to be effective, and that it was important to get the Iraqis to commit to a political timeline." Casey, *Strategic Reflections*, 124.

94. In the fall of 2005, Andrew Krepinevich published an article recommending an "oil-spot" strategy, making him among the first people to publicly endorse an alternative. In June 2006, two staff members at the National Security Council arranged for President Bush to meet with four critics of the war at Camp David, though the meeting reportedly had little effect on the president's thinking. Andrew F. Krepinevich Jr., "How to Win in Iraq," *Foreign Affairs* 84, no. 5 (September–October 2005): 87–104; Ricks, *The Gamble*, 41–4.

95. Ricks, *The Gamble*, 50; Russell, *Innovation, Transformation, and War*, 57; Feaver, "Anatomy of the Surge," 26.

96. Ricks, *The Gamble*, 58.

97. Feaver, "Right to Be Right," 103; Colonel Joel D. Rayburn and Colonel Frank K. Sobchak, eds., *The U.S. Army in the Iraq War*, vol. 2: *Surge and Withdrawal, 2007–2011*, Carlisle, PA: Strategic Studies Institute and US Army War College Press, January 2019, 6–7.

98. In early fall, retired army general Jack Keane suggested to the chairman of the Joint Chiefs of Staff that he charter a Council of Colonels to help provide input to the administration's strategy review. Though this group recommended adopting a smaller but sustainable US military presence, Colonel H. R. McMaster—who pioneered the counterinsurgency approach in Tal Afar discussed in Chapter 7—drafted a five-page minority report that recommended deploying additional US brigades to Iraq. Keane, who played a critical role in developing the idea of the

surge, as discussed later, would almost certainly have been aware of McMaster's recommendations. Rayburn and Sobchak, *U.S. Army*, vol. 2, 12–3, 17–8.

99. Cloud and Jaffe, *The Fourth Star*, 243–4; Casey, *Strategic Reflections*, 138.

100. Sheryl Gay Stolberg and Jim Rutenberg, "Rumsfeld Resigns as Defense Secretary after Big Election Gains for Democrats," *New York Times*, November 8, 2006.

101. Rayburn and Sobchak, *U.S. Army*, vol. 2, 9–12.

102. Woodward, *The War Within*, 42, 214–5.

103. Casey, *Strategic Reflections*, 136.

104. Woodward, *The War Within*, 215.

105. These included Abizaid; Pace; Lieutenant General Peter Chiarelli, who commanded the Multi-National Corps in Iraq (and reported to Casey); and General Peter Schoomaker, the army chief of staff. Ricks, *The Gamble*, 92–3, 114; Ucko, *New Counterinsurgency Era*, 206; Cloud and Jaffe, *The Fourth Star*, 246–7.

106. James A. Baker and Lee H. Hamilton, Co-Chairs, *The Iraq Study Group Report*, New York: Vintage Books, 2006, 6.

107. For more on the AEI study, and especially the three-day exercise that refined its ideas, see the detailed accounts in Woodward, *The War Within*, and in Ricks, *The Gamble*. Barno, who retired from the military in May 2006, participated in the exercise as an adviser.

108. Ricks, *The Gamble*, 99–100.

109. Feaver, "Right to Be Right," 111. See a similar analysis in Rayburn and Sobchak, *U.S. Army*, vol. 2, 23.

110. Rayburn and Sobchak, *U.S. Army*, vol. 2, 23–4.

111. Casey, *Strategic Reflections*, 142–3.

112. Casey, *Strategic Reflections*, 143–4.

113. Feaver, "Right to Be Right," 107–9; Rayburn and Sobchak, *U.S. Army*, vol. 2, 24.

114. Woodward, *The War Within*, 287.

115. Woodward, *The War Within*, 288.

116. Cloud and Jaffe, *The Fourth Star*, 247.

117. Feaver, "Right to Be Right," 109. Lieutenant General Raymond Odierno, Casey's subordinate corps commander, had also been trying to convince Casey to support a surge of five brigades but encountered the same resistance. Feaver, "Right to Be Right," 102; Ricks, *The Gamble*, 91–2, 112–3; Woodward, *The War Within*, 296; Rayburn and Sobchak, *U.S. Army*, vol. 2, 27–31.

118. Rayburn and Sobchak, *U.S. Army*, vol. 2, 97.

119. Cloud and Jaffe, *The Fourth Star*, 248. See also Ricks, *The Gamble*, 104; Woodward, *The War Within*, 295.

120. Woodward, *The War Within*, 295; Cloud and Jaffe, *The Fourth Star*, 248; Ricks, *The Gamble*, 104; Malkasian, "Counterinsurgency in Iraq," 258.

121. Michael R. Gordon and Thom Shanker, "Bush to Name a New General to Oversee Iraq," *New York Times*, January 5, 2007.

122. White House, Office of the Press Secretary, "President's Address to the Nation," White House Library, January 10, 2007. Bush announced that the surge would consist of approximately 20,000 additional US troops, though the final number ended up being closer to 30,000.

123. For the United States, World War II lasted 45 months (December 1941 to August 1945); 46 months elapsed between the initial invasion of Iraq in March 2003 and the announcement of the surge strategy in January 2007.

124. Woodward, *The War Within*, 328.

125. Cloud and Jaffe, *The Fourth Star*, 262; Kagan, 32.

126. The charts only show data through the end of 2009 because the numbers remained largely constant from that time until the full withdrawal of US forces at the end of 2011.

127. Stephen Biddle, Jeffrey A. Friedman, and Jacob N. Shapiro, "Testing the Surge: Why Did Violence Decline in Iraq in 2007?," *International Security* 37, no. 1 (Summer 2012), especially 18–9.

128. Cloud and Jaffe, *The Fourth Star*, 255. See also Andrew F. Krepinevich, Jr., *The Army and Vietnam*, Baltimore: Johns Hopkins University Press, 1986, especially 268–71; Kenneth J. Campbell, "Once Burned, Twice Cautious: Explaining the Weinberger-Powell Doctrine," *Armed Forces & Society* 24, no. 3 (Spring 1998): 357–74.

129. Nora Bensahel, "The Coalition Paradox: The Politics of Military Cooperation," PhD diss., Stanford University, 1999, 207–17.

130. Paul Yingling, "A Failure in Generalship," *Armed Forces Journal*, May 1, 2007.

131. Ricks, *The Gamble*, 104.

132. Yingling, "A Failure in Generalship."

133. See, for example, Mackubin Thomas Owens, "Promoting Failure," *National Review*, February 21, 2007; Lawrence F. Kaplan, "War at Home," *New Republic*, March 19, 2007; Tom Bowman, "Army Chief of Staff Defends Iraq Decisions," *All Things Considered*, National Public Radio, January 15, 2009; Williamson Murray, *America and the Future of War*, Stanford, CA: Hoover Institution Press, 2017, 107–9.

134. As discussed later, McKiernan also served as the NATO commander of the International Security Assistance Force in Afghanistan. Gates had the authority to remove McKiernan from the US chain of command, but only the North Atlantic Council (NAC) could remove McKiernan from the NATO chain of command. Though Gates did not consult with the NAC before making this decision, he notified the NAC shortly before the public announcement, and it readily accepted his decision, which effectively terminated McKiernan from the NATO command as well.

135. Rajiv Chandrasekaran, "Pentagon Worries Led to Command Change," *Washington Post*, August 17, 2009; Robert M. Gates, *Duty: Memoirs of a Secretary at War*, New York: Alfred A. Knopf, 2014, 346.

136. US Department of Defense, "Press Conference with Secretary Gates and Adm. Mullen on Leadership Changes in Afghanistan from the Pentagon," May 11, 2009.

137. Gates, *Duty*, 344.

138. Before August 2003, individual NATO members served as the lead nation for ISAF—first the United Kingdom, then Turkey, and then Germany and the Netherlands jointly—but NATO as an organization had not been involved. See Sten Rynning, *NATO in Afghanistan: The Liberal Disconnect*, Stanford, CA: Stanford University Press, 2012, 44–5, 83–7.

139. For a good summary of NATO's expansion throughout Afghanistan, see Vincent Morelli and Paul Belkin, "NATO in Afghanistan: A Test of the Transatlantic Alliance," RL33627, Congressional Research Service, December 3, 2009, especially 9–10, 17–8.

140. Rynning, *NATO in Afghanistan*, 109–10; Sten Rynning, "ISAF and NATO: Campaign Innovation and Organizational Adaptation," in Theo Farrell, Frans Osinga, and James A. Russell, eds., *Military Adaptation in Afghanistan*, Stanford, CA: Stanford University Press, 2013, 88–91. NATO described its operations in Afghanistan as "the comprehensive approach." That term had no agreed meaning, however, which meant that national contributions to the operation varied widely.

141. See Jason H. Campbell and Jeremy Shapiro, "Afghanistan Index: Tracking Variables of Reconstruction & Security in Post-9/11 Afghanistan," Brookings Institution, June 24, 2009, 4–6, 18. IED data are only available starting in January 2007. The peak number of attacks and casualties usually occurs during the summer, because of the winter lull in fighting.

142. Rynning, *NATO in Afghanistan*, 161; Gates, *Duty*, 217.

143. McKiernan later stated that he had approached Casey about the position: "I knew that Dan McNeill was going to need a replacement. I approached the Army Chief, George Casey, and said, 'I would like to go back to the fight and I'd be more than willing to be successor to Dan McNeill.'" See "An Exit Interview with General David D. McKiernan," Senior Leader Debriefing Program, Carlisle Barracks, PA: US Army Military History Institute, 2010, 1.

144. McKiernan's military biography is available in "An Exit Interview with General David D. McKiernan," ii–iii.

145. There has been some speculation that army leaders nominated McKiernan for the ISAF position because they wanted to reward his success in the 2003 campaign with a fourth star. For the full list of units serving under McKiernan during the invasion of Iraq, see COL Gregory Fontenot, US Army, Retired, LTC E. J. Degen, US Army, and LTC David Tohn, US Army, *On Point: The United States Army in Operation IRAQI FREEDOM*, Fort Leavenworth, KS: Combat Studies Institute Press, 2004, 441–96.

146. Chandrasekaran, "Pentagon Worries."

147. Gates, *Duty*, 344.

148. Gates, *Duty*, 217.

149. Josh White, "U.S. Deaths Rise in Afghanistan," *Washington Post*, July 2, 2008.

150. Rynning, *NATO in Afghanistan*, 161, 171.

151. Gates, *Duty*, 205–6. This odd arrangement resulted from the inactivation of the senior US and coalition headquarters called Combined Forces Command–Afghanistan in late 2006. Most US forces were shifted to ISAF, including the five regional commands, but there were still significant US forces operating as part of Operation Enduring Freedom. The OEF forces reported to the US division commander in the eastern region, because he was dual-hatted with a direct US chain of command in addition to his ISAF chain of command. Those were the only forces in Afghanistan directly controlled by the United States. In September 2008, 15,000 US troops were part of ISAF, while 19,000 were part of OEF. Jason H. Campbell and Jeremy Shapiro, "Afghanistan Index: Tracking Variables of Reconstruction & Security in Post-9/11 Afghanistan," Brookings Institution, November 18, 2008, 10.

152. Rynning, *NATO in Afghanistan*, 165; David P. Auerswald and Stephen M. Saideman, *NATO in Afghanistan: Fighting Together, Fighting Alone*, Princeton, NJ: Princeton University Press, 2014, 99. McKiernan later emphasized that this change also fixed a problem with the Title 10 authorities that applied to all US troops in Afghanistan. See "An Exit Interview with General David D. McKiernan," 52–60.

153. "An Exit Interview with General David D. McKiernan," 11–2, 45–6.

154. "An Exit Interview with General David D. McKiernan," 15.

155. Carlotta Gall, "U.S. Killed 90, Including 60 Children, in Afghan Village, U.N. Finds," *New York Times*, August 26, 2008; John F. Burns, "Afghan Officials Say Airstrike Killed Civilians," *New York Times*, October 16, 2008.

156. Jim Garamone, "Directive Aimed at Reducing Civilian Casualties," American Forces Press Service, September 16, 2008; Burns, "Afghan Officials."

157. Campbell and Shapiro, "Afghanistan Index," June 24, 2009, 5.

158. Auerswald and Saideman, *NATO in Afghanistan*, 99.

159. "An Exit Interview with General David D. McKiernan," 18, 98.

160. General David Petraeus asked for McKiernan's counterinsurgency guidance to be entered into the *Congressional Record* alongside his testimony in April 2009. The reprinted guidance is not dated, and Petraeus only describes it as "recently issued." Yet McKiernan recalled that it was "a natural follow on" from the September tactical directive, which suggests it was issued earlier than the date of Petraeus's testimony suggests. "An Exit Interview with General David D. McKiernan," 18; US Senate, Committee on Armed Services, "United States Policy toward Afghanistan and Pakistan," April 7, 2009, 13–7.

161. Chandrasekaran, "Pentagon Worries."

162. Gates, *Duty*, 344.

163. Mark Mazzetti and Eric Schmitt, "U.S. Study Is Said to Warn of Crisis in Afghanistan," *New York Times*, October 8, 2008.

164. Petraeus was then a two-star general commanding the 101st Airborne Division, which reported to McKiernan's three-star land component headquarters.

165. Auerswald and Saideman, *NATO in Afghanistan*, 100.

166. McKiernan later described the tensions between NATO and the United States as follows: "If you're in NATO you're subject to the political pressures from 26 member states, and a mandate and OPLAN [operational plan] that doesn't include words such as counterinsurgency, counterterrorism, war. And if you're in CENTCOM you're saying I don't have time to mince around with other terminology, we've got to go in there and fight Taliban and others and fight with the Afghans." "An Exit Interview with General David D. McKiernan," 2–3.

167. Chandrasekaran, "Pentagon Worries."

168. Elisabeth Bumiller and Thom Shanker, "Commander's Ouster Is Tied to Shift in Afghan War," *New York Times*, May 11, 2009; Eric Schmitt and Mark Mazzetti, "Switch Signals New Path for Afghan War," *New York Times*, May 12, 2009.

169. "One-on-One with General David McKiernan," *NBC Nightly News with Brian Williams*, June 16, 2008; "Transcript: General David McKiernan Speaks at Council's Commanders Series," Atlantic Council, November 19, 2008.

170. "An Exit Interview with General David D. McKiernan," 76.

171. US Department of Defense, "DoD News Briefing with Gen. McKiernan from the Pentagon," October 1, 2008. See also Gates, *Duty*, 217; "An Exit Interview with General David D. McKiernan," 61.

172. O'Hanlon and Livingston, "Iraq Index," 18.

173. Kenneth Katzman, "Afghanistan: Post-Taliban Governance, Security, and U.S. Policy," RL30588, Congressional Research Service, June 8, 2009, 31. See also Secretary of Defense Robert M. Gates, testimony to the US Senate, Committee on Armed Services, September 23, 2008.

174. Campbell and Shapiro, "Afghanistan Index," November 18, 2008.

175. Chandrasekaran, "Pentagon Worries."

176. The election was later postponed until August. Gates, *Duty*, 338–40; Bob Woodward, *Obama's Wars*, New York: Simon & Schuster, 2010, 94–7.

177. Chandrasekaran, "Pentagon Worries"; Matthew C. Brand, Colonel, USAF, "General McChrystal's Strategic Assessment: Evaluating the Operating Environment in Afghanistan in the Summer of 2009," Research Paper 2011-1, Air Force Research Institute, July 2011, 34.

178. Chandrasekaran, "Pentagon Worries."

179. These were Multi-National Force–Iraq and Multi-National Corps–Iraq, respectively. See Catherine Dale, "Operation Iraqi Freedom: Strategies, Approaches,

Results, and Issues for Congress," RL34387, Congressional Research Service, March 28, 2008.

180. Barno commanded CFC-A from 2003 to 2005. See Chapter 5.

181. Lieutenant General David W. Barno, US Army, Retired, "Fighting 'The Other War': Counterinsurgency Strategy in Afghanistan, 2003–2005," *Military Review* 87, no. 5 (September–October 2007): 34. For more on CFC-A (which Barno commanded from 2003 to 2005), see Chapter 6.

182. Gates, *Duty*, 344–5.

183. Rynning, *NATO in Afghanistan*, 166. McKiernan's mandate as the overall US commander did include many of these tasks, however. For example, US Forces–Afghanistan supervised the training of all Afghan national security forces through the Combined Security Transition Assistance Command–Afghanistan.

184. Chandrasekaran, "Pentagon Worries."

185. Complex attacks involved more than one method of attack, or involved more than 20 insurgents. See Joseph H. Felter and Jacob N. Shapiro, "Limiting Civilian Casualties as Part of a Winning Strategy: The Case of Courageous Restraint," *Daedalus* 146, no. 1 (Winter 2017): 46.

186. See, for example, the transcript of the first presidential debate between Barack Obama and John McCain, September 26, 2008, available at debates.org; and "Obama's Remarks on Iraq and Afghanistan," *New York Times*, July 15, 2008.

187. Three separate reviews of US policy toward Afghanistan were either underway at the time or had been recently completed: one led by Lieutenant General Douglas Lute at the National Security Council, a second by Mullen and the Joint Staff in the Pentagon, and a third by Petraeus at CENTCOM. Obama reportedly felt that there was no coherent strategy among them, and wanted a fresh look by someone he had appointed. The 60-day timeline reflected the president's desire to announce his new strategy before the NATO summit scheduled for April 4. Woodward, *Obama's Wars*, 80, 88; Rynning, *NATO in Afghanistan*, 179–81.

188. Woodward, *Obama's Wars*, 94–8; Gates, *Duty*, 338–40.

189. Chandrasekaran, "Pentagon Worries." See also Woodward, *Obama's Wars*, 99, 117–8.

190. Gates, *Duty*, 345.

191. Gates, *Duty*, 345.

192. JSOC conducted extensive counterterrorist strikes throughout Iraq and Afghanistan both during and after McChrystal's tenure. Naylor, *Relentless Strike*.

193. For more on his background, see Elisabeth Bumiller and Mark Mazzetti, "A General Steps from the Shadows," *New York Times*, May 12, 2009.

194. Chandrasekaran, "Pentagon Worries."

195. Chandrasekaran, "Pentagon Worries"; Thomas E. Ricks, *The Generals*, New York: Penguin, 2012.

196. Chandrasekaran, "Pentagon Worries." Emphasis added.

197. Casey, whose track record in Iraq had been so poor, "argued strenuously" against firing McKiernan and called it a "rotten" thing to do, according to Gates. See *Duty*, 345.

198. Schmitt and Mazzetti, "Switch Signals New Path"; Chandrasekaran, "Pentagon Worries"; Gates, *Duty*, 345.

199. Chandrasekaran, "Pentagon Worries."

200. Mullen was representing Gates's views as well. For more on the proper relationship between generals and their civilian leaders in war, see Eliot A. Cohen, *Supreme Command*, New York: Free Press, 2002.

201. McKiernan later chose to submit a letter of resignation to Gates instead of being officially terminated. Chandrasekaran, "Pentagon Worries"; Bumiller and Shanker, "Commander's Ouster"; Gates, *Duty*, 346.

202. "Press Conference with Secretary Gates and Adm. Mullen on Leadership Changes in Afghanistan from the Pentagon."

203. See, for example, Bumiller and Shanker, "Commander's Ouster"; Ann Scott Tyson, "Top U.S. Commander in Afghanistan Is Fired," *Washington Post*, May 12, 2009.

204. As noted earlier, Gates did not consult with the North Atlantic Council about his decision to fire McKiernan, but did notify the NAC before the announcement.

205. Gates, *Duty*, 345.

206. Gates, *Duty*, 345.

207. Chandrasekaran, "Pentagon Worries."

208. "An Exit Interview with General David D. McKiernan," 99–100.

209. "An Exit Interview with General David D. McKiernan," 43.

210. "An Exit Interview with General David D. McKiernan," 75.

211. Chandrasekaran, "Pentagon Worries."

212. Gates, *Duty*, 344. Emphasis added.

CHAPTER 9

1. For more on their work, see Michael Lewis, *The Undoing Project*, New York: Norton, 2017. Many of their findings are discussed in Daniel Kahneman, *Thinking, Fast and Slow*, New York: Farrar, Straus and Giroux, 2011.

2. Tversky, who died in 1996, was not eligible to win the award because Nobel Prizes may not be awarded posthumously. Occasional exceptions have been made for those who have died after the award has been announced but before it has been awarded, and in one recent case, for an awardee who had died three days before the announcement, unbeknownst to the selection committee. See Macy Halford, "Should Death Stop the Nobel?," *New Yorker*, October 3, 2011.

3. See Amos Tversky and Daniel Kahneman, "Availability: A Heuristic for Judging Frequency and Probability," *Cognitive Psychology* 5, no. 2 (September 1973): 207–32.

4. David Barno and Nora Bensahel, "Mirages of War: Six Illusions from Our Recent Conflicts," *War on the Rocks*, April 11, 2017.

5. Joseph J. Collins, "Initial Planning and Execution in Afghanistan and Iraq," in Richard D. Hooker Jr. and Joseph J. Collins, eds., *Lessons Encountered: Learning from the Long War*, Washington, DC: National Defense University Press, September 2015, 24.

6. Francis Fukuyama, "The End of History?," *National Interest* 16 (Summer 1989): 3–18; Charles Krauthammer, "The Unipolar Moment," *Foreign Affairs* 70, no. 1 (1990–1991): 23–33.

7. Defense spending, for example, declined by approximately 30 percent between 1990 and 1998. Army active-duty strength fell from 770,000 in 1989 to 510,000 in 1995, and shrank from 18 active divisions to 10. Theo Farrell, Sten Rynning, and Terry Terriff, *Transforming Military Power since the Cold War: Britain, France, and the United States, 1991–2012*, New York: Cambridge University Press, 2013, 19.

8. See Joint Chiefs of Staff, *Joint Doctrine for Military Operations Other Than War*, Joint Publication 3-07, June 16, 1995.

9. For a terrific overview of this decade, see Derek Chollet and James Goldgeier, *America between the Wars: From 11/9 to 9/11*, New York: PublicAffairs, 2008.

10. The text of the president's speech to the nation on the evening of September 11, 2001, is available at http://edition.cnn.com/2001/US/09/11/bush.speech.text/.

11. World Bank, "China Overview," April 8, 2019, available at https://www.world-bank.org/en/country/china/overview. See also Wayne M. Morrison, "China's Economic Rise: History, Trends, Challenges, and Implications for the United States," RL33534, Congressional Research Service, February 5, 2018.

12. World Economic Outlook, "Gross Domestic Product, Current Dollars, Purchasing Power Parity," International Monetary Fund, April 2019.

13. Defense Intelligence Agency, *China Military Power: Modernizing a Force to Fight and Win*, January 2019; and International Institute for Strategic Studies, *The Military Balance 2018*, 5.

14. See, for example, "The Battle for Digital Supremacy," *The Economist*, March 15, 2018.

15. David W. Barno, "Silicon, Iron, and Shadow: Three Wars That Will Define America's Future," *Foreign Policy*, March 19, 2013.

16. In 2017, China's military included more than 2 million active-duty forces and 500,000 reserve forces; the US military included almost 1.4 million active-duty forces and more than 850,000 reserve forces. International Institute for Strategic Studies, *The Military Balance 2018*, 46, 250.

17. Nora Bensahel, "Darker Shades of Gray: Why Gray Zone Conflicts Will Become More Frequent and Complex," Foreign Policy Research Institute E-Notes, February 13, 2017.

18. See Defense Intelligence Agency, *Russia Military Power: Building a Military to Support Great Power Aspirations*, 2017, 12–13.

19. Lee Willett, "Back on the World Stage: Russian Naval Power in 2013 and Beyond," *Jane's Navy International*, December 7, 2012; Dmitri Trenin, "The Revival of the Russian Military: How Moscow Reloaded," *Foreign Affairs* 95, no. 3 (May–June 2016): 23–9.

20. For more on gray zone conflicts, see David Barno and Nora Bensahel, "Fighting and Winning in the 'Gray Zone,'" *War on the Rocks*, May 19, 2015; Bensahel, "Darker Shades of Gray."

21. US Senate, Select Committee on Intelligence, "The Intelligence Community Assessment: Assessing Russian Activities and Intentions in Recent U.S. Elections, Summary of Initial Findings," July 3, 2018; Office of the Director of National Intelligence, "Background to 'Assessing Russian Activities and Intentions in Recent US Elections': The Analytic Process and Cyber Incident Attribution," January 6, 2017, 2–3.

22. Frank Hoffman, "Conflict in the 21st Century: The Rise of Hybrid Wars," Potomac Institute for Policy Studies, December 2007; Michael J. Mazarr, *Mastering the Gray Zone: Understanding a Changing Era of Conflict*, Carlisle, PA: Strategic Studies Institute and US Army War College Press, December 2015.

23. Michael C. McCarthy, Matthew A. Moyer, and Brett H. Venable, *Deterring Russia in the Gray Zone*, Carlisle, PA: Strategic Studies Institute, US Army War College, March 2019.

24. Commander T. J. Gilmore, "Iran Owns the Gray Zone," *Proceedings* (US Naval Institute), 144, no. 3 (March 2018): 48–53.

25. Koh Swee Lean Collin, "Is There Any Way to Counter China's Gray Zone Tactics in the South China Sea?," *National Interest*, September 13, 2017.

26. In 2017, the Defense Intelligence Agency estimated that North Korea had between 30 and 60 nuclear weapons, though other analysts believe that the number was smaller. Joby Warrick, Ellen Nakashima, and Anna Fifield, "North Korea Now Making Missile-Ready Nuclear Weapons, U.S. Analysts Say," *Washington Post*, August 8, 2017.

27. Mark Landler, "Trump Abandons Nuclear Deal He Long Scorned," *New York Times*, May 8, 2018.

28. See, for example, Stephen Hummel, "The Islamic State and WMD: Assessing the Future Threat," *CTC Sentinel* 9, no. 1 (January 2016): 18–21.

29. NASA's data are available at https://earthobservatory.nasa.gov/world-of-change/DecadalTemp.

30. David Barno and Nora Bensahel, "The Anti-Access Challenge You're Not Thinking About," *War on the Rocks*, May 5, 2015; Michon Scott and Kathryn Hansen, "Sea Ice," NASA Earth Observatory, September 16, 2016.

31. Most studies show that the Arctic will be seasonally ice-free sometime between 2030 and 2050, though newer studies predict that this will happen in the earlier

part of that time period. American Geophysical Union, "Ice-Free Arctic Summers Could Happen on Earlier Side of Predictions," *Phys.org*, February 27, 2019.

32. Barno and Bensahel, "Anti-Access Challenge"; Stephanie Pezard, Abbie Tingstad, and Alexandra Hall, "The Future of Arctic Cooperation in a Changing Strategic Environment," PE-268-RC, Rand Europe, 2018; Rachael Gosnell, "Caution in the High North: Geopolitical and Economic Challenges of the Arctic Maritime Environment," *War on the Rocks*, June 25, 2018; Mathieu Boulègue, "Russia's Military Posture in the Arctic: Managing Hard Power in a 'Low Tension' Environment," Chatham House, June 2019.

33. See, for example, Union of Concerned Scientists, "Fact Sheet: The Science Connecting Extreme Weather to Climate Change," June 2018; US Global Change Research Program, *Fourth National Climate Assessment*, vol. 2: *Impacts, Risks, and Adaptation in the United States*, 2018; Umair Irfan, "How Antarctica's Melting Ice Could Change Weather around the World," *Vox.com*, February 6, 2019.

34. United Nations Department of Economic and Social Affairs, Population Division, *The World's Cities in 2018—Data Booklet*, ST/ESA/SER.A/417, 2018, 2.

35. Chief of Staff of the Army, Strategic Studies Group, "Megacities and the United States Army: Preparing for a Complex and Uncertain Future," June 2014; Joe Lacdan, "Warfare in Megacities: A New Frontier in Military Operations," Army News Service, May 24, 2018.

36. See United Nations Population Fund, "World Population Trends," available at https://www.unfpa.org/world-population-trends.

37. These problems could be alleviated through greater immigration, though this option tends to be unpopular and can exacerbate domestic social and political divisions. They might also be alleviated through increased automation and robotics, but the social and political consequences of such changes in the nature of work remain unknown. See the subsequent discussion of the Fourth Industrial Revolution.

38. Facundo Alvaredo et al., "World Inequality Report 2018: Executive Summary," 9, available at https://wir2018.wid.world/files/download/wir2018-summary-english.pdf.

39. Rupert Neate, "Richest 1% Own Half the World's Wealth, Study Finds," *The Guardian*, November 14, 2017.

40. Michael Savage, "Richest 1% on Target to Own Two-Thirds of All Wealth by 2030," *The Guardian*, April 7, 2018.

41. Over 1 million asylum seekers and migrants arrived in Europe in 2015, although as of mid-2017, 85 percent of the world's refugees were hosted in developing regions. See European Parliament, "EU Migrant Crisis: Facts and Figures," June 30, 2017.

42. Steven Levitsky and Daniel Ziblatt, *How Democracies Die*, New York: Crown, 2018; "Democracy Continues Its Disturbing Retreat," *The Economist*, January 31, 2018.

43. Jane Perlez, "China Creates a World Bank of Its Own, and the U.S. Balks," *New York Times*, December 4, 2015; Tamar Gutner, "AIIB: Is the Chinese-Led

Development Bank a Role Model?," Council on Foreign Relations, June 25, 2018; Enda Curran, "China's Growing Clout Triggers Economic Arms Race with Old Order," *Bloomberg*, November 1, 2018.

44. Thomas Wright, "Trump's 19th-Century Foreign Policy," *Politico*, January 20, 2016; "Transcript: Donald Trump Expounds on His Foreign Policy Views," *New York Times*, March 26, 2016.

45. "Donald Trump Is Undermining the Rules-Based International Order," *The Economist*, June 7, 2018.

46. That said, from both a military and commercial standpoint, the world's space architecture is vulnerable to cyber disruption and exploitation as much as are land-, sea-, or air-based systems today. All orbital systems have terrestrial command-and-control nodes that can be disrupted. In October 2008, a cyberattack attributed to China reportedly "achieved all steps required to command" a NASA earth observation satellite "but did not issue commands." US-China Economic and Security Review Commission, *2011 Report to Congress*, November 2011, 215–6.

47. See Paul Tullis, "The World Economy Runs on GPS. It Needs a Backup Plan," *Bloomberg Businessweek*, July 25, 2018.

48. The text of the 1967 treaty (officially called Treaty on Principles Governing the Activities of States in the Exploration and Use of Outer Space, Including the Moon and Other Celestial Bodies) and a list of its signatories are available at https://www.state.gov/t/isn/5181.htm.

49. Joe Pappalardo, "No Treaty Will Stop Space Weapons," *Popular Mechanics*, January 25, 2018; Ryan Browne and Barbara Starr, "U.S. General: Russia and China Building Space Weapons to Target US Satellites," *CNN.com*, December 2, 2017; T. X. Hammes, *Deglobalization and International Security*, Amherst, NY: Cambria Press, 2019, 182–3.

50. The Stuxnet worm, which significantly damaged Iranian nuclear facilities, may be the best-known example of a cyberattack. Other notable examples include Russian cyberattacks against Estonia in 2007 and Georgia in 2008; hacks against Sony Pictures in 2014 and against the British national health system in 2017, both attributed to North Korea; and major hacks of the US government, to include the Office of Personnel Management in 2015 and the Democratic National Committee in 2016, believed to originate with China and Russia, respectively. See Kim Zetter, *Countdown to Zero Day: Stuxnet and the Launch of the World's First Digital Weapon*, New York: Crown Publishers, 2014; US House of Representatives, Committee on Oversight and Government Reform, "The OPM Data Breach: How the Government Jeopardized Our National Security for More Than a Generation," September 7, 2016; Office of the Director of National Intelligence, "Background to Assessing Russian Activities," 2–3.

51. Sarah P. White, "Understanding Cyberwarfare: Lessons from the Russia-Georgia War," Modern War Institute at West Point, March 20, 2018.

52. Key elements of this section are drawn from David E. Sanger, *The Perfect Weapon: War, Sabotage, and Fear in the Cyber Age*, New York: Crown Publishers, 2018; Gary Brown and Kurt Sanger, "Mattis Faces a Challenge in Equipping US to Engage in Cyberwarfare," *The Hill*, February 6, 2017.

53. See the infographic "A Guide to the Internet of Things," available at https://www.intel.com/content/dam/www/public/us/en/images/iot/guide-to-iot-infographic.png.

54. The International Telecommunications Union defined the internet of things as "a global infrastructure for the information society, enabling advanced services by interconnecting (physical and virtual) things based on existing and evolving interoperable information and communication technologies." In other words, it integrates unrelated, physical, noncomputer objects into a wider communications network. International Telecommunications Union, "Overview of the Internet of Things," Recommendation ITU-T Y.2060, June 2012, 1.

55. See, for example, Nicole Perlroth and David E. Sanger, "Cyberattacks Put Russian Fingers on the Switch at Power Plants, U.S. Says," *New York Times*, March 15, 2018; Sanger, *The Perfect Weapon*.

56. A zero-day vulnerability is a previously unsuspected software vulnerability for which the software developer has not yet released a patch. All systems operating that software are temporarily vulnerable to exploitation during the window between discovery and patching. See "How Do Zero-Day Vulnerabilities Work," Norton by Symantec, available at https://us.norton.com/internetsecurity-emerging-threats-how-do-zero-day-vulnerabilities-work-30sectech.html.

57. For a superb novel that details a near-future war exhibiting all manner of cyber exploitation of US military systems, see P. W. Singer and August Cole, *Ghost Fleet*, Boston: Houghton Mifflin Harcourt, 2015. Unusually for a work of fiction, all of the technologies discussed are buttressed by detailed footnotes.

58. Threats to both military and commercial logistics and communications extend far beyond the cyber domain. See, for example, US Department of Defense, Office of the Under Secretary of Defense for Research and Engineering, *Defense Science Board Task Force on Survivable Logistics: Executive Summary*, November 2018; David Barno and Nora Bensahel, "The U.S. Military's Dangerous Embedded Assumptions," *War on the Rocks*, April 17, 2018.

59. Much of this section draws on David Barno and Nora Bensahel, "War in the Fourth Industrial Revolution," *War on the Rocks*, June 19, 2018.

60. Klaus Schwab, *The Fourth Industrial Revolution*, New York: Crown Business, 2016.

61. For more on the ways in which these technologies will affect the future of warfare, see Paul Scharre, *Army of None: Autonomous Weapons and the Future of War*, New York: Norton, 2018.

62. T. X. Hammes, "Technological Change and the Fourth Industrial Revolution," in George P. Shultz, Jim Hoagland, and James Timbie, eds., *Beyond Disruption:*

Technology's Challenge to Governance, Stanford, CA: Hoover Institution Press, 2018, 37–73; Hammes, *Deglobalization*, 191-3.

63. Additive manufacturing (also known as 3D printing) is a manufacturing process that constructs digitally modeled objects in a "bottom-up" layer-by-layer manner using computer-operated tools. It permits a level of precision and complexity unattainable through traditional casting, forging, or machining techniques, as well as a highly streamlined production process. The potential military applications of additive manufacturing include radically simplifying the logistics chain by permitting the manufacture of spare parts on site, as in the navy's "Print the Fleet" project; or the rapid in-theater manufacture of large numbers of missile-armed drones envisioned by DARPA. See Syed A. M. Tofail et al., "Additive Manufacturing: Scientific and Technological Challenges, Market Uptake and Opportunities," *Materials Today* 21, no. 1 (January–February 2018): 22–37; information about the US Navy's "Print the Fleet" initiative, available at https://www.navsea.navy.mil/Portals/103/ Documents/NSWC_Dahlgren/WhatWeDo/PrinttheFleet/InHouse_DNeck_ Print_the_Fleet_Trifold.pdf; Kyle Mizokami, "Drone 'Factory in a Can' Would Change Air War Forever," *Popular Mechanics*, September 7, 2017; Matthew J. Louis, Tom Seymour, and Jim Joyce, "3D Opportunity in the Department of Defense: Additive Manufacturing Fires Up," Deloitte Consulting, 2014, 3–6, available at https://www2.deloitte.com/content/dam/insights/us/articles/additive-manufacturing-defense-3d-printing/DUP_1064-3D-Opportunity-DoD_ MASTER1.pdf.

64. Hammes, "Technological Change," 56.

65. For a good overview of some of these developing technologies, see Hammes, *Deglobalization*, especially Chapter 5; R. Jeffrey Smith, "Hypersonic Missiles Are Unstoppable. And They Are Starting a New Global Arms Race," *New York Times Magazine*, June 19, 2019.

66. See, for example, Ellen Nakashima and Paul Sonne, "China Hacked a Navy Contractor and Secured a Trove of Highly Sensitive Data on Submarine Warfare," *Washington Post*, June 8, 2018.

CHAPTER 10

1. For a detailed explanation, see Headquarters, Department of the Army, *Doctrine Primer*, Army Doctrine Publication 1-01, September 2014, 2-2 to 2-3.

2. Joint Chiefs of Staff, *Joint Operations*, Joint Publication 3-0, January 17, 2017, Incorporating Change 1, October 22, 2018, i.

3. Joint Chiefs of Staff, *Doctrine for the Armed Forces of the United States*, Joint Publication 1, March 25, 2013, Incorporating Change 1, July 12, 2017.

4. US Marine Corps, *Marine Corps Operations*, Marine Corps Doctrinal Publication 1-0 (with Change 1), July 26, 2017; Headquarters, Department of the Army, *Army Leadership*, Army Doctrine Reference Publication 6-22, August 1, 2012.

5. The US military has sought to improve adaptability before the battle through a model called "operational design" in its deliberate planning system. It aims to encourage creativity, foresight, and critical thinking rather than mechanical adherence to methodical planning processes. Although beneficial, this initiative does not help promote adaptability *during* combat operations. See Joint Staff, J-7, *Planner's Handbook for Operational Design,* Version 1.0, October 7, 2011; Joint Chiefs of Staff, *Joint Planning,* Joint Publication 5-0, June 16, 2017, IV-1 to IV-39.

6. Joint Chiefs of Staff, *Joint Doctrine Development Process,* Chairman of the Joint Chiefs of Staff Manual 5120.01A, December 29, 2014, especially B-12 and B-24.

7. See Marine Corps Order 3504.1, "Marine Corps Lessons Learned Program (MCLLP) and the Marine Corps Center for Lessons Learned (MCCLL)," July 31, 2006.

8. Jon T. Thomas and Douglas L. Schultz, "Lessons about Lessons: Growing the Joint Lessons Learned Program," *Joint Force Quarterly* 79 (4th Quarter 2015): 113–20.

9. David Barno and Nora Bensahel, "The U.S. Military's Dangerous Embedded Assumptions," *War on the Rocks,* April 17, 2018.

10. Ian W. Toll, *Pacific Crucible: War at Sea in the Pacific, 1941–1942,* New York: Norton, 2012, 380–3.

11. US Joint Forces Command was established in 1999 as the successor to US Atlantic Command. It was disestablished in 2011 and its functions were transferred to other parts of the Joint Staff.

12. Nicholas Kristof, "How We Won the War," *New York Times,* September 6, 2002.

13. Captain Dale C. Rielage, "An Open Letter to the U.S. Navy from Red," *Proceedings* (US Naval Institute) 143, no. 6 (June 2017): 28–31.

14. Barno heard this argument frequently, both explicitly and implicitly, while serving on the army staff in 2005 and 2006.

15. See J. Ronald Fox, *Defense Acquisition Reform, 1960–2009: An Elusive Goal,* Washington, DC: Center of Military History, US Army, 2011; Moshe Schwartz, "Defense Acquisition Reform: Background, Analysis, and Issues for Congress," R43566, Congressional Research Service, May 23, 2014.

16. Wolters Kluwer Editorial Staff, *CCH Federal Acquisition Regulation as of January 1, 2019,* n.p.: Kluwer Law International, 2019; Wolters Kluwer Editorial Staff, *CCH Department of Defense FAR Supplement as of January 1, 2019,* n.p.: Kluwer Law International, 2019.

17. Steven Brill, "Donald Trump, Palantir, and the Crazy Battle to Clean Up a Multi-Billion-Dollar Procurement Swamp," *Fortune,* April 1, 2017.

18. Moshe Schwartz, "Defense Acquisitions: How DOD Acquires Weapon Systems and Recent Efforts to Reform the Process," RL34026, Congressional Research Service, May 23, 2014.

19. Colonel Timothy D. Chyma, "Rapid Acquisition," Senior Service College Fellowship Civilian Research Project, US Army War College, February 2010, 13.

20. Colonel Joseph W. Roberts, "Agile Acquisition: How Does the Army Capitalize on Success?," Strategy Research Project, US Army War College, January 2017, 12–4; Army Rapid Equipping Force, About Us, available at http://www.ref.army.mil/AboutUs.

21. Paul Wolfowitz, "Joint Rapid Acquisition Cell Initiative," memo to Secretary Rumsfeld, November 16, 2004; Robert L. Buhrkuhl, "When the Warfighter Needs It Now," *Defense AT&L*, November–December 2006, 29.

22. Julie Ann Mattocks, "Institutionalizing the Marine Corps' Rapid Acquisition Capability: Equipping, Training and Sustaining the Nation's Expeditionary Force to Win," Marine Corps University, 2011, 8–12.

23. As noted in Chapter 6, the head of the REF responded to a unit's request for Palantir by saying, "While I don't disagree with your need, I cannot buy Palantir anymore without involving the senior leadership of the Army, and they are very resistant." Ken Dilanian, "Top Army Brass Defend Troubled Intelligence System," Associated Press, July 10, 2014.

24. See US House of Representatives, Committee on Armed Services, Subcommittee on Oversight & Investigations, "The Joint Improvised Explosive Device Defeat Organization: DOD's Fight against IEDs Today and Tomorrow," Committee Print 110-11, 45–137, November 2008; Ian Duncan, "Elite Army Unit at Fort Meade Searching for Ways to Fight ISIS," *Baltimore Sun*, July 20, 2015; Kris Osborn, "How the U.S. Military Is Using a New Special Attack Force to Kill Its Enemies," *National Interest*, May 2, 2019.

25. Authors' conversation with Colonel (Ret.) George Topic, USA, deputy director of the Center for Joint and Strategic Logistics, National Defense University, June 2019.

26. Secretary of the Army Memorandum, "Army Directive 2017-33 (Enabling the Army Modernization Task Force)," November 7, 2017, available at https://army-pubs.army.mil/epubs/DR_pubs/DR_a/pdf/web/ARN6391_AD2017-33_Web_Final.pdf.

27. More specifically, Futures Command has established eight cross-functional teams that align with the army's six modernization priorities: long-range precision fires; the next-generation combat vehicle; future vertical lift; network command, control, communication, and intelligence; assured positioning, navigation, and timing; air and missile defense; soldier lethality; and synthetic training environment. See Andrew Feickert, "Army Futures Command (AFC)," IN10889, Congressional Research Service, updated September 10, 2018; COL Daniel S. Roper, USA, Ret., and LTC Jessica Grassetti, USA, "Seizing the High Ground—United States Army Futures Command," *ILW Spotlight* 18-4, Institute of Land Warfare, Association of the US Army, October 2018; Sean Kimmons, "In First Year, Futures Command Grows From 12 to 24,000 Personnel," Army News Service, July 19, 2019.

28. David Barno and Nora Bensahel, "Three Things the Army Chief of Staff Wants You to Know," *War on the Rocks*, May 23, 2017.

29. C. Todd Lopez, "Milley: On Cusp of Profound, Fundamental Change," Army News Service, October 6, 2016.

30. See, for example, Katherine W. Phillips, "How Diversity Makes Us Smarter," *Scientific American*, October 1, 2014; David Rock and Heidi Grant, "Why Diverse Teams Are Smarter," *Harvard Business Review*, November 4, 2016; Alison Reynolds and David Lewis, "Teams Solve Problems Faster When They're More Cognitively Diverse," *Harvard Business Review*, March 30, 2017; Scott E. Page, *The Diversity Bonus: How Great Teams Pay Off in the Knowledge Economy*, Princeton, NJ: Princeton University Press, 2017; Vivian Hunt et al., *Delivering through Diversity*, McKinsey & Company, January 2018; Rocío Lorenzo et al., "How Diverse Leadership Teams Boost Innovation," Boston Consulting Group, January 23, 2018.

31. The US military is also nearly unique in the world for the extensive noncommissioned officer (NCO) professional education program it conducts as well. American NCOs often attend more leader development courses in their career than do many nations' commissioned officers. This investment is a significant factor in the remarkable capabilities that US military NCOs contribute to the force. Yet many of the observations and shortcomings we identify for officer PME are shortcomings in the NCO PME program as well. We focus here on officer PME programs, since officers who rise to the senior ranks directly affect theater and strategic issues in ways that even the most senior NCOs do not.

32. US Department of Defense, *Summary of the 2018 National Defense Strategy of the United States of America*, January 2018, 8.

33. Col. Francis J. H. Park, "A Rigorous Education for an Uncertain Future," *Military Review* 96, no. 3 (May–June 2016): 75.

34. Barno's personal experience as a student at the US Army War College 1994 to 1995, and subsequent observations by both authors over many years.

35. Usually, PME students are dismissed from their programs only if they are involved in academic misconduct, such as cheating. In civilian graduate programs, students must maintain a minimum GPA in order to continue in and graduate from their programs. The same substandard student would nearly always be allowed to graduate from a PME program.

36. Authors' observations.

37. B. A. Friedman, "The End of the Fighting General," *Foreign Policy*, September 12, 2018.

38. Barno's personal experience working with promotion boards and officer promotion discussions.

39. Everett S. P. Spain, J. D. Mohundo, and Bernard B. Banks, "Intellectual Capital: A Case for Cultural Change," *Parameters* 45, no. 2 (Summer 2015): 88.

40. In July 2019, the army announced that all captains attending the career course would be required to take the GRE, with results reported to the service. This policy is primarily designed to help the army improve its talent management efforts, but the army has said it will also use the scores in assessments for competitive PME slots. Army Talent Management Task Force, "Army Captains to Take GRE to Assess Talent," July 29, 2019; Jennifer H. Svan, "With New Army Policy, Officers Take Standardized GRE Test during Captains Career Course," *Stars and Stripes*, August 9, 2019.

41. In 2015, the authors offered to teach a semester-long elective on military adaptation at a senior service college. We had taught (and continue to teach) this course to civilian graduate students, with highly positive reviews. The senior service college told us that the writing and reading requirements in our civilian syllabus were too demanding for the military students, and we were encouraged to significantly reduce them. We cut both by more than half and submitted a syllabus with approximately the same reading requirements and far fewer writing requirements than a different undergraduate course that we taught. Even so, only one student signed up and the course was canceled.

42. Most PME institutions include a sizable number of foreign military officers. They also typically include a small number of civilian students from across the US government. These civilians do bring some different experiences into the classroom, but their impact on diverse thinking remains limited because they come from backgrounds related to US national security and have frequently worked extensively with the US military throughout their careers.

43. Stephen J. Gerras and Leonard Wong, *Changing Minds in the Army: Why It Is So Difficult and What to Do about It*, Carlisle, PA: Strategic Studies Institute, US Army War College, October 2013, 8. Emphasis added.

44. Gerras and Wong, *Changing Minds*, 9.

45. See Chapter 8.

46. See Headquarters, Department of the Army, *Mission Command*, Army Doctrine Publication 6-0, May 2012.

47. Gen. Stephen Townsend, US Army, Maj. Gen. Douglas Crissman, US Army, and Maj. Kelly McCoy, US Army, "Reinvigorating the Army's Approach to Mission Command: It's Okay to Run with Scissors (Part I)," *Military Review*, April 2019, 4.

48. All army regulations are available at https://armypubs.army.mil/ProductMaps/PubForm/AR.aspx.

49. Examples often include medical readiness, dental readiness, and completion of such closely watched mandatory training classes as Sexual Harassment and Rape Prevention (SHARP).

50. Leonard Wong, "Strategic Insights: Letting the Millennials Drive," Strategic Studies Institute, US Army War College, May 2, 2016.

51. David Barno and Nora Bensahel, "The Future of the Army: Today, Tomorrow, and the Day after Tomorrow," Atlantic Council, September 2016, 20.

52. In 2002, company commanders averaged 297 mandatory days of training—even though there are only 256 training days in a calendar year. Leonard Wong and Stephen J. Gerras, *Lying to Ourselves: Dishonesty in the Army Profession*, Carlisle, PA: Strategic Studies Institute, US Army War College, February 2015, 4.

53. Wong and Gerras, *Lying to Ourselves*. See also David Barno and Nora Bensahel, "Lying to Ourselves: The Demise of Military Integrity," *War on the Rocks*, March 10, 2015.

54. Chad Garland, "Army Secretary: PT Belts Aren't Needed in Daylight," *Stars and Stripes*, January 12, 2019; Meghann Myers, "No PT Belts Needed during the Day, on Running Tracks, Army Secretary Says," *Army Times*, January 14, 2019.

55. See, for example, Jonathan J. Vaccaro, "The Next Surge: Counterbureaucracy," *New York Times*, December 8, 2009.

56. "Townsend: Doctrine Critical to a 'Winning Army,'" Association of the United States Army, January 9, 2019.

57. "Townsend: Doctrine Critical to a 'Winning Army.'"

58. Peter W. Singer, "Tactical Generals: Leaders, Technology, and the Perils," Brookings Institution, July 7, 2009.

59. These investigations are conducted under Army Regulation 15-6, "Procedures for Investigating Officers and Boards of Officers." It governs formal and informal fact-finding procedures and investigations, and is commonly used to investigate incidents when no criminal activities are suspected.

CHAPTER 11

1. Williamson Murray and Allan R. Millett, *A War to Be Won: Fighting the Second World War*, Cambridge, MA: Belknap Press of Harvard University Press, 2000, 18; Williamson Murray, *Military Adaptation in War: With Fear of Change*, New York: Cambridge University Press, 2011, 77.

2. See, for example, US Army, *2019 Army Modernization Strategy: Investing in the Future*, 2019; John H. Pendleton, "Army Readiness: Progress and Challenges in Rebuilding Personnel, Equipping, and Training," Testimony before the Subcommittee on Readiness and Management Support, Committee on Armed Services, US Senate, US Government Accountability Office, GAO-19-367T, February 6, 2019; Maj. Brett Lea, "1st Stryker Brigade Combat Team Converts to Armored Brigade," Army News Service, June 20, 2019.

3. Joint Chiefs of Staff, *Joint Operations*, Joint Publication 3-0, January 17, 2017, Incorporating Change 1, October 22, 2018, ix. Many of these nine principles were codified by the US War Department as early as 1921 and mirrored the British Field Service Regulations of the same era. John I. Alger, "The Origins and Adaptation

of the Principles of War," US Army Command and General Staff College, June 6, 1975, especially Appendixes 1, 6, 13.

4. Department of the Navy, US Marine Corps, "Commandant's Planning Guidance: 38th Commandant of the Marine Corps," July 2019, 19.

5. Berger has also strongly advocated for this approach. Department of the Navy, US Marine Corps, "Commandant's Planning Guidance," 19.

6. We are grateful to Colonel (Ret.) Jim Shufelt, USA, for many of the insights in this section.

7. The Marine Corps, for example, is making a number of changes in its force structure to improve capability for independent operations at the infantry platoon and company levels. These include adding drone and counterdrone operators, logisticians, and an intelligence cell at the company level. Sydney J. Freedberg Jr., "Marines Reorganize Infantry for High-Tech War: Fewer Riflemen, More Drones," *Breaking Defense*, May 4, 2018.

8. This section draws on David Barno and Nora Bensahel, "The Future of the Army: Today, Tomorrow, and the Day after Tomorrow," Atlantic Council, September 2016, 30–1.

9. The total number of deaths, including those from nonhostile causes, is approximately 6,900. Data current as of January 2019. See https://dod.defense.gov/News/Casualty-Status/.

10. "Research Starters: Worldwide Deaths in World War II," National World War II Museum, available at https://www.nationalww2museum.org/students-teachers/student-resources/research-starters/research-starters-worldwide-deaths-world-war.

11. The total number of casualties was more than 58,000. Data from the US National Archives, available at https://www.archives.gov/research/military/vietnam-war/casualty-statistics.

12. The Marine Corps is taking a particularly interesting approach through its Marine Maker initiative to incorporating 3D printing. Marines from a wide variety of occupational specialties can apply to participate in a week-long Marine Makers Course to learn skills in additive manufacturing, coding, and adaptive problem-solving. Maker Units deploy with 3D printers so they can create capabilities needed to accomplish their missions. And Maker Labs at various marine installations offer training and equipment to help familiarize marines with their capabilities. See Captain Alexander Morrow, US Marine Corps, "The Corps Challenges Makers to Make Their Future," January 4, 2017, available at https://www.secnav.navy.mil/innovation/Pages/2017/01/MarineMaker.aspx; "Marine Maker, Innovation Labs," First Marine Logistics Group, March 19, 2018, available at https://www.1stmlg.marines.mil/News/Article/1469796/marine-maker-innovation-labs/.

13. It will be interesting to see how well the army's new Futures Command is able to address these challenges in the coming years. See Chapter 10.

14. See "Asymmetric Warfare Group," Information Papers, 2008 Army Posture Statement, available at https://www.army.mil/aps/08/information_papers/prepare/Army_Asymmetric_Warfare_Group.html; "Asymmetric Warfare Group Command Brief," https://www.awg.army.mil/About-Us/Command-Overview/.

15. Ash Carter, *Inside the Five-Sided Box: Lessons from a Lifetime of Leadership at the Pentagon*, New York: Dutton, 2019, 325.

16. See, for example, Dan Lamothe, "Veil of Secrecy Lifted on Pentagon Office Planning 'Avatar' Fighters and Drone Swarms," *Washington Post*, March 8, 2016.

17. Carter, *Five-Sided Box*, 325–6.

18. This move means that the director of SCO will report to the director of DARPA, who reports to the deputy undersecretary for research and engineering, who reports to the undersecretary for research and engineering, who reports to the secretary of defense. Colin Clark, "Top DoD Official Shank Resigns; SCO Moving to DARPA," *Breaking Defense*, June 17, 2019; Aaron Mehta, "Griffin Makes Case for Why SCO Should Live under DARPA—and Why Its Director Had to Go," *Defense News*, August 8, 2019.

19. The term "graceful degradation" comes from the field of psychology, which defines it as "a property of cognitive networks in which damage to a portion of the network produces relatively little damage to overall performance, because performance is distributed across the units in the network and no one unit is solely responsible for any aspect of processing." The term is now widely used to describe a similar property of computer networks. See "Graceful Degradation," *APA Dictionary of Psychology*, American Psychological Association, available at https://dictionary.apa.org/graceful-degradation.

20. The term "skunkworks" originated in 1943, when the US Army Air Tactical Service Command asked Lockheed Martin to develop a fighter jet as quickly as possible. Though "Skunk Works" is now a registered trademark of Lockheed Martin, it is now commonly used to refer to a small group working outside of normal procedures, and appears in many dictionaries. See "Skunk Works® Origin Story," available at https://www.lockheedmartin.com/en-us/who-we-are/business-areas/aeronautics/skunkworks/skunk-works-origin-story.html.

21. Fitness reports are sometimes issued at other times as well, such as when changing jobs.

22. In the army, for example, these include measures of character (such as discipline and adherence to army values); presence (such as military bearing and fitness); intellect (such as mental agility, sound judgment, and expertise); leadership (such as the ability to build trust and lead by example); development (such as the ability to develop others and steward the profession), and achievement (getting results). Examples from the US Army Officer Evaluation Report for company grade officers, DA Form 67-10-1, March 2019.

23. Service secretaries issue written guidance to every promotion board that outlines the most important traits the board should consider when selecting officers or NCOs for advancement. After the board finishes its work, this guidance becomes part of the public record, available for those who aspire to future promotion.

24. For more on how the frozen middle hinders innovation in all types of organizations, see, for example, Eamon Barrett, "In Business, the 'Frozen Middle' Blocks Design Thinking. Here's How to Fix It," *Fortune*, March 6, 2019; Maj. Paul L. Stokes, USMC (Ret) and CWO4 Brian D. Bethke, "Fanning the Flames of Innovation: Thawing Out the Frozen Middle," *Marine Corps Gazette*, 103, no. 4 (April 2019): 61–4.

25. See "Around the Air Force: The Frozen Middle/Servant Leadership," available at https://www.af.mil/News/Article-Display/Article/1791452/around-the-air-force-the-frozen-middle-servant-leadership/; "Thawing the Middle," *Airman Magazine*, August 6, 2018, available at https://airman.dodlive.mil/2018/08/06/thawing-the-middle/.

26. Meghann Myers, "The Army Just Dumped a Bunch of Mandatory Training to Free Up Soldiers' Time," *Army Times*, April 24, 2018; Meghann Myers, "Good News, Soldiers: The Army Has Slashed Even More Mandatory Training Requirements," *Army Times*, June 5, 2018; Chad Garland, "Army Secretary: PT Belts Aren't Needed in Daylight," *Stars and Stripes*, January 12, 2019.

27. See Headquarters, Department of the Army, *Evaluation Reporting System*, Army Regulation 623-3, June 14, 2019, 50–3; Department of the Navy, US Marine Corps, "Commandant's Planning Guidance," 8.

28. Personal communication with the authors.

29. Such partnering would benefit the civilian students as well and be a small step toward bridging the ever-increasing civil-military gap. See Rosa Brooks, "Civil-Military Paradoxes," in Kori Schake and Jim Mattis, eds., *Warriors & Citizens: American Views of Our Military*, Stanford, CA: Hoover Institution Press, 2016, 21–68; David Barno and Nora Bensahel, "The Deepest Obligation of Citizenship: Looking beyond the Warrior Caste," *War on the Rocks*, May 15, 2018.

30. Stephen J. Gerras and Leonard Wong, *Changing Minds in the Army: Why It Is So Difficult and What to Do about It*, Carlisle, PA: Strategic Studies Institute, US Army War College, October 2013.

31. See Headquarters, Department of the Army, *General Douglas MacArthur Leadership Award Program*, Army Regulation 600-89, September 22, 2017.

32. The Force of the Future initiative was a sweeping approach to reforming DOD's personnel policies and processes advanced by Secretary Carter in an effort to revitalize the recruiting, development, and retention of the force. The services largely opposed the effort, causing its failure. However, a number of Carter's initiatives have subsequently been enacted in law in the 2019 National Defense Authorization Act and are now available to the services to implement as they choose. David Barno

and Nora Bensahel, "Can the U.S. Military Halt Its Brain Drain?," *The Atlantic*, November 5, 2015; Leo Shane III, "Congress Is Giving the Military Promotion System a Massive Overhaul," *Military Times*, July 25, 2018; Blaise Misztal, Jack Rametta, and Mary Farrell, "Personnel Reform Lives, but Don't Call It 'Force of the Future,'" *War on the Rocks*, August 9, 2018.

33. General McConville served as the army's G-1, or head of army personnel, for three years before serving as the army vice chief of staff and then the army chief of staff. See Kyle Rempfer, "Army Chief: New Talent Management Will Start with Officers, Then Go to Enlisted," *Army Times*, October 15, 2019.

34. For decades, officer assignments were centrally managed through the army's Human Resources Command. The new Army Talent Alignment Process uses software that enables officers to include their individual skills and areas of expertise alongside their Officer Record Briefs, and which lists position descriptions written by the units that have vacancies. Officers and units are encouraged to contact each other when they see a potential fit, and then each lists their preference for each position. Of the 15,000 officers who participated in the first run of the system, about 45 percent received a one-to-one match—meaning that the officer and the receiving unit had listed each other as their first preference. See Maj. Gen. J. P. McGee and Ryan Evans, "The Army's New Approach to People," *War on the Rocks*, December 16, 2019; Kyle Rempfer, "Almost Half of Officers Match to Their Top Job Choice under New Army System," *Army Times*, December 13, 2019.

35. Rempfer, "Army Chief."

36. McGee and Evans, "The Army's New Approach."

37. See, for example, Leonard Wong and Stephen Gerras, "Army Talent Management Reform: The Culture Problem," *War on the Rocks*, February 22, 2019.

38. See the subsequent discussion of commander's initiative groups and commander's action groups.

39. These committees are authorized by the Federal Advisory Committee Act (FACA) of 1972 (5 U.S.C., Appendix, as amended) and 41 CFR 102-3-50(d).

40. As of this writing, both authors serve on the Reserve Forces Policy Board, a FACA board that serves as "the principal policy advisor to the Secretary of Defense on matters relating to the Reserve components." See US Department of Defense, Reserve Forces Policy Board, Mission & History, available at https://rfpb.defense.gov/Mission-History/.

41. In 2016, Secretary of Defense Ash Carter chartered the Defense Innovation Board (DIB), chaired by Eric Schmidt, the former executive chairman of Alphabet. It provides recommendations on organizational structure and process, business and functional concepts, and technology applications. By contrast, a Defense Adaptation Board would focus specifically on gaps in the US military's ability to adapt rapidly in wartime, especially in the areas of doctrine, technology, and leadership. More information on its role is available at https://innovation.defense.gov/.

42. Nora Bensahel, "The Coalition Paradox: The Politics of Military Cooperation," PhD diss., Stanford University, 1999, Chapter 3.

43. These include the Center for Army Lessons Learned and the Marine Corps Center for Lessons Learned.

44. Usefully, the teams could also offer a redundant means of communications to battlefield commanders to whom they are attached.

45. Corelli Barnett, *The Swordbearers: Supreme Command in the First World War*, New York: Morrow, 1964, 11.

Bibliography

Abouseada, Hamdy Sobhy, "The Crossing of the Suez Canal, October 6, 1973," Strategy Research Project, US Army War College, April 2000.

Ackerman, Spencer, "Taliban Suicide Attacks Have Held Steady for Most of the War," *Wired*, February 24, 2012.

Adamsky, Dmitry (Dima), and Kjell Inge Bjerga, "Conclusion: Military Innovation between Anticipation and Adaptation," in Dmitry (Dima) Adamsky and Kjell Inge Bjerga, eds., *Contemporary Military Adaptation: Between Anticipation and Adaptation*, London: Routledge, 2012, 188–93.

Adkin, Mark, *Urgent Fury: The Battle for Grenada*, New York: Lexington Books, 1989.

"After Action Report, 735th Tank Bn, July–Nov 44, Feb–Mar 45," available through the Ike Skelton Combined Arms Research Digital Library, http://cgsc.cdmhost. com/cdm/singleitem/collection/p4013coll8/id/3596/rec/8.

"After Action Report, 747th Tank Battalion, June thru December 1944, January thru April 1945," available through the Ike Skelton Combined Arms Research Digital Library, http://cgsc.cdmhost.com/cdm/ref/collection/p4013coll8/id/ 3499.

Al-Ansary, Khalid, and Ali Adeeb, "Most Tribes in Anbar Agree to Unite against Insurgents," *New York Times*, September 18, 2006.

Alexander, Martin S., "In Defence of the Maginot Line: Security Policy, Domestic Politics and the Economic Depression in France," in Robert Boyce, ed., *French Foreign and Defence Policy, 1918–1940: The Decline and Fall of a Great Power*, New York: Routledge, 1998, 164–94.

Alexander, Martin S., *The Republic in Danger: General Maurice Gamelin and the Politics of French Defence, 1933–1940*, New York: Cambridge University Press, 1992.

Alford, LtCol Julian D., and Maj Edwin O. Rueda, "Winning in Iraq," *Marine Corps Gazette 90*, no. 6 (June 2006): 29–30.

Alger, John I., "The Origins and Adaptation of the Principles of War," US Army Command and General Staff College, June 6, 1975.

Allison, Graham T., *Essence of Decision*, Boston: Little, Brown, 1971.

Altman, Howard, "Saddled with Challenges," *Tampa Bay Times*, January 6, 2018.

Ambrose, Stephen, *D-Day, June 6, 1944: The Climactic Battle of World War II*, New York: Simon & Schuster, 1994.

American Geophysical Union, "Ice-Free Arctic Summers Could Happen on Earlier Side of Predictions," *Phys.org*, February 27, 2019.

Andrade, Dale, "Westmoreland Was Right: Learning the Wrong Lessons from the Vietnam War," *Small Wars & Insurgencies* 19, no. 2 (June 2008): 145–81.

"An Exit Interview with General David D. McKiernan," Senior Leader Debriefing Program, Carlisle Barracks, PA: US Army Military History Institute, 2010.

"A Profile of Iraq's Anbar Province," *Talk of the Nation*, National Public Radio, September 20, 2007.

Army Talent Management Task Force, "Army Captains to Take GRE to Assess Talent," July 29, 2019.

Army Test and Evaluation Command, "Palantir Operational Assessment Report," April 2012.

Army Test and Evaluation Command, "Palantir Operational Assessment Report," May 2012.

Asher, Dani, *The Egyptian Strategy for the Yom Kippur War: An Analysis*, trans. Moshe Tlamim, Jefferson, NC: McFarland, 2009.

Auerswald, David P., and Stephen M. Saideman, *NATO in Afghanistan: Fighting Together, Fighting Alone*, Princeton, NJ: Princeton University Press, 2014.

Avant, Deborah D., *Political Institutions and Military Change: Lessons from Peripheral Wars*, Ithaca, NY: Cornell University Press, 1994.

Baker, James A., and Lee H. Hamilton, Co-Chairs, *The Iraq Study Group Report*, New York: Vintage Books, 2006.

Baldor, Lolita C., "Congress Eyes Probe into Army Program," *San Diego Union-Tribune*, July 24, 2012.

Barnett, Corelli, *The Swordbearers: Supreme Command in the First World War*, New York: Morrow, 1964.

Barno, David W., "Military Adaptation in Complex Operations," *Prism* 1, no. 1 (December 2009): 27–36.

Barno, David W., "Silicon, Iron, and Shadow: Three Wars That Will Define America's Future," *Foreign Policy*, March 19, 2013.

Barno, David W., "Theater Command in Afghanistan: Taking Charge of 'The Other War' in 2003–2005," in Michael A. Sheehan and Erich Marquardt, eds., *Counterterrorism & Irregular Warfare*, United States Military Academy Combating Terrorism Center, West Point, NY, 2019.

Barno, Lieutenant General David W., US Army, Retired, "Fighting 'The Other War'": Counterinsurgency Strategy in Afghanistan, 2003–2005," *Military Review* 87, no. 5 (September–October 2007): 32–44.

Barno, David, and Nora Bensahel, "A New Generation of Unrestricted Warfare," *War on the Rocks*, April 19, 2016.

Barno, David, and Nora Bensahel, "Can the U.S. Military Halt Its Brain Drain?," *The Atlantic*, November 5, 2015.

Barno, David, and Nora Bensahel, "Fighting and Winning in the 'Gray Zone,'" *War on the Rocks*, May 19, 2015.

Barno, David, and Nora Bensahel, "Lying to Ourselves: The Demise of Military Integrity," *War on the Rocks*, March 10, 2015.

Barno, David, and Nora Bensahel, "Mirages of War: Six Illusions from Our Recent Conflicts," *War on the Rocks*, April 11, 2017.

Barno, David, and Nora Bensahel, "The Anti-Access Challenge You're Not Thinking About," *War on the Rocks*, May 5, 2015.

Barno, David, and Nora Bensahel, "The Deepest Obligation of Citizenship: Looking Beyond the Warrior Caste," *War on the Rocks*, May 15, 2018.

Barno, David, and Nora Bensahel, "The Future of the Army: Today, Tomorrow, and the Day after Tomorrow," Atlantic Council, September 2016.

Barno, David, and Nora Bensahel, "The U.S. Military's Dangerous Embedded Assumptions," *War on the Rocks*, April 17, 2018.

Barno, David, and Nora Bensahel, "Three Things the Army Chief of Staff Wants You to Know," *War on the Rocks*, May 23, 2017.

Barno, David, and Nora Bensahel, "War in the Fourth Industrial Revolution," *War on the Rocks*, June 19, 2018.

Barno, David, and Nora Bensahel, "What We Saw in *War Machine*," *War on the Rocks*, June 13, 2017.

Barratto, Colonel David J., "Special Forces in the 1980s: A Strategic Reorientation," *Military Review* 63, no. 3 (March 1983): 2–14.

Barrett, Eamon, "In Business, the 'Frozen Middle' Blocks Design Thinking. Here's How to Fix It," *Fortune*, March 6, 2019.

Barry, John, " 'Hillbilly Armor,' " *Newsweek*, December 19, 2004.

Bender, Bryan, "Panel on Iraq Bombings Grows to $3B Effort," *Boston Globe*, June 25, 2006.

Bennett, John T., "What Next for U.S. Joint Anti-IED Efforts?," *Defense News*, September 17, 2007.

Bensahel, Nora, "Darker Shades of Gray: Why Gray Zone Conflicts Will Become More Frequent and Complex," Foreign Policy Research Institute E-Notes, February 13, 2017.

Bensahel, Nora, "Humanitarian Relief and Nation Building in Somalia," in Robert J. Art and Patrick M. Cronin, eds., *The United States and Coercive Diplomacy*, Washington, DC: United States Institute of Peace Press, 2003, 21–56.

Bensahel, Nora, "Securing Iraq: The Mismatch of Demand and Supply," in Markus E. Bouillon, David M. Malone, and Ben Rowswell, eds., *Iraq: Preventing a New Generation of Conflict*, Boulder, CO: Lynne Rienner, 2007, 227–43.

Bensahel, Nora, "The Coalition Paradox: The Politics of Military Cooperation," PhD diss., Stanford University, 1999.

Bensahel, Nora, et al., *After Saddam: Prewar Planning and the Occupation of Iraq*, MG-642-A, Santa Monica, CA: Rand Corporation, 2008.

Bernstein, Jonathan, *AH-64 Apache Units of Operations* Enduring Freedom *and* Iraqi Freedom, Osprey Combat Aircraft 57, Oxford: Osprey Publishing, 2005.

Beyerchen, Alan, "Clausewitz, Nonlinearity, and the Unpredictability of War," *International Security* 17, no. 3 (Winter 1992–1993): 59–90.

Biddle, Stephen, *Afghanistan and the Future of Warfare: Implications for Army and Defense Policy*, Carlisle, PA: Strategic Studies Institute, US Army War College, November 2002.

Biddle, Stephen, *Military Power*, Princeton, NJ: Princeton University Press, 2004.

Biddle, Stephen, Jeffrey A. Friedman, and Jacob N. Shapiro, "Testing the Surge: Why Did Violence Decline in Iraq in 2007?," *International Security* 37, no. 1 (Summer 2012): 7–40.

Bishop, Chris, *Apache AH-64 Boeing (McDonnell Douglas), 1976–2005*, Oxford: Osprey Publishing, 2005.

Bloch, Marc, *Strange Defeat: A Statement of Evidence Written in 1940*, trans. Gerard Hopkins, New York: Octagon Books, 1968.

Blumenson, Martin, *Breakout and Pursuit*, Washington, DC: Center of Military History, US Army, 1993.

Boal, Matthew A., "Field Expedient Armor Modifications to US Armored Vehicles," master's thesis, US Army Command and General Staff College, June 16, 2006.

Bond, Brian, "Slim and Fourteenth Army in Burma," in Brian Bond and Kyoichi Tachikawa, eds., *British and Japanese Military Leadership in the Far Eastern War, 1941–45*, New York: Routledge, 2004, 38–52.

Boulègue, Mathieu, "Russia's Military Posture in the Arctic: Managing Hard Power in a 'Low Tension' Environment," Chatham House, June 2019.

Bowman, Tom, "Army Chief of Staff Defends Iraq Decisions," *All Things Considered*, National Public Radio, January 15, 2009.

Brand, Matthew C., Colonel, USAF, "General McChrystal's Strategic Assessment: Evaluating the Operating Environment in Afghanistan in the Summer of 2009," Research Paper 2011-1, Air Force Research Institute, July 2011.

Branigin, William, "Bush, Rumsfeld Pledge to Protect Troops," *Washington Post*, December 9, 2004.

Brill, Steven, "Donald Trump, Palantir, and the Crazy Battle to Clean Up a Multi-Billion-Dollar Procurement Swamp," *Fortune*, April 1, 2017.

Briscoe, Charles H., et al., *Weapon of Choice: U.S. Army Special Operations Forces in Afghanistan*, Fort Leavenworth, KS: Combat Studies Institute Press, 2003.

Brody, Ben, "Bones, Fobbits and Green Beans: The Definitive Glossary of Modern U.S. Military Slang," *Salon.com*, December 4, 2013.

Brooks, Drew, "The True Story of '12 Strong,' Horse Soldiers Were First to Enter Afghanistan," *Fayetteville Observer*, January 21, 2018.

Brooks, Risa A., and Elizabeth Stanley-Mitchell, eds., *Creating Military Power*, Stanford, CA: Stanford University Press, 2007.

Brooks, Rosa, "Civil-Military Paradoxes," in Kori Schake and Jim Mattis, eds., *Warriors & Citizens: American Views of Our Military*, Stanford, CA: Hoover Institution Press, 2016, 21–68.

Brown, Gary, and Kurt Sanger, "Mattis Faces a Challenge in Equipping US to Engage in Cyberwarfare," *The Hill*, February 6, 2017.

Browne, Ryan, and Barbara Starr, "U.S. General: Russia and China Building Space Weapons to Target US Satellites," *CNN.com*, December 2, 2017.

Buhrkuhl, Robert L., "When the Warfighter Needs It Now," *Defense AT&L*, November–December 2006, 28–31.

Builder, Carl H., *The Masks of War: American Military Styles in Strategy and Analysis*, Baltimore: Johns Hopkins University Press, 1989.

Bumiller, Elisabeth, and Mark Mazzetti, "A General Steps from the Shadows," *New York Times*, May 12, 2009.

Bumiller, Elisabeth, and Thom Shanker, "Commander's Ouster Is Tied to Shift in Afghan War," *New York Times*, May 11, 2009.

Burns, John F., "Afghan Officials Say Airstrike Killed Civilians," *New York Times*, October 16, 2008.

Burton, Brian, and John Nagl, "Learning as We Go: The US Army Adapts to Counterinsurgency in Iraq, July 2004–December 2006," *Small Wars & Insurgencies* 19, no. 3 (September 2008): 303–27.

Buss, Major Darren W., "Evolution of Army Attack Aviation: A Chaotic Coupled Pendulums Analogy," School of Advanced Military Studies, US Army Command and General Staff College, Fort Leavenworth, KS, 2013.

Butler, Howard K., *Army Aviation and Logistics in Vietnam*, St. Louis, MO: US Army Aviation Command Historical Office, 1985.

Campbell, Jason H., and Jeremy Shapiro, "Afghanistan Index: Tracking Variables of Reconstruction & Security in Post-9/11 Afghanistan," Brookings Institution, November 18, 2008.

Campbell, Jason H., and Jeremy Shapiro, "Afghanistan Index: Tracking Variables of Reconstruction & Security in Post-9/11 Afghanistan," Brookings Institution, June 24, 2009.

Campbell, Kenneth J., "Once Burned, Twice Cautious: Explaining the Weinberger-Powell Doctrine," *Armed Forces & Society* 24, no. 3 (Spring 1998): 357–74.

Carter, Ash, *Inside the Five-Sided Box: Lessons from a Lifetime of Leadership at the Pentagon*, New York: Dutton, 2019.

Carter, Sara, "'Matter of Life and Limb': The Congressman Who's Going to Battle with the Army Over a Software Program," *The Blaze*, April 22, 2014.

Carter, Sara, "The Software You've Likely Never Heard about That Could Have Saved Soldiers' Lives—So Why Did the Army Stop It from Being Used?," *The Blaze*, April 21, 2014.

Casey, George W., Jr., *Strategic Reflections: Operation* Iraqi Freedom *July 2004–February 2007*, Washington, DC: National Defense University Press, October 2012.

Catignani, Sergio, "Coping with Knowledge: Organizational Learning in the British Army?," *Journal of Strategic Studies* 37, no. 1 (January 2014): 30–64.

Catignani, Sergio, "'Getting COIN' at the Tactical Level in Afghanistan: Reassessing Counter-Insurgency Adaptation in the British Army," *Journal of Strategic Studies* 35, no. 4 (August 2012): 513–39.

Censer, Marjorie, "Army Report: Military Has Spent $32 Billion since '95 on Abandoned Weapons Programs," *Washington Post*, May 27, 2011.

Chandrasekaran, Rajiv, "Pentagon Worries Led to Command Change," *Washington Post*, August 17, 2009.

Cheng, Roger, "By 2020, More People Will Own a Phone Than Have Electricity," *Cnet.com*, February 3, 2016.

Chief of Staff of the Army, Strategic Studies Group, "Megacities and the United States Army: Preparing for a Complex and Uncertain Future," June 2014.

Chollet, Derek, and James Goldgeier, *America between the Wars: From 11/9 to 9/11*, New York: PublicAffairs, 2008.

Chyma, Colonel Timothy D., "Rapid Acquisition," Senior Service College Fellowship Civilian Research Project, US Army War College, February 2010.

Citino, Robert M., *The Path to Blitzkrieg: Doctrine and Training in the German Army, 1920–1939*, Boulder, CO: Lynne Rienner Publishers, 1999.

Clancy, Tom, with John Gresham, *Special Forces: A Guided Tour of U.S. Army Special Forces*, New York: Berkley Books, 2001.

Clark, Alan, *Barbarossa: The Russian-German Conflict, 1941–1945*, New York: Perennial, 2002.

Clark, Colin, "Top DoD Official Shank Resigns; SCO Moving to DARPA," *Breaking Defense*, June 17, 2019.

Clausewitz, Carl von, *On War*, ed. and trans. Michael Howard and Peter Paret, Princeton, NJ: Princeton University Press, 1976.

Clausewitz, Carl von, *Principles of War*, ed. and trans. Hans W. Gatzke, Harrisburg, PA: Stackpole Company, 1960.

Cloud, David, and Greg Jaffe, *The Fourth Star: Four Generals and the Epic Struggle for the U.S. Army*, New York: Crown Publishers, 2009.

Cohen, Eliot A., *Supreme Command*, New York: Free Press, 2002.

Cohen, Eliot A., Lieutenant Colonel Conrad Crane, US Army, Retired, Lieutenant Colonel Jan Horvath, US Army, and Lieutenant Colonel John Nagl, US Army, "Principles, Imperatives, and Paradoxes of Counterinsurgency," *Military Review* 86, no. 2 (March–April 2006): 49–53.

Cohen, Eliot A., and John Gooch, *Military Misfortunes*, New York: Anchor Books, 2003.

Cole, Ronald H., *Operation Urgent Fury: The Planning and Execution of Joint Operations in Grenada, 12 October–2 November 1983*, Washington, DC: Joint History Office, Office of the Chairman of the Joint Chiefs of Staff, 1997.

Collin, Koh Swee Lean, "Is There Any Way to Counter China's Gray Zone Tactics in the South China Sea?," *National Interest*, September 13, 2017.

Collins, Joseph J., "Initial Planning and Execution in Afghanistan and Iraq," in Richard D. Hooker Jr. and Joseph J. Collins, eds., *Lessons Encountered: Learning from the Long War*, Washington, DC: National Defense University Press, September 2015, 21–88.

Cordesman, Anthony, and Abraham Wagner, *The Lessons of Modern War*, vol. 1: *The Arab-Israeli Conflicts, 1973–1989*, Boulder, CO: Westview Press, 1990.

Crane, Conrad C., *Avoiding Vietnam: The U.S. Army's Response to Defeat in Southeast Asia*, Carlisle, PA: Strategic Studies Institute, US Army War College, September 2002.

Crane, Conrad C., *Cassandra in Oz: Counterinsurgency and Future War*, Annapolis, MD: Naval Institute Press, 2016.

Croke, Vicki Constantine, *Elephant Company: The Inspiring Story of an Unlikely Hero and the Animals Who Helped Him Save Lives in World War II*, New York: Random House, 2014.

Curran, Enda, "China's Growing Clout Triggers Economic Arms Race with Old Order," *Bloomberg*, November 1, 2018.

Currey, Richard, "Waiting for Justice: The Saga of Army Lt. Julian Goodrum, PTSD, Hillbilly Armor, and Whistle-Blowing," *VVA Veteran*, March–April 2005.

Cyert, Richard M., and James G. March, *A Behavioral Theory of the Firm*, Englewood Cliffs, NJ: Prentice-Hall, 1963.

Daddis, Gregory A., *No Sure Victory: Measuring U.S. Army Effectiveness and Progress in the Vietnam War*, New York: Oxford University Press, 2011.

Dale, Catherine, "Operation Iraqi Freedom: Strategies, Approaches, Results, and Issues for Congress," RL34387, Congressional Research Service, March 28, 2008.

Davidson, Janine, *Lifting the Fog of Peace: How Americans Learned to Fight Modern War*, Ann Arbor: University of Michigan Press, 2010.

Defense Intelligence Agency, *China Military Power: Modernizing a Force to Fight and Win*, January 2019.

Defense Intelligence Agency, *Russia Military Power: Building a Military to Support Great Power Aspirations*, 2017.

"Democracy Continues Its Disturbing Retreat," *The Economist*, January 31, 2018.

Department of the Navy, US Marine Corps, "Commandant's Planning Guidance: 38th Commandant of the Marine Corps," July 2019.

Dilanian, Ken, "Top Army Brass Defend Troubled Intelligence System," Associated Press, July 10, 2014.

Dobbins, James F., *Foreign Service: Five Decades on the Frontlines of American Diplomacy*, Santa Monica, CA: Rand Corporation; Washington, DC: Brookings Institution Press, 2017.

Dobbs, Richard, James Manyika, and Jonathan Woetzel, *No Ordinary Disruption: The Four Global Forces Breaking All the Trends*, New York: PublicAffairs, 2015.

"Donald Trump Is Undermining the Rules-Based International Order," *The Economist*, June 7, 2018.

Doubler, Captain Michael D., *Busting the Bocage: American Combined Arms Operations in France, 6 June–31 July 1944*, Fort Leavenworth, KS: Combat Studies Institute, US Army Command and General Staff College, November 1988.

Doughty, Robert A., *Breaking Point: Sedan and the Fall of France, 1940*, Hamden, CT: Archon Books, 1990.

Doughty, Robert A., *Pyrrhic Victory: French Strategy and Operations in the Great War*, Cambridge, MA: Belknap Press of Harvard University Press, 2005.

Doughty, Robert A., *The Seeds of Disaster: The Development of French Army Doctrine, 1919–1939*, Hamden, CT: Archon Books, 1985.

Downie, Richard Duncan, *Learning from Conflict: The U.S. Military in Vietnam, El Salvador, and the Drug War*, Westport, CT: Praeger, 1998.

Draper, Robert, "Boondoggle Goes Boom," *New Republic*, June 19, 2013.

Duncan, Ian, "Elite Army Unit at Fort Meade Searching for Ways to Fight ISIS," *Baltimore Sun*, July 20, 2015.

Dupuy, Colonel T. N., USA, Ret., *Elusive Victory: The Arab Israeli Wars, 1947–1974*, Fairfax, VA: Hero Books, 1984.

Eisenstadt, Michael, and Kenneth Pollack, "Armies of Snow and Armies of Sand: The Impact of Soviet Military Doctrine on Arab Armies," *Middle East Journal* 55, no. 4 (Autumn 2001): 549–78.

Eisler, Peter, Blake Morrison, and Tom Vanden Brook, "Pentagon Balked at Pleas from Officers in Field for Safer Vehicles," *USA Today*, July 16, 2007.

El Shazly, Saad, *The Crossing of the Suez*, San Francisco: American Mideast Research, 2003.

Etherington, Darrell, "Pokémon Go Estimated at over 75M Downloads Worldwide," *TechCrunch*, July 25, 2016.

European Parliament, "EU Migrant Crisis: Facts and Figures," June 30, 2017.

Evans, Michael, "Forward from the Past: The Development of Australian Army Doctrine, 1972–Present," Study Paper No. 301, Land Warfare Studies Centre, September 1999.

Evans, Rob, "Afghanistan War Logs: How the IED Became the Taliban's Weapon of Choice," *The Guardian*, July 25, 2010.

Farrell, Theo, "Culture and Military Power," *Review of International Studies* 24, no. 3 (July 1998): 407–16.

Farrell, Theo, "Improving in War: Military Adaptation and the British in Helmand Province, Afghanistan, 2006–2009," *Journal of Strategic Studies* 33, no. 4 (August 2010): 567–94.

Farrell, Theo, "Introduction: Military Adaptation in War," in Theo Farrell, Frans Osinga, and James A. Russell, eds., *Military Adaptation in Afghanistan*, Stanford, CA: Stanford University Press, 2013, 1–23.

Farrell, Theo, Sten Rynning, and Terry Terriff, *Transforming Military Power since the Cold War: Britain, France, and the United States, 1991–2012*, New York: Cambridge University Press, 2013.

Farrell, Theo, and Terry Terriff, "The Sources of Military Change," in Theo Farrell and Terry Terriff, eds., *The Sources of Military Change*, Boulder, CO: Lynne Rienner Publishers, 2002, 3–20.

Feaver, Peter D., "Anatomy of the Surge," *Commentary* 125, no. 4 (April 2008): 24–8.

Feaver, Peter D., "The Right to Be Right: Civil-Military Relations and the Iraq Surge Decision," *International Security* 35, no. 4 (Spring 2011): 87–125.

Feickert, Andrew, "Army Futures Command (AFC)," IN10889, Congressional Research Service, updated September 10, 2018.

Feickert, Andrew, "Joint Light Tactical Vehicle (JLTV): Background and Issues for Congress," RS22942, Congressional Research Service, updated June 24, 2019.

Felter, Joseph H., and Jacob N. Shapiro, "Limiting Civilian Casualties as Part of a Winning Strategy: The Case of Courageous Restraint," *Daedalus* 146, no. 1 (Winter 2017): 44–58.

Finch, M. P. M., "*Outre-Mer* and *Métropole*: French Officers' Reflections on the Use of the Tank in the 1920s," *War in History* 15, no. 3 (July 2008): 294–313.

Finkel, Meir, *On Flexibility: Recovery from Technological and Doctrinal Surprise on the Battlefield*, Stanford, CA: Stanford University Press, 2011.

Fitzgerald, David, *Learning to Forget: US Army Counterinsurgency Doctrine and Practice from Vietnam to Iraq*, Stanford, CA: Stanford University Press, 2013.

Flavin, William, "Civil-Military Operations: Afghanistan," US Army Peacekeeping and Stability Operations Institute, March 23, 2004.

Flynn, Major General Michael T., USA, Captain Matt Pottinger, USMC, and Paul D. Batchelor, DIA, "Fixing Intel: A Road Map for Making Intelligence Relevant in Afghanistan," Center for a New American Security, January 2010.

Fontenot, COL Gregory, US Army, Retired, LTC E. J. Degen, US Army, and LTC David Tohn, US Army, *On Point: The United States Army in Operation IRAQI FREEDOM*, Fort Leavenworth, KS: Combat Studies Institute Press, 2004.

Fox, J. Ronald, *Defense Acquisition Reform, 1960–2009: An Elusive Goal*, Washington, DC: Center of Military History, US Army, 2011.

Freedberg, Sydney J., Jr., "Abizaid of Arabia," *Atlantic Monthly* 292, no. 5 (December 2003): 32–6.

Freedberg, Sydney J., Jr., "Marines Reorganize Infantry for High-Tech War: Fewer Riflemen, More Drones," *Breaking Defense*, May 4, 2018.

Friedman, B. A., "The End of the Fighting General," *Foreign Policy*, September 12, 2018.

Friedman, Thomas L., *Thank You for Being Late: An Optimist's Guide to Thriving in the Age of Accelerations*, New York: Farrar, Straus and Giroux, 2016.

Fukuyama, Francis, "The End of History?," *National Interest* 16 (Summer 1989): 3–18.

Fuller, J. F. C., *The Foundations of the Science of War*, London: Hutchinson, 1926.

Gabel, Christopher R., *Seek, Strike, and Destroy: U.S. Army Tank Destroyer Doctrine in World War II*, Leavenworth Papers No. 12, Fort Leavenworth, KS: US Army Command and General Staff College, 1985.

Gale, Tim, "'A Charming Toy': The Surprisingly Long Life of the Renault Light Tank, 1917–1940," in Alaric Searle, ed., *Genesis, Employment, Aftermath: First World War Tanks and the New Warfare, 1900–1945*, Solihull, West Midlands, England: Helion, 2015, 191–209.

Gale, Tim, *The French Army's Tank Force and Armoured Warfare in the Great War: The Artillerie Spéciale*, Burlington, VT: Ashgate, 2013.

Gall, Carlotta, "U.S. Killed 90, Including 60 Children, in Afghan Village, U.N. Finds," *New York Times*, August 26, 2008.

Galloway, Joseph L., "Vietnam Story," *U.S. News & World Report*, October 29, 1990.

Galvin, John R., "Uncomfortable Wars: Toward a New Paradigm," *Parameters* 16, no. 4 (Winter 1986): 2–8.

Gansler, Jacques S., William Lucyshyn, and William Varettoni, "Acquisition of Mine-Resistant, Ambush-Protected (MRAP) Vehicles: A Case Study," Center for Public Policy and Private Enterprise, School of Public Policy, University of Maryland, March 2010.

Garamone, Jim, "Directive Aimed at Reducing Civilian Casualties," American Forces Press Service, September 16, 2008.

Garland, Chad, "Army Secretary: PT Belts Aren't Needed in Daylight," *Stars and Stripes*, January 12, 2019.

Gartenberg, Chaim, "We Compared Hardware Specs for Every iPhone Ever Made," *TheVerge.com*, June 28, 2017.

Gaskell, Stephanie, "House Panel Probes Army IED Review," *Politico*, August 1, 2012.

Gates, Robert M., *Duty: Memoirs of a Secretary at War*, New York: Alfred A. Knopf, 2014.

Gentile, Colonel Gian, *Wrong Turn: America's Deadly Embrace of Counterinsurgency*, New York: New Press, 2013.

Gentile, Gian P., "A Strategy of Tactics: Population-centric COIN and the Army," *Parameters* 39, no. 3 (Autumn 2009): 5–17.

Gerras, Stephen J., and Leonard Wong, *Changing Minds in the Army: Why It Is So Difficult and What to Do about It*, Carlisle, PA: Strategic Studies Institute, US Army War College, October 2013.

Gerring, John, *Social Science Methodology: A Criterial Framework*, Cambridge: Cambridge University Press, 2001.

Gilmore, Commander T. J., "Iran Owns the Gray Zone," *Proceedings* (US Naval Institute) 144, no. 3 (March 2018): 48–53.

Gittler, Juliana, "'Skunk Werks' Armor Shop Helps Soldiers through Better Protection for U.S. Vehicles," *Stars and Stripes*, October 31, 2004.

Gordon, John, IV, Bruce Nardulli, and Walter L. Perry, "The Operational Challenges of Task Force Hawk," *Joint Force Quarterly* 29 (Autumn–Winter 2001–2002): 52–7.

Gordon, Michael R., and Thom Shanker, "Bush to Name a New General to Oversee Iraq," *New York Times*, January 5, 2007.

Gordon, Michael R., and General Bernard E. Trainor, *Cobra II: The Inside Story of the Invasion and Occupation of Iraq*, New York: Pantheon Books, 2006.

Gordon, Michael R., and General Bernard E. Trainor, *The Endgame: The Inside Story of the Struggle for Iraq, from George W. Bush to Barack Obama*, New York: Pantheon Books, 2012.

Gordon, Michael R., and General Bernard E. Trainor, *The Generals' War: The Inside Story of the Conflict in the Gulf*, Boston: Little, Brown, 1995.

Gosnell, Rachael, "Caution in the High North: Geopolitical and Economic Challenges of the Arctic Maritime Environment," *War on the Rocks*, June 25, 2018.

Grant, Ian, and Kazuo Tamayama, *Burma 1942: The Japanese Invasion*, Chichester, England: Zampi Press, 1999.

Gray, Colin, "Strategic Culture as Context: The First Generation of Theory Strikes Back," *Review of International Studies* 25, no. 1 (January 1999): 49–69.

Greenhalgh, Elizabeth, "Technology Development in Coalition: The Case of the First World War Tank," *International History Review* 22, no. 4 (December 2000): 806–36.

Greenhalgh, Elizabeth, *The French Army and the First World War*, Cambridge: Cambridge University Press, 2014.

Grissom, Adam, "The Future of Military Innovation Studies," *Journal of Strategic Studies* 29, no. 5 (2006): 905–34.

Guderian, Heinz, *Achtung-Panzer! The Development of Tank Warfare*, trans. Christopher Duffy, London: Cassell, 1999.

Gutner, Tamar, "AIIB: Is the Chinese-Led Development Bank a Role Model?," Council on Foreign Relations, June 25, 2018.

Halford, Macy, "Should Death Stop the Nobel?," *New Yorker*, October 3, 2011.

Hammes, T. X., *Deglobalization and International Security*, Amherst, NY: Cambria Press, 2019.

Hammes, T. X., "Technological Change and the Fourth Industrial Revolution," in George P. Shultz, Jim Hoagland, and James Timbie, eds., *Beyond Disruption: Technology's Challenge to Governance*, Stanford, CA: Hoover Institution Press, 2018, 37–73.

Hannan, Michael T., and John Freeman, "Structural Inertia and Organizational Change," *American Sociological Review* 49, no. 2 (April 1984): 149–64.

Harris, Shane, "Palantir Wins Competition to Build Army Intelligence System," *Washington Post*, March 26, 2019.

Harrison, Gordon A., *Cross-Channel Attack,* Washington, DC: Center of Military History, US Army, 1993.

Hartung, Frank E., "The Sociology of Positivism," *Science & Society* 8, no. 4 (Fall 1944): 328–41.

Headquarters, Department of the Army, *Army Leadership*, Army Doctrine Publication 6-22, August 1, 2012.

Headquarters, Department of the Army, *Army Leadership*, Army Doctrine Reference Publication 6-22, August 1, 2012.

Headquarters, Department of the Army, *Army Leadership*, Field Manual 6-22, October 2006.

Headquarters, Department of the Army, *Civil Affairs Operations*, Field Manual 3-05.40, September 2006.

Headquarters, Department of the Army, *Civil Affairs Operations*, Field Manual 41-10, January 11, 1993.

Headquarters, Department of the Army, *Civil Affairs Operations*, Field Manual 41-10, February 2000.

Headquarters, Department of the Army, *Counterinsurgency Operations*, Field Manual-Interim 3-07.22, October 2004.

Headquarters, Department of the Army, *Doctrine Primer*, Army Doctrine Publication 1-01, September 2014.

Headquarters, Department of the Army, *Evaluation Reporting System*, Army Regulation 623-3, June 14, 2019.

Headquarters, Department of the Army, *General Douglas MacArthur Leadership Award Program*, Army Regulation 600-89, September 22, 2017.

Headquarters, Department of the Army, *Leader Development*, Field Manual 6-22, June 2015.

Headquarters, Department of the Army, *Mission Command*, Army Doctrine Publication 6-0, May 2012.

Headquarters, Department of the Army, *Mission Command*, Army Doctrine Reference Publication 6-0, May 2012.

Headquarters, Department of the Army, *Operations*, Field Manual 3-0, October 2017.

Headquarters, Department of the Army, *The Army*, Army Doctrine Publication 1, September 17, 2012.

Headquarters, Department of the Army / Headquarters, Marine Corps Combat Development Command, *Counterinsurgency*, Field Manual 3-24 / Marine Corps Warfighting Publication 3-33.5, December 15, 2006.

Hendren, John, "Unarmored Military Vehicles to Be Restricted," *Los Angeles Times*, February 13, 2005.

Hendrix, John T., "The Interwar Army and Mechanization: The American Approach," *Journal of Strategic Studies* 16, no. 1 (March 1993): 75–108.

Henry, Dallas J., "The Battle at LZ Albany," *Infantry* 103, no. 2 (April–June 2014): 20–3.

Herbert, Major Paul H., *Deciding What Has to Be Done: General William E. DePuy and the 1976 Edition of FM 100-5, Operations*, Leavenworth Papers Number 16, Fort Leavenworth, KS: Combat Studies Institute, US Army Command and General Staff College, 1988.

Hoehn, Andrew R., and Sarah Harting, *Risking NATO: Testing the Limits of the Alliance in Afghanistan*, MG-974, Santa Monica, CA: Rand Corporation, 2010.

Hoffman, Frank, "A Second Look at the Powell Doctrine," *War on the Rocks*, February 20, 2014.

Hoffman, Frank, "Changing Tires on the Fly: The Marines and Postconflict Stability Ops," Foreign Policy Research Institute E-Notes, September 2, 2006.

Hoffman, Frank, "Conflict in the 21st Century: The Rise of Hybrid Wars," Potomac Institute for Policy Studies, December 2007.

Høiback, Harald, "What Is Doctrine?," *Journal of Strategic Studies* 34, no. 6 (December 2011): 879–900.

Holmes, Richard, ed., *The Oxford Companion to Military History*, Oxford: Oxford University Press, 2001.

Horne, Alistair, *To Lose a Battle: France 1940*, London: Macmillan, 1969.

Hoskinson, Charles, "Computer Bug Hurts Army Ops," *Politico*, June 29, 2011.

Howard, Michael, "Military Science in an Age of Peace," *RUSI Journal* 119, no. 1 (March 1974): 3–11.

Howard, Michael, "The Use and Abuse of Military History," in Michael Howard, *The Causes of War*, Cambridge, MA: Harvard University Press, 1983.

Hummel, Stephen, "The Islamic State and WMD: Assessing the Future Threat," *CTC Sentinel* 9, no. 1 (January 2016): 18–21.

Hunt, Vivian, et al., *Delivering through Diversity*, McKinsey & Company, January 2018.

Hutcheson, Joshua, "'Ready First' Brigade Takes Responsibility for Tal Afar Area," *Stars and Stripes*, February 22, 2006.

Ingersoll, Ralph, *The Battle Is the Pay-off*, New York: Harcourt, Brace, 1943.

International Telecommunications Union, "Overview of the Internet of Things," Recommendation ITU-T Y.2060, June 2012.

Irfan, Umair, "How Antarctica's Melting Ice Could Change Weather around the World," *Vox.com*, February 6, 2019.

Jensen, Benjamin M., *Forging the Sword: Doctrinal Change in the U.S. Army*, Stanford, CA: Stanford University Press, 2016.

Johnson, David E., *Fast Tanks and Heavy Bombers*, Ithaca, NY: Cornell University Press, 1998.

Johnson, Hubert C., *Breakthrough! Tactics, Technology, and the Search for Victory on the Western Front in World War I*, Novato, CA: Presidio Press, 1994.

Johnson, Jeannie L., *The Marines, Counterinsurgency, and Strategic Culture: Lessons Learned and Lost in America's Wars*, Washington, DC: Georgetown University Press, 2018.

Johnston, Alastair Iain, "Thinking about Strategic Culture," *International Security* 19, no. 4 (Spring 1995): 32–64.

Johnston, Paul, "Doctrine Is Not Enough: The Effect of Doctrine on the Behavior of Armies," *Parameters* 30, no. 3 (Autumn 2000): 30–9.

Joint Chiefs of Staff, *Civil-Military Operations*, Joint Publication 3-57, July 8, 2008.

Joint Chiefs of Staff, *Close Air Support*, Joint Publication 3-09.3, November 25, 2014.

Joint Chiefs of Staff, *Doctrine for the Armed Forces of the United States*, Joint Publication 1, March 25, 2013.

Joint Chiefs of Staff, *Doctrine for the Armed Forces of the United States*, Joint Publication 1, March 25, 2013, Incorporating Change 1, July 12, 2017.

Joint Chiefs of Staff, *Doctrine for Joint Operations*, Joint Publication 3-0, September 10, 2001.

Joint Chiefs of Staff, *DOD Dictionary of Military and Associated Terms*, Joint Publication 1-02, November 8, 2010, as amended through February 15, 2016.

Joint Chiefs of Staff, *Interagency Coordination during Joint Operations*, vol. 1, Joint Publication 3-08, October 9, 1996.

Joint Chiefs of Staff, *Joint Doctrine Development Process*, Chairman of the Joint Chiefs of Staff Manual 5120.01A, December 29, 2014.

Joint Chiefs of Staff, *Joint Doctrine for Civil-Military Operations*, Joint Publication 3-57, June 21, 1995.

Joint Chiefs of Staff, *Joint Doctrine for Civil-Military Operations*, Joint Publication 3-57, February 8, 2001.

Joint Chiefs of Staff, *Joint Doctrine for Military Operations Other Than War*, Joint Publication 3-07, June 16, 1995.

Joint Chiefs of Staff, *Joint Operations*, Joint Publication 3-0, January 17, 2017.

Joint Chiefs of Staff, *Joint Operations*, Joint Publication 3-0, January 17, 2017, Incorporating Change 1, October 22, 2018.

Joint Chiefs of Staff, *Joint Planning*, Joint Publication 5-0, June 16, 2017.

Joint Staff, J-7, *Planner's Handbook for Operational Design,* Version 1.0, October 7, 2011.

Judson, Jen, "Army Releases RFP for DCGS-A Increment 2," *Army Times*, December 23, 2015.

Judson, Jen, "Palantir—Who Successfully Sued the Army—Has Won a Major Army Contract," *Defense News*, March 29, 2019.

Judson, Jen, "The Army Turns to a Former Legal Opponent to Fix Its Intel Analysis System," *C4ISRNET*, March 9, 2018.

Kagan, Frederick W., "Choosing Victory: A Plan for Success in Iraq—Phase I Report," A Report of the Iraq Planning Group at the American Enterprise Institute, January 2007.

Kagan, Kimberly, "The Anbar Awakening: Displacing al Qaeda from Its Stronghold in Western Iraq," Institute for the Study of War and WeeklyStandard.com, August 21, 2006–March 30, 2007.

Kahneman, Daniel, *Thinking, Fast and Slow*, New York: Farrar, Straus and Giroux, 2011.

Kaplan, Fred, *The Insurgents: David Petraeus and the Plot to Change the American Way of War*, New York: Simon & Schuster, 2013.

Kaplan, Lawrence F., "War at Home," *New Republic*, March 19, 2007.

Katzman, Kenneth, "Afghanistan: Post-Taliban Governance, Security, and U.S. Policy," RL30588, Congressional Research Service, June 8, 2009.

Kennedy, Major Edwin L., "Failure of Israeli Armored Tactical Doctrine, Sinai, 6-8 October 1973," *Armor* 99, no. 6 (November–December 1990): 28–31.

Kier, Elizabeth, "Culture and Military Doctrine: France between the Wars," *International Security* 19, no. 4 (Spring 1995): 65–93.

Kier, Elizabeth, *Imagining War*, Princeton, NJ: Princeton University Press, 1997.

Kilcullen, David, "Anatomy of a Tribal Revolt," *Small Wars Journal*, August 30, 2007.

Kimmons, Sean, "In First Year, Futures Command Grows from 12 to 24,000 Personnel," Army News Service, July 19, 2019.

Kollars, Nina, "Military Innovation's Dialectic: Gun Trucks and Rapid Acquisition," *Security Studies* 23, no. 4 (October-December 2014): 787–813.

Kollars, Nina, "Organising Adaptation in War," *Survival* 57, no. 6 (December 2015–January 2016): 111–26.

Kollars, Nina A., "War's Horizon: Soldier-Led Adaptation in Iraq and Vietnam," *Journal of Strategic Studies* 38, no. 4 (June 2015): 529–53.

Komer, R. W., *Bureaucracy Does Its Thing: Institutional Constraints on U.S.-GVN Performance in Vietnam*, R-967-ARPA, Santa Monica, CA: Rand Corporation, August 1972.

Krauthammer, Charles, "The Unipolar Moment," *Foreign Affairs* 70, no. 1 (1990–1991): 23–33.

Krepinevich, Andrew F., Jr., "How to Win in Iraq," *Foreign Affairs* 84, no. 5 (September–October 2005): 87–104.

Krepinevich, Andrew F., Jr., "Recovery from Defeat: The U.S. Army and Vietnam," in George J. Andreopolis and Harold E. Selesky, eds., *The Aftermath of Defeat: Societies, Armed Forces, and the Challenge of Recovery*, New Haven, CT: Yale University Press, 1994, 124–42.

Krepinevich, Andrew F., Jr., *The Army and Vietnam*, Baltimore: Johns Hopkins University Press, 1986.

Kristof, Nicholas, "How We Won the War," *New York Times*, September 6, 2002.

Kukis, Mark, "Turning Iraq's Tribes against Al-Qaeda," *Time*, December 26, 2006.

Lacdan, Joe, "Warfare in Megacities: A New Frontier in Military Operations," Army News Service, May 24, 2018.

Lahaie, Olivier, "The Development of Tank Warfare on the Western Front," in Alaric Searle, ed., *Genesis, Employment, Aftermath: First World War Tanks and the New Warfare, 1900–1945,* Solihull, West Midlands, England: Helion, 2015, 57–79.

Lamb, Christopher J., Matthew J. Schmidt, and Berit G. Fitzsimmons, "MRAPs, Irregular Warfare, and Pentagon Reform," *Joint Force Quarterly* 55 (4th Quarter 2009): 76–85.

Lamothe, Dan, "Veil of Secrecy Lifted on Pentagon Office Planning 'Avatar' Fighters and Drone Swarms," *Washington Post*, March 8, 2016.

Landler, Mark, "Trump Abandons Nuclear Deal He Long Scorned," *New York Times*, May 8, 2018.

Landon, James J., "CIMIC: Civil Military Cooperation," in Larry Wentz, ed., *Lessons from Bosnia: The IFOR Experience*, Washington, DC: Institute for National Strategic Studies, National Defense University / CCRP, 1997, 119–38.

Lea, Maj. Brett, "1st Stryker Brigade Combat Team Converts to Armored Brigade," Army News Service, June 20, 2019.

Leonard, Captain Steven M., OD, "Steel Curtain: The Guns on the Ia Drang," *Field Artillery*, July–August 1998, 17–20.

Levitsky, Steven, and Daniel Ziblatt, *How Democracies Die*, New York: Crown, 2018.

Lewin, Ronald, *Slim: The Standardbearer*, London: Cooper, 1976.

Lewis, Michael, *The Undoing Project*, New York: Norton, 2017.

Liewer, Steve, "Sandbags Become Makeshift Vehicle Armor," *Stars and Stripes*, April 26, 2004.

Lindblom, Charles E., "The Science of 'Muddling Through,'" *Public Administration Review* 19, no. 2 (Spring 1959): 79–88.

Linn, Brian McAllister, *Elvis's Army: Cold War GIs and the Atomic Battlefield*, Cambridge, MA: Harvard University Press, 2016.

Livingston, Ian S., Heather L. Messera, and Michael O'Hanlon, "Afghanistan Index: Tracking Variables of Reconstruction & Security in Post-9/11 Afghanistan," Brookings Institution, January 29, 2010.

Livingston, Ian S., and Michael O'Hanlon, "Afghanistan Index: Tracking Variables of Reconstruction & Security in Post-9/11 Afghanistan," Brookings Institution, September 29, 2017.

Lock-Pullan, Richard, "How to Rethink War: Conceptual Innovation and AirLand Battle Doctrine," *Journal of Strategic Studies* 28, no. 4 (August 2005): 679–702.

Long, Austin, *Doctrine of Eternal Recurrence: The U.S. Military and Counterinsurgency Doctrine, 1960–1970 and 2003–2006*, Rand Counterinsurgency Study Paper 6, Santa Monica, CA: Rand Corporation, 2008.

Long, Austin, "The Anbar Awakening," *Survival* 50, no. 2 (April–May 2008): 67–94.

Long, Jeffrey W., *The Evolution of U.S. Army Doctrine: From Active Defense to AirLand Battle and Beyond*, Fort Leavenworth, KS: US Army Command and General Staff College, 1991.

Lopez, C. Todd, "Milley: On Cusp of Profound, Fundamental Change," Army News Service, October 6, 2016.

Lorenzo, Rocío, et al., "How Diverse Leadership Teams Boost Innovation," Boston Consulting Group, January 23, 2018.

Louis, Matthew J., Tom Seymour, and Jim Joyce, "3D Opportunity in the Department of Defense: Additive Manufacturing Fires Up," Deloitte Consulting, 2014.

Lovell, John P., "Vietnam and the U.S. Army: Learning to Cope with Failure," in George K. Osborn et al., eds., *Democracy, Strategy, and Vietnam: Implications for American Policymaking*, Lexington, MA: Lexington Books, 1987, 121–54.

Lucas, Pascal Marie Henri, *The Evolution of Tactical Ideas in France and Germany during the War of 1914–1918*, trans. P. V. Kieffer, Paris: Berger-Leorault, 1925.

Lupfer, Timothy T., "The Challenge of Military Reform," in Asa A. Clark IV et al., eds., *The Defense Reform Debate*, Baltimore: Johns Hopkins University Press, 1984, 23–32.

Lupfer, Timothy T., *The Dynamics of Doctrine: The Changes in German Tactical Doctrine during the First World War*, Fort Leavenworth, KS: US Army Command and General Staff College, 1981.

Luttwak, Edward N., "The Operational Level of War," *International Security* 5, no. 3 (Winter 1980–1981): 61–79.

Lynch, Marc, "Explaining the Awakening: Engagement, Publicity, and the Transformation of Iraqi Sunni Political Attitudes," *Security Studies* 20, no. 1 (January-March 2011): 36–72.

Maass, Peter, "Professor Nagl's War," *New York Times Magazine*, January 11, 2004.

MacFarland, Colonel Sean, "Addendum: Anbar Awakens," *Military Review* 88, no. 3 (May–June 2008): 2–3.

MacKenzie, Donald, and Judy Wajcman, *The Social Shaping of Technology*, 2nd ed., Buckingham, England: Open University Press, 1999.

Malkasian, Carter, "Counterinsurgency in Iraq, May 2003–January 2007," in Daniel Marston and Carter Malkasian, eds., *Counterinsurgency in Modern Warfare*, Oxford: Osprey Publishing, 2008, 241–59.

Malkasian, Carter, "Did the Coalition Need More Forces in Iraq? Evidence from Al Anbar," *Joint Force Quarterly* 46 (3rd Quarter 2007): 120–6.

Malkasian, Carter, *Illusions of Victory: The Anbar Awakening and the Rise of the Islamic State*, New York: Oxford University Press, 2017.

Malkasian, Carter, "Signaling Resolve, Democratization, and the First Battle of Fallujah," *Journal of Strategic Studies* 29, no. 3 (June 2006): 423–52.

Manea, Octavian, "Reflections on the 'Counterinsurgency Decade': Small Wars Journal Interview with General David H. Petraeus," *Small Wars Journal*, September 1, 2013.

Mansoor, Peter R., *Baghdad at Sunrise: A Brigade Commander's War in Iraq*, New Haven, CT: Yale University Press, 2008.

March, James G., "Footnotes to Organizational Change," *Administrative Science Quarterly* 26, no. 4 (December 1981): 563–77.

Marine Corps Order 3504.1, "Marine Corps Lessons Learned Program (MCLLP) and the Marine Corps Center for Lessons Learned (MCCLL)," July 31, 2006.

Mattocks, Julie Ann, "Institutionalizing the Marine Corps' Rapid Acquisition Capability: Equipping, Training and Sustaining the Nation's Expeditionary Force to Win," Marine Corps University, 2011.

Mayo, Lida, *The Ordnance Department: On Beachhead and Battlefront*, Washington, DC: Office of the Chief of Military History, US Army, 1968.

Mazarr, Michael J., *Light Forces and the Future of U.S. Military Strategy*, McLean, VA: Brassey's, 1990.

Mazarr, Michael J., *Mastering the Gray Zone: Understanding a Changing Era of Conflict*, Carlisle, PA: Strategic Studies Institute and US Army War College Press, December 2015.

Mazzetti, Mark, and Eric Schmitt, "U.S. Study Is Said to Warn of Crisis in Afghanistan," *New York Times*, October 8, 2008.

McCarthy, Michael C., Matthew A. Moyer, and Brett H. Venable, *Deterring Russia in the Gray Zone*, Carlisle, PA: Strategic Studies Institute, US Army War College, March 2019.

McCary, John A., "The Anbar Awakening: An Alliance of Incentives," *Washington Quarterly* 32, no. 1 (Winter 2008–2009): 43–59.

McGee, Maj. Gen. J. P., and Ryan Evans, "The Army's New Approach to People," *War on the Rocks*, December 16, 2019.

McGlone, Ashly, "Hunter, IEDs, and Campaign Contributions," *Morning Call*, August 3, 2012.

McNerney, Michael J., "Stabilization and Reconstruction in Afghanistan: Are PRTs a Model or a Muddle?," *Parameters* 35, no. 4 (Winter 2005–2006): 32–46.

McWilliams, Chief Warrant Officer-4 Timothy S., and Lieutenant Colonel Kurtis P. Wheeler, *Al-Anbar Awakening*, vol. 1: *American Perspectives: U.S. Marines and Counterinsurgency in Iraq, 2004–2009*, Quantico, VA: Marine Corps University Press, 2009.

Mehta, Aaron, "Griffin Makes Case for Why SCO Should Live under DARPA—and Why Its Director Had to Go," *Defense News*, August 8, 2019.

Michaels, Jim, *A Chance in Hell*, New York: St. Martin's Press, 2010.

Millett, Allan R., and Williamson Murray, eds., *Military Effectiveness*, Boston: Allan & Unwin, 1988.

Misztal, Blaise, Jack Rametta, and Mary Farrell, "Personnel Reform Lives, but Don't Call It 'Force of the Future,'" *War on the Rocks*, August 9, 2018.

Mizokami, Kyle, "Drone 'Factory in a Can' Would Change Air War Forever," *Popular Mechanics*, September 7, 2017.

Mockaitis, Thomas R., *Civil-Military Cooperation in Peace Operations: The Case of Kosovo*, Carlisle, PA: Strategic Studies Institute, US Army War College, October 2004.

Moore, Harold G., and Joseph L. Galloway, *We Were Soldiers Once . . . and Young*, New York: Open Road Media, 2012.

Moore, Solomon, and Julian Barnes, "Many Iraqi Troops Are No-Shows in Baghdad," *Los Angeles Times*, September 23, 2006.

Moran, Michael, "Frantically, the Army Tries to Armor Humvees," *MSNBC.com*, April 15, 2004.

Morelli, Vincent, and Paul Belkin, "NATO in Afghanistan: A Test of the Transatlantic Alliance," RL33627, Congressional Research Service, December 3, 2009.

Moreman, T. R., *The Jungle, the Japanese and the British Commonwealth Armies at War, 1941–45*, New York: Frank Cass, 2005.

Morin, Monte, "'Raider Brigade' Takes Over Ramadi," *Stars and Stripes*, February 19, 2007.

Morrison, Wayne M., "China's Economic Rise: History, Trends, Challenges, and Implications for the United States," RL33534, Congressional Research Service, February 5, 2018.

Murray, Williamson, *America and the Future of War*, Stanford, CA: Hoover Institution Press, 2017.

Murray, Williamson, "Armored Warfare: The British, French, and German Experiences," in Williamson Murray and Allan R. Millett, eds., *Military Innovation in the Interwar Period*, Cambridge; New York: Cambridge University Press, 1996, 6–49.

Murray, Williamson, "Does Military Culture Matter?," *Orbis* 43, no. 1 (Winter 1999): 27–42.

Murray, Williamson, "Innovation: Past and Future," in Williamson Murray and Allan R. Millett, eds., *Military Innovation in the Interwar Period*, Cambridge; New York: Cambridge University Press, 1996, 300–28.

Murray, Williamson, *Military Adaptation in War: With Fear of Change*, New York: Cambridge University Press, 2011.

Murray, Williamson, and Allan R. Millett, *A War to Be Won: Fighting the Second World War*, Cambridge, MA: Belknap Press of Harvard University Press, 2000.

Murray, Williamson, and Allan R. Millett, "Introduction: Military Effectiveness Twenty Years Later," in Allan R. Millett and Williamson Murray, eds., *Military Effectiveness*, vol. 1: *The First World War*, new ed., Cambridge; New York: Cambridge University Press, 2010, xiii–xxi.

Myers, Meghann, "Good News, Soldiers: The Army Has Slashed Even More Mandatory Training Requirements," *Army Times*, June 5, 2018.

Myers, Meghann, "No PT Belts Needed during the Day, on Running Tracks, Army Secretary Says," *Army Times*, January 14, 2019.

Myers, Meghann, "The Army Just Dumped a Bunch of Mandatory Training to Free Up Soldiers' Time," *Army Times*, April 24, 2018.

Myre, Greg, "'12 Strong': When the Afghan War Looked Like a Quick, Stirring Victory," *Morning Edition*, National Public Radio, January 18, 2018.

Nagl, John A., "Constructing the Legacy of Field Manual 3-24," *Joint Force Quarterly* 58 (3rd Quarter 2010): 118–20.

Nagl, John A., "Learning and Adapting to Win," *Joint Force Quarterly* 58 (3rd Quarter 2010): 123–4.

Nagl, John A., *Learning to Eat Soup with a Knife: Counterinsurgency Lessons from Malaya and Vietnam*, Westport, CT: Praeger, 2002.

Nakashima, Ellen, and Paul Sonne, "China Hacked a Navy Contractor and Secured a Trove of Highly Sensitive Data on Submarine Warfare," *Washington Post*, June 8, 2018.

National Commission on Terrorist Attacks upon the United States, *The 9/11 Commission Report*, New York: Norton, 2004.

National Intelligence Council, *Global Trends: Paradox of Progress*, January 9, 2017.

Naylor, Sean, *Not a Good Day to Die: The Untold Story of Operation Anaconda*, New York: Berkley Books, 2005.

Naylor, Sean, *Relentless Strike: The Secret History of Joint Special Operations Command*, New York: St. Martin's Press, 2015.

Neate, Rupert, "Richest 1% Own Half the World's Wealth, Study Finds," *The Guardian*, November 14, 2017.

Neumann, Brian, Lisa Mundy, and Jon Mikolashek, *Operation ENDURING FREEDOM: The United States Army in Afghanistan, March 2002–April 2005*, Washington, DC: US Army Center for Military History, [2013?].

Nielsen, Suzanne C., *An Army Transformed: The U.S. Army's Post-Vietnam Recovery and the Dynamics of Change in Military Organizations*, Carlisle, PA: Strategic Studies Institute, US Army War College, September 2010.

North Atlantic Treaty Organization, *ISAF Provincial Reconstruction Team (PRT) Handbook*, 4th ed. [n.d.].

Nuzum, Henry, *Shades of CORDS in the Kush: The False Hope of "Unity of Effort" in American Counterinsurgency*, Carlisle, PA: Strategic Studies Institute, US Army War College, April 2010.

O'Ballance, Edgar, *No Victor, No Vanquished: The Yom Kippur War*, San Rafael, CA: Presidio Press, 1978.

"Obama's Remarks on Iraq and Afghanistan," *New York Times*, July 15, 2008.

Office of the Director of National Intelligence, "Background to 'Assessing Russian Activities and Intentions in Recent US Elections': The Analytic Process and Cyber Incident Attribution," January 6, 2017.

Office of the Special Investigator General for Iraqi Reconstruction, "Challenges in Obtaining Reliable and Useful Data on Iraqi Security Forces Continue," SIGIR-09-002, October 21, 2008.

Office of the Special Inspector General for Iraqi Reconstruction, "Interim Analysis of Iraqi Security Force Information Provided by the Department of Defense Report, *Measuring Stability and Security in Iraq*," SIGIR-08-015, April 25, 2008.

O'Hanlon, Michael E., and Jason Campbell, "Iraq Index: Tracking Variables of Reconstruction & Security in Post-Saddam Iraq," Brookings Institution, January 28, 2008.

O'Hanlon, Michael E., and Jason Campbell, "Iraq Index: Tracking Variables of Reconstruction & Security in Post-Saddam Iraq," Brookings Institution, July 28, 2008.

O'Hanlon, Michael E., and Nina Kamp, "Iraq Index: Tracking Variables of Reconstruction & Security in Post-Saddam Iraq," Brookings Institution, April 27, 2006.

O'Hanlon, Michael E., and Ian Livingston, "Iraq Index: Tracking Variables of Reconstruction & Security in Post-Saddam Iraq," Brookings Institution, January 31, 2011.

Oliker, Olga, *Iraqi Security Forces: Defining Challenges and Assessing Progress*, CT-277, Santa Monica, CA: Rand Corporation, March 2007.

Oliker, Olga, et al., *Aid during Conflict: Interaction between Civilian Assistance Providers in Afghanistan, September 2001–June 2002*, MG-212, Santa Monica, CA: Rand Corporation, 2004.

"One-on-One with General David McKiernan," *NBC Nightly News with Brian Williams*, June 16, 2008.

Osborn, Kris, "How the U.S. Military Is Using a New Special Attack Force to Kill Its Enemies," *National Interest*, May 2, 2019.

Oudshoorn, Nelly, and Trevor Pinch, eds., *How Users Matter: The Co-Construction of Users and Technologies*, Cambridge, MA: MIT Press, 2003.

Owens, Mackubin Thomas, "Promoting Failure," *National Review*, February 21, 2007.

Packer, George, "The Lesson of Tal Afar," *New Yorker*, April 10, 2006.

Page, Scott E., *The Diversity Bonus: How Great Teams Pay Off in the Knowledge Economy*, Princeton, NJ: Princeton University Press, 2017.

Pappalardo, Joe, "No Treaty Will Stop Space Weapons," *Popular Mechanics*, January 25, 2018.

Paret, Peter, *Clausewitz and the State*, New York: Oxford University Press, 1976.

Park, Colonel Francis J. H., "A Rigorous Education for an Uncertain Future," *Military Review* 96, no. 3 (May–June 2016): 70–7.

Partlow, Joshua, "Sheikhs Help Curb Violence in Iraq's West, U.S. Says," *Washington Post*, January 27, 2007.

Paschall, Rod, "The 27-Day War," *MHQ: The Quarterly Journal of Military History* 24, no. 3 (Spring 2012): 54–65.

Pendleton, John H., "Army Readiness: Progress and Challenges in Rebuilding Personnel, Equipping, and Training," Testimony before the Subcommittee on Readiness and Management Support, Committee on Armed Services, US Senate, US Government Accountability Office, GAO-19-367T, February 6, 2019.

Perito, Robert M., "The U.S. Experience with Provincial Reconstruction Teams in Afghanistan," Special Report 152, United States Institute of Peace, October 2005.

Perlez, Jane, "China Creates a World Bank of Its Own, and the U.S. Balks," *New York Times*, December 4, 2015.

Perlroth, Nicole, and David E. Sanger, "Cyberattacks Put Russian Fingers on the Switch at Power Plants, U.S. Says," *New York Times*, March 15, 2018.

Perry, Mark, *Four Stars*, Boston: Houghton Mifflin, 1989.

Perry, Walter L., and David Kassing, *Toppling the Taliban: Air-Ground Operations in Afghanistan, October 2001–June 2002*, RR-381, Santa Monica, CA: Rand Corporation, 2015.

Pezard, Stephanie, Abbie Tingstad, and Alexandra Hall, "The Future of Arctic Cooperation in a Changing Strategic Environment," PE-268-RC, Rand Europe, 2018.

Phillips, Katherine W., "How Diversity Makes Us Smarter," *Scientific American*, October 1, 2014.

Porch, Douglas, *Counterinsurgency: Exposing the Myths of the New Way of War*, New York: Cambridge University Press, 2013.

Posen, Barry R., "Foreword: Military Doctrine and the Management of Uncertainty," *Journal of Strategic Studies* 39, no. 2 (2016): 159–73.

Posen, Barry R., *The Sources of Military Doctrine: France, Britain, and Germany between the World Wars*, Ithaca, NY: Cornell University Press, 1984.

Pribbenow, Merle L., "The Fog of War: The Vietnamese View of the Ia Drang Battle," *Military Review* 81, no. 1 (January–February 2001): 93–7.

Project on National Security Reform, *Forging a New Shield*, Arlington, VA: Project on National Security Reform, November 2008.

"Provincial Reconstruction Teams in Afghanistan: An Interagency Assessment," Office of the Coordinator for Stabilization and Reconstruction, Department of State; Joint Center for Operational Analysis, United States Joint Forces Command; and Bureau of Policy and Program Coordination, United States Agency for International Development, April 5, 2006.

Raines, Edgar F., Jr., *The Rucksack War: U.S. Operational Logistics in Grenada, 1983*, Washington, DC: Center of Military History, US Army, 2010.

Rayburn, Colonel Joel D., and Colonel Frank K. Sobchak, eds., *The U.S. Army in the Iraq War*, vol. 1: *Invasion, Insurgency, Civil War, 2003–2006*, Carlisle, PA: Strategic Studies Institute and US Army War College Press, January 2019.

Rayburn, Colonel Joel D., and Colonel Frank K. Sobchak, eds., *The U.S. Army in the Iraq War*, vol. 2: *Surge and Withdrawal, 2007–2011*, Carlisle, PA: Strategic Studies Institute and US Army War College Press, January 2019.

Reed, Dan, "Shootout at LZ Albany: The 7th Cavalry's Second Fight for Survival at the Battle of Ia Drang," *Vietnam* 28, no. 4 (December 2015): 28–37.

Rempfer, Kyle, "Almost Half of Officers Match to Their Top Job Choice under New Army System," *Army Times*, December 13, 2019.

Rempfer, Kyle, "Army Chief: New Talent Management Will Start with Officers, Then Go to Enlisted," *Army Times*, October 15, 2019.

Reynolds, Alison, and David Lewis, "Teams Solve Problems Faster When They're More Cognitively Diverse," *Harvard Business Review*, March 30, 2017.

Ricks, Thomas E., *Fiasco*, New York: Penguin, 2006.

Ricks, Thomas E., *The Gamble*, New York: Penguin, 2009.

Ricks, Thomas E., *The Generals*, New York: Penguin, 2012.

Rielage, Captain Dale C., "An Open Letter to the U.S. Navy from Red," *Proceedings* (US Naval Institute) 143, no. 6 (June 2017): 28–31.

Roberts, Colonel Joseph W., "Agile Acquisition: How Does the Army Capitalize on Success?," Strategy Research Project, US Army War College, January 2017.

Robson, Seth, "2nd ID Vehicles Are Upgraded to Survive Rough Traveling in Iraq," *Stars and Stripes*, August 20, 2004.

Rock, David, and Heidi Grant, "Why Diverse Teams Are Smarter," *Harvard Business Review*, November 4, 2016.

Rodman, David, "Combined Arms Warfare: The Israeli Experience in the 1973 Yom Kippur War," *Defence Studies* 15, no. 2 (April 2015): 161–74.

Romjue, John L., "From Active Defense to AirLand Battle: The Development of Army Doctrine, 1973–1982," Fort Monroe, VA: United States Training and Doctrine Command Historical Office, June 1984.

Roper, COL Daniel S., USA, Ret., and LTC Jessica Grassetti, USA, "Seizing the High Ground—United States Army Futures Command," *ILW Spotlight* 18-4, Institute of Land Warfare, Association of the US Army, October 2018.

Rosen, Stephen Peter, *Winning the Next War: Innovation and the Modern Military*, Ithaca, NY: Cornell University Press, 1991.

Rumbaugh, Russell, "What We Bought: Defense Procurement from FY01 to FY10," Stimson Center, October 2011.

Russakoff, Dale, "Lessons of Might and Right," *Washington Post*, September 9, 2001.

Russell, James A., *Innovation, Transformation, and War: Counterinsurgency in Anbar and Ninewa Provinces, Iraq, 2005–2007*, Stanford, CA: Stanford University Press, 2011.

Russen, Major David M., "Combat History: 102nd Cavalry Reconnaissance Squadron (MECZ), World War II (D Day—June 6, 1944 through VE Day—May 8, 1945)," available at http://njcavalryandarmorassociation.org/History%20of%20the%20102nd.pdf.

Rynning, Sten, "ISAF and NATO: Campaign Innovation and Organizational Adaptation," in Theo Farrell, Frans Osinga, and James A. Russell, eds., *Military Adaptation in Afghanistan*, Stanford, CA: Stanford University Press, 2013, 83–107.

Rynning, Sten, *NATO in Afghanistan: The Liberal Disconnect*, Stanford, CA: Stanford University Press, 2012.

Safire, William, "Whack-a-Mole," *New York Times Magazine*, October 29, 2006.

Sanger, David E., *The Perfect Weapon: War, Sabotage, and Fear in the Cyber Age*, New York: Crown Publishers, 2018.

Sanger, David E., David D. Kirkpatrick, and Nicole Perlroth, "The World Once Laughed at North Korean Cyberpower. No More," *New York Times*, October 15, 2017.

Savage, Michael, "Richest 1% on Target to Own Two-Thirds of All Wealth by 2030," *The Guardian*, April 7, 2018.

Scales, Robert H., Jr., *Certain Victory: The US Army in the Gulf War*, Washington, DC: Office of the Chief of Staff, United States Army, 1993.

Scarborough, Rowan, "Army Mulls Funding for Controversial Intel Network," *Washington Times*, February 14, 2014.

Scarborough, Rowan, "Problems with Army's Battlefield Intel System Unresolved after Two Years," *Washington Times*, May 1, 2014.

Scarborough, Rowan, "Rep. Duncan Hunter Takes Army to Task over Requests for IED Finder," *Washington Times*, March 28, 2013.

Schadlow, Nadia, "Peace and War: The Space Between," *War on the Rocks*, August 18, 2014.

Scharre, Paul, *Army of None: Autonomous Weapons and the Future of War*, New York: Norton, 2018.

Schmitt, Eric, "Iraq-Bound Troops Confront Rumsfeld over Lack of Armor," *New York Times*, December 8, 2004.

Schmitt, Eric, and Mark Mazzetti, "Switch Signals New Path for Afghan War," *New York Times*, May 12, 2009.

Schroen, Gary, *First In: An Insider's Account of How the CIA Spearheaded the War on Terror in Afghanistan*, New York, Ballantine Books, 2005.

Schwab, Klaus, *The Fourth Industrial Revolution*, New York: Crown Business, 2016.

Schwartz, Moshe, "Defense Acquisition Reform: Background, Analysis, and Issues for Congress," R43566, Congressional Research Service, May 23, 2014.

Schwartz, Moshe, "Defense Acquisitions: How DOD Acquires Weapon Systems and Recent Efforts to Reform the Process," RL34026, Congressional Research Service, May 23, 2014.

Scott, Michon, and Kathryn Hansen, "Sea Ice," NASA Earth Observatory, September 16, 2016.

Seiple, Chris, *The U.S. Military/NGO Relationship in Humanitarian Interventions*, Carlisle, PA: Peacekeeping Institute, Center for Strategic Leadership, US Army War College, 1996.

Sellers, Cameron S., "Provincial Reconstruction Teams: Improving Effectiveness," master's thesis, Naval Postgraduate School, September 2007.

Sepp, Kalev I., "Best Practices in Counterinsurgency," *Military Review* 85, no. 3 (May–June 2005): 8–12.

Shachtman, Noah, "Brain, Damaged: Army Says Its Software Mind Is 'Not Survivable,'" *Wired*, August 8, 2012.

Shachtman, Noah, "No Spy Software Scandal Here, Army Claims," *Wired*, November 30, 2012.

Shachtman, Noah, "Spy Chief Called Silicon Valley Stooge in Army Software Civil War," *Wired*, August 1, 2012.

Shane, Leo, III, "Congress Is Giving the Military Promotion System a Massive Overhaul," *Military Times*, July 25, 2018.

Shanker, Thom, and Eric Schmitt, "Armor Scarce for Heavy Trucks Transporting U.S. Cargo in Iraq," *New York Times*, December 10, 2004.

Shirer, William L., *The Collapse of the Third Republic: An Inquiry Into the Fall of France*, New York: Simon & Schuster, 1969.

Sinclair, Captain Wayne A., "Answering the Landmine," *Marine Corps Gazette* 80, no. 7 (July 1996): 37–40.

Singer, P. W., and August Cole, *Ghost Fleet*, Boston: Houghton Mifflin Harcourt, 2015.

Singer, Peter W., "Tactical Generals: Leaders, Technology, and the Perils," Brookings Institution, July 7, 2009.

Slim, Field Marshal Viscount, *Defeat into Victory*, New York: Cooper Square Press, 2000.

Smith, Andrew, *Improvised Explosive Devices in Iraq, 2003–09: A Case Study of Operational Surprise and Institutional Response*, Carlisle, PA: Strategic Studies Institute, US Army War College, April 2011.

Smith, Major Niel, US Army, and Colonel Sean MacFarland, US Army, "Anbar Awakens: The Tipping Point," *Military Review* 88, no. 2 (March–April 2008): 41–52.

Smith, R. Jeffrey, "Hypersonic Missiles Are Unstoppable. And They Are Starting a New Global Arms Race," *New York Times Magazine*, June 19, 2019.

"Soldiers Must Rely on 'Hillbilly Armor' for Protection," *ABC News*, December 8, 2004.

Sorley, Lewis, *A Better War: The Unexamined Victories and Final Tragedies of America's Last Years in Vietnam*, New York: Harcourt Brace, 1999.

Sorley, Lewis, *Westmoreland: The General Who Lost Vietnam*, Boston: Houghton Mifflin Harcourt, 2011.

South, Todd, "Here's When the Marine Corps' Newest Tactical Vehicle Hits the Rest of the Fleet," *Marine Corps Times*, August 16, 2019.

Spain, Everett S. P., J. D. Mohundo, and Bernard B. Banks, "Intellectual Capital: A Case for Cultural Change," *Parameters* 45, no. 2 (Summer 2015): 77–91.

"Special Forces, Marines Embrace Palantir Software," *Military.com*, July 1, 2013.

Spencer, John, "What Is Army Doctrine?," Modern War Institute, March 21, 2016.

Stanton, Doug, *Horse Soldiers: The Extraordinary Story of a Band of U.S. Soldiers Who Rode to Victory in Afghanistan*, New York: Scribner, 2009.

Stanton, Shelby L., *The Rise and Fall of an American Army: U.S. Ground Forces in Vietnam, 1965–1973*, Novato, CA: Presidio, 1985.

Stapleton, Barbara J., "A British Agency's Afghanistan Group Briefing Paper on the Development of Joint Regional Teams in Afghanistan," British Agencies Afghanistan Group, January 2003.

Stewart, Richard W., *Operation ENDURING FREEDOM: The United States Army in Afghanistan, October 2001–March 2002*, CMH Pub 70-83-1, US Army Center for Military History, [2004].

Stokes, Maj Paul L., USMC (Ret) and CWO4 Brian D. Bethke, "Fanning the Flames of Innovation: Thawing Out the Frozen Middle," *Marine Corps Gazette*, 103, no. 4 (April 2019): 61–4.

Stolberg, Sheryl Gay, and Jim Rutenberg, "Rumsfeld Resigns as Defense Secretary after Big Election Gains for Democrats," *New York Times*, November 8, 2006.

Sullivan, General Gordon R., US Army, "Doctrine: A Guide to the Future," *Military Review* 72, no. 2 (February 1992): 2–9.

Summers, Harry G., Jr., *On Strategy: A Critical Analysis of the Vietnam War*, Novato, CA: Presidio Press, 1982.

Svan, Jennifer H., "With New Army Policy, Officers Take Standardized GRE Test during Captains Career Course," *Stars and Stripes*, August 9, 2019.

Tetlock, Philip E., *Expert Political Judgment: How Good Is It? How Can We Know?*, Princeton, NJ: Princeton University Press, 2005.

Tetlock, Philip E., and Dan Gardner, *Superforecasting: The Art and Science of Prediction*, New York: Crown Publishers, 2015.

"The Battle for Digital Supremacy," *The Economist*, March 15, 2018.

Thomas, Evan, "General McChrystal's Plan for Afghanistan," *Newsweek*, September 25, 2009.

Thomas, Jon T., and Douglas L. Schultz, "Lessons about Lessons: Growing the Joint Lessons Learned Program," *Joint Force Quarterly* 79 (4th Quarter 2015): 113–20.

Thompson, Mark, "How Safe Are Our Troops?," *Time*, December 13, 2004.

Tingley, Kim, "The Loyal Engineers Steering NASA's Voyager Probes across the Universe," *New York Times Magazine*, August 3, 2017.

Tofail, Syed A. M., et al., "Additive Manufacturing: Scientific and Technological Challenges, Market Uptake and Opportunities," *Materials Today* 21, no. 1 (January–February 2018): 22–37.

Toll, Ian W., *Pacific Crucible: War at Sea in the Pacific, 1941–1942*, New York: Norton, 2012.

"Townsend: Doctrine Critical to a 'Winning Army,'" Association of the US Army, January 9, 2019.

Townsend, Gen. Stephen, US Army, Maj. Gen. Douglas Crissman, US Army, and Maj. Kelly McCoy, US Army, "Reinvigorating the Army's Approach to Mission Command: It's Okay to Run with Scissors (Part I)," *Military Review*, April 2019.

"Transcript: Donald Trump Expounds on His Foreign Policy Views," *New York Times*, March 26, 2016.

"Transcript: General David McKiernan Speaks at Council's Commanders Series," Atlantic Council, November 19, 2008.

Trenin, Dmitri, "The Revival of the Russian Military: How Moscow Reloaded," *Foreign Affairs* 95, no. 3 (May–June 2016): 23–9.

Tucker, Patrick, "The War over Soon-to-Be-Outdated Army Intelligence Systems," *Defense One*, July 5, 2016.

Tullis, Paul, "The World Economy Runs on GPS. It Needs a Backup Plan," *Bloomberg Businessweek*, July 25, 2018.

Tversky, Amos, and Daniel Kahneman, "Availability: A Heuristic for Judging Frequency and Probability," *Cognitive Psychology* 5, no. 2 (September 1973): 207–32.

Tyson, Ann Scott, "A Sober Assessment of Afghanistan," *Washington Post*, June 15, 2008.

Tyson, Ann Scott, "Top U.S. Commander in Afghanistan Is Fired," *Washington Post*, May 12, 2009.

Ucko, David, "Militias, Tribes and Insurgents: The Challenge of Political Reintegration in Iraq," *Conflict, Security & Development* 8, no. 3 (October 2008): 341–73.

Ucko, David H., "Critics Gone Wild: Counterinsurgency as the Root of All Evil," *Small Wars & Insurgencies* 25, no. 1 (2014): 161–79.

Ucko, David H., *The New Counterinsurgency Era: Transforming the U.S. Military for Modern Wars*, Washington, DC: Georgetown University Press, 2009.

Union of Concerned Scientists, "Fact Sheet: The Science Connecting Extreme Weather to Climate Change," June 2018.

United Nations Department of Economic and Social Affairs, Population Division, *The World's Cities in 2018—Data Booklet*, ST/ESA/SER.A/417, 2018.

US-China Economic and Security Review Commission, *2011 Report to Congress*, November 2011.

US Census Bureau, *Statistical Abstract of the United States: 2012*, Washington, DC: Government Printing Office, 2012.

US Department of Defense, "Department of Defense Press Briefing by Secretary Carter on Force of the Future Reforms in the Pentagon Press Briefing Room," January 28, 2016.

US Department of Defense, "DoD News Briefing with Gen. McKiernan from the Pentagon," October 1, 2008.

US Department of Defense, *Fiscal Year 2009 Supplemental Request*, April 2009.

US Department of Defense, *Joint Improvised Explosive Device Defeat Organization Annual Report 2010*, 2010.

US Department of Defense, *Measuring Stability and Security in Iraq*, October 2005.

US Department of Defense, "Press Conference with Secretary Gates and Adm. Mullen on Leadership Changes in Afghanistan from the Pentagon," May 11, 2009.

US Department of Defense, "Secretary of Defense Speech, United States Military Academy (West Point, NY)," February 25, 2011.

US Department of Defense, *Summary of the 2018 National Defense Strategy of the United States of America*, January 2018.

US Department of Defense, Office of the Assistant Secretary of Defense, "Special Defense Department Briefing on Uparmoring HMMWV," December 15, 2004.

US Department of Defense, Office of the Under Secretary of Defense for Acquisition, Technology, and Logistics, *Report of the Defense Science Board Task Force on the Fulfillment of Urgent Operational Needs*, July 2009.

US Department of Defense, Office of the Under Secretary of Defense for Research and Engineering, *Defense Science Board Task Force on Survivable Logistics: Executive Summary*, November 2018.

US Department of Defense, Press Operations, "News Briefing with Secretary of Defense Donald Rumsfeld and Gen. Peter Pace," November 29, 2005.

US Department of Defense, Press Operations, "Secretary Rumsfeld Media Availability with Jay Garner," June 18, 2003.

US General Accounting Office, "Operation Desert Storm: Apache Helicopter Was Considered Effective in Combat, but Reliability Problems Persist," GAO/NSIAD-92-146, April 1992.

US Global Change Research Program, *Fourth National Climate Assessment*, vol. 2: *Impacts, Risks, and Adaptation in the United States*, 2018.

US Government Accountability Office, "Defense Major Automated Information Systems: Cost and Schedule Commitments Need to Be Established Earlier," GAO-15-282, February 2015.

US House of Representatives, Committee on Armed Services, Subcommittee on Oversight & Investigations, "Agency Stovepipes vs. Strategic Agility: Lessons We Need to Learn from Provincial Reconstruction Teams in Iraq and Afghanistan," September 2008.

US House of Representatives, Committee on Armed Services, Subcommittee on Oversight & Investigations, "The Joint Improvised Explosive Device Defeat Organization: DOD's Fight against IEDs Today and Tomorrow," Committee Print 110-11, 45-137, November 2008.

US House of Representatives, Committee on Oversight and Government Reform, "The OPM Data Breach: How the Government Jeopardized Our National Security for More Than a Generation," September 7, 2016.

US Marine Corps, *Marine Corps Operations*, Marine Corps Doctrinal Publication 1-0 (with Change 1), July 26, 2017.

US Senate, Committee on Armed Services, "Allegations of Mistreatment of Iraqi Prisoners," May 19, 2004.

US Senate, Committee on Armed Services, "United States Policy toward Afghanistan and Pakistan," April 7, 2009.

US Senate, Select Committee on Intelligence, "The Intelligence Community Assessment: Assessing Russian Activities and Intentions in Recent U.S. Elections, Summary of Initial Findings," July 3, 2018.

Vaccaro, Jonathan J., "The Next Surge: Counterbureaucracy," *New York Times*, December 8, 2009.

Vanden Brook, Tom, "New Vehicles Protect Marines in 300 Attacks in Iraq Province," *USA Today*, April 19, 2007.

Van Evera, Stephen, *Guide to Methods for Students of Political Science*, Ithaca, NY: Cornell University Press, 1997.

Vought, Lieutenant Colonel Donald B., "Preparing for the Wrong War?," *Military Review* 57, no. 5 (May 1977): 16–34.

Waghelstein, Colonel John D., "Post-Vietnam Counterinsurgency Doctrine," *Military Review* 65, no. 5 (May 1985): 42–9.

Ward, Orlando, "Foreword to Original Edition," in Russell A. Gugeler, *Combat Actions in Korea*, Washington, DC: Office of the Chief of Military History, US Army, 1970.

Warrick, Joby, Ellen Nakashima, and Anna Fifield, "North Korea Now Making Missile-Ready Nuclear Weapons, U.S. Analysts Say," *Washington Post*, August 8, 2017.

Watts, Barry, and Williamson Murray, "Military Innovation in Peacetime," in Williamson Murray and Allan R. Millett, eds., *Military Innovation in the Interwar Period*, Cambridge; New York: Cambridge University Press, 1996, 369–415.

Weber, Max, *From Max Weber: Essays in Sociology*, ed. H. H. Gerth and C. Wright Mills, New York: Oxford University Press, 1946.

Weigley, Russell F., *The American Way of War: A History of United States Military Strategy and Policy*, Bloomington: University of Indiana Press, 1973.

Weinberger, Sharon, and Noah Shachtman, "Military Dragged Feet on Bomb-Proof Vehicles (Updated Again)," *Wired*, May 22, 2007.

Weiner, Sharon K., "Organizational Interests versus Battlefield Needs: The U.S. Military and Mine-Resistant Ambush Protected Vehicles in Iraq," *Polity* 42, no. 4 (October 2010): 461–82.

West, Bing, *No True Glory: A Frontline Account of the Battle for Fallujah*, New York: Bantam Books, 2005.

White, Josh, "U.S. Deaths Rise in Afghanistan," *Washington Post*, July 2, 2008.

White, Sarah P., "Understanding Cyberwarfare: Lessons from the Russia-Georgia War," Modern War Institute at West Point, March 20, 2018.

White House, Office of the Press Secretary, "President Addresses Nation, Discusses Iraq, War on Terror," Fort Bragg, North Carolina, June 28, 2005.

White House, Office of the Press Secretary, "Presidential Address to the Nation," Treaty Room, October 7, 2001.

White House, Office of the Press Secretary, "President Outlines Strategy for Victory in Iraq," United States Naval Academy, Annapolis, Maryland, November 30, 2005.

White House, Office of the Press Secretary, "President's Address to the Nation," White House Library, January 10, 2007.

White House, Office of the Press Secretary, "Remarks of President Barack Obama—State of the Union Address as Delivered," January 13, 2016.

Widder, Major General Werner, "Auftragstaktik and Innere Führung: Trademarks of German Leadership," *Military Review* 82, no. 5 (September–October 2002): 3–9.

Wilkins, Major Aaron L., "The Civil Military Operations Center (CMOC) in Operation *Uphold Democracy* (Haiti)," paper presented to the Research Department of the Air Command and Staff College in partial fulfilment of graduation requirements, March 1997.

Willett, Lee, "Back on the World Stage: Russian Naval Power in 2013 and Beyond," *Jane's Navy International*, December 7, 2012.

Williams, Brian Glyn, "General Dostum and the Mazar i Sharif Campaign: New Light on the Role of Northern Alliance Warlords in Operation Enduring Freedom," *Small Wars & Insurgencies* 21, no. 4 (December 2010): 610–32.

Wolters Kluwer Editorial Staff, *CCH Department of Defense FAR Supplement as of January 1, 2019*, n.p.: Kluwer Law International, 2019.

Wolters Kluwer Editorial Staff, *CCH Federal Acquisition Regulation as of January 1, 2019*, n.p.: Kluwer Law International, 2019.

Wong, Leonard, "Strategic Insights: Letting the Millennials Drive," Strategic Studies Institute, US Army War College, May 2, 2016.

Wong, Leonard, and Stephen Gerras, "Army Talent Management Reform: The Culture Problem," *War on the Rocks*, February 22, 2019.

Wong, Leonard, and Stephen J. Gerras, *Lying to Ourselves: Dishonesty in the Army Profession*, Carlisle, PA: Strategic Studies Institute, US Army War College, February 2015.

Woodward, Bob, *Bush at War*, New York: Simon & Schuster, 2002.

Woodward, Bob, *Obama's Wars*, New York: Simon & Schuster, 2010.

Woodward, Bob, *The War Within: A Secret White House History, 2006–2008*, New York: Simon & Schuster, 2008.

World Bank, "China Overview," April 8, 2019.

World Economic Outlook, "Gross Domestic Product, Current Dollars, Purchasing Power Parity," International Monetary Fund, April 2019.

Wright, Austin, "Hunter Battles Army on Intel," *Politico*, May 1, 2013.

Wright, Donald P., et al., *A Different Kind of War: Operation ENDURING FREEDOM, October 2001–September 2005*, Fort Leavenworth, KS: Combat Studies Institute Press, US Army Combined Arms Center, 2010.

Wright, Thomas, "Trump's 19th-Century Foreign Policy," *Politico*, January 20, 2016.

Yingling, Paul, "A Failure in Generalship," *Armed Forces Journal*, May 1, 2007.

Zabecki, David T., ed., *World War II in Europe: An Encyclopedia*, New York: Routledge, 2015.

Zaloga, Steven, *Armored Attack 1944: U.S. Army Tank Combat in the European Theater from D-Day to the Battle of the Bulge*, Mechanicsburg, PA: Stackpole Books, 2011.

Zaloga, Steven, *Armored Thunderbolt: The U.S. Army Sherman in World War II*, Mechanicsburg, PA: Stackpole Books, 2008.

Zaloga, Steven J., *Sherman Medium Tank, 1942–1945*, Oxford: Osprey Publishing, 1978.

Zegart, Amy B., "Agency Design and Evolution," in Robert F. Durant, ed., *The Oxford Handbook of American Bureaucracy*, Oxford: Oxford University Press, 2010, 207–30.

Zegart, Amy B., *Spying Blind: The CIA, the FBI, and the Origins of 9/11*, Princeton, NJ: Princeton University Press, 2007.

Zetter, Kim, *Countdown to Zero Day: Stuxnet and the Launch of the World's First Digital Weapon*, New York: Crown Publishers, 2014.

Zisk, Kimberly Marten, *Engaging the Enemy*, Princeton, NJ: Princeton University Press, 1993.

Index

For the benefit of digital users, indexed terms that span two pages (e.g., 52–53) may, on occasion, appear on only one of those pages.